Engaging the Atom

Energy and Society
Brian Black, Series Editor

OTHER TITLES IN THE SERIES

Transportation and the Culture of Climate Change: Accelerating Ride to Global Crisis
Edited by Tatiana Prorokova-Konrad

Energy Culture: Art and Theory on Oil and Beyond
Edited by Imre Szeman and Jeff Diamanti

On Petrocultures: Globalization, Culture, and Energy
Imre Szeman

Oil and Urbanization on the Pacific Coast: Ralph Bramel Lloyd and the Shaping of the Urban West
Michael R. Adamson

Oil and Nation: A History of Bolivia's Petroleum Sector
Stephen C. Cote

Engaging The Atom

THE HISTORY OF NUCLEAR ENERGY AND SOCIETY IN EUROPE FROM THE 1950s TO THE PRESENT

EDITED BY
ARNE KAIJSER, MARKKU LEHTONEN,
JAN-HENRIK MEYER, AND MAR RUBIO-VARAS

West Virginia University Press / Morgantown

First edition published 2021 by West Virginia University Press
Printed in the United States of America

ISBN 978-1-952271-31-1 (cloth) / 978-1-952271-32-8 (paperback) /
978-1-952271-33-5 (ebook)

Library of Congress Control Number: 2021037030

Book and cover design by Than Saffel / WVU Press
Cover photograph by G_O_S / Shutterstock.com

Contents

Preface

Every book has a moment when the idea for writing it is born. For this book, it occurred at a seaside restaurant in Barcelona in September 2017. We were fifteen researchers from Europe and the United States, enjoying dinner after a long day of teaching at a summer school organized as part of our joint research project. This project, called HoNESt, explored the "History of Nuclear Energy and Society" in Europe, the United States, and the Soviet Union/Russia. We talked about our common research agenda and all the remaining deliverables that we would need to prepare and deliver according to the very elaborate research plan that HoNESt—like any European Commission–funded project—would have to submit to its funders. It struck us that the plan did not include an overarching presentation of our research findings, independent of the scrutiny of the project officers, who took their role so seriously that it seemed to border on censorship. Then someone suddenly said, "Why don't we write an edited book, outside of the plan?" This idea was met with enthusiasm and immediately led to a lively discussion on the content of such a book. We continued to sketch out the contents the next evening at another restaurant, after yet another day of teaching. At the end of the second dinner, we had a basic idea about what the book could look like and were excited to take on the new challenge.

As a transnational and interdisciplinary group, we had some additional assets. These included, first, a unique and quickly growing knowledge base about public engagement with nuclear energy during the postwar era in twenty countries, jointly produced by all project partners. Second, we could rely on the mutual trust that had gradually built up among the HoNESt partners during several years of collaboration—first when writing a research proposal together, and then when implementing it in close, often daily interaction.

In the subsequent two and a half years, this book gradually took shape. The four of us who volunteered to serve as editors depended on thirteen other HoNESt project colleagues who were willing to contribute chapters. We met at two intensive workshops at the University of Central Lancashire in Preston, UK (at the time still part of the European Union), and at the Leibniz Center for Contemporary History in Potsdam, Germany, to discuss first a book proposal and then our draft chapters. We thank both institutions for their hospitality. The West Virginia University Press series on Energy and Society seemed to us to be a perfect fit for the book. Luckily, the editors shared our view and accepted the proposal.

This book has seventeen authors, but many more have contributed indirectly. In fact, the book could not have been written without all the reports produced by the more than sixty researchers who worked on the HoNESt project at different times. In particular, the project leadership has been crucial in the production of all this material: the project coordinator Albert Presas i Puig, who brought and held the consortium together in good spirit and led the project to successful completion; the work package leaders Matthew Cotton, Josep Espluga, Wilfried Konrad, Jan-Henrik Meyer, Karl-Erik Michelsen, Mar Rubio-Varas, and John Whitton; and many dedicated administrators, not least Josep Niubò i Garcia, who diligently managed project finances and administration. Without these people, our book would never have come into being. We are grateful for the assistance of our skilled (and fast!) copyeditor, Martin Rodden, in Berlin, and diligent copyediting by Rachel Fudge. We would also like to extend our gratitude to the two anonymous reviewers of our book proposal, the external reviewers of the intermediate versions of the Short Country Reports, the project's advisory committee, all the eyewitnesses who shared their knowledge and memories with us over the past few years, and the archivists across Europe who dug up valuable and often hitherto-unexplored source material. Last but not least, the HoNESt project would never have seen the light of day without the personal commitment of Bruno Schmitz during his last term as the head of unit for fission energy at the European Commission's Directorate-General for Research and Innovation. He firmly defended the idea that the history of the relationship between nuclear and society had to be written, but not by a nuclear-sector insider. Instead, this would be a joint task for trained social scientists and historians, free from conflicts of interest and direct implication in nuclear-sector institutions. His enlightened commitment made possible this joint effort at bridging the gap between nuclear and social sciences. We hope this book honors his plea. Most of the research for this volume was possible thanks to funding by the European Commission Euratom/Horizon2020 research and training program 2014–2018 for the research project *History of Nuclear Energy and Society (HoNESt)* (grant agreement no. 662268). The work by Markku Lehtonen was also supported by the European Commission Marie Skłodowska-Curie Individual Fellowships (IF) grant number 794697-TENUMECA.

The journey from the dinner in Barcelona to the final production of the book has at times meant hard work, but the friendship and mutual trust among all involved has made it a pleasant endeavor.

Arne Kaijser, Markku Lehtonen, Jan-Henrik Meyer, Mar Rubio-Varas
Stockholm, Barcelona, Potsdam, Pamplona, March 2021

INTRODUCTION

Nuclear Energy and Society in Postwar Europe

Arne Kaijser, Markku Lehtonen, Jan-Henrik Meyer, and Mar Rubio-Varas

Introduction

Public engagement with nuclear energy in Europe has changed dramatically over time. It has also varied considerably between—and even within—countries and societies, ranging from promotion to questioning or outright rejection and protest, from silent acceptance to public controversy. Such engagement has been characterized by varying perceptions and discourses, between hope and promise on the one hand, and fear and risk on the other. Public engagement has involved not only different types of activities and perceptions but also different kinds of societal actors: at first, these included primarily scientists and engineers, industrialists, managers of utilities, policymakers, and public relations experts, but increasingly they were also citizens of various backgrounds, from farmers and housewives to pastors and scientists acting as counter-experts. Eventually, antinuclear activists—many of them women—formed social movements to organize citizen-initiated public engagement.

Public engagement first emerged in the mid-1950s, when something of an atomic euphoria swept over the world following President Eisenhower's "Atoms for Peace" talk at the United Nations (UN) in 1953. In this speech, Eisenhower announced his intention to promote international collaboration to develop peaceful uses of atomic power. Many European countries soon launched programs with the goal of building nuclear power plants of their own. In parallel and aided by the United States (US) government, European governments, industry, and research bodies launched campaigns to enlighten the general public about the benefits of the new technology's peaceful uses. These efforts served to counterbalance the concerns and fears that the general public was presumed to hold toward a technology that was initially developed for military purposes and had demonstrated its destructive power in Hiroshima and Nagasaki.[1]

The Netherlands may serve as an example of such activities.[2] In the summer of 1957, an exhibition called *The Atom* took place in a new hangar building at the Schiphol Airport outside Amsterdam. Its main attraction was a small, swimming pool–type nuclear reactor with a 10 kW capacity. Most of the 750,000 visitors seemed fascinated when they saw the blue glimmer of the reactor. The exhibition was planned and sponsored by a range of institutions and government bodies, including national ministries, the municipality of Amsterdam, industrial companies like Philips, nuclear scientists, and representatives from nuclear organizations in other countries, in particular the US. Their goal was to make the Dutch public "atomic minded." The exhibition also displayed other symbols of technological progress that promised to facilitate the challenges of modern life, including novel electric kitchen appliances and a large electronic calculator. The official visitors' guide argued: "It is up to us, the generation of 1957, to determine the future of tomorrow's world. Our future: atomic energy!" One week after the pompous opening of the exhibition, the Dutch government presented a memorandum on nuclear energy to the Dutch parliament, proposing a swift transition in the Netherlands toward nuclear energy.

The atom exhibition at Schiphol was by no means unique. In most European countries on both sides of the Iron Curtain, the same kind of actors organized similar exhibitions with the aim of persuading the general public of the benefits of nuclear energy for modern consumer societies.[3] The biggest of these was the 1958 world exhibition in Brussels, visited by no less than forty-two million people. Its main attraction was the Atomium, a 102-meter-high structure of nine spheres representing an iron crystal. Its main purpose was to celebrate the coming of the Atomic Age. In their exhibition pavilions, countries such as the US, the Soviet Union and the United Kingdom (UK) all showed models of their nuclear reactors.[4]

In the Soviet Union, the establishment of the Obninsk nuclear power plant in 1954 was equally presented as a source of national pride. Located just 100 kilometers from Moscow, it was the first reactor in the world to be connected to the power grid. Aimed at demonstrating the USSR's scientific and technical superiority, Obninsk was a Cold War propaganda coup and was widely covered in the mass media throughout the Soviet Union and by its allies in Eastern Europe.

On both sides of the Iron Curtain, the engagement with the atom in these early years was orchestrated from above by the nuclear sector at large, encompassing government, industry, and academia. The exhibitions and highly enthusiastic and optimistic media coverage probably helped to generate broad enthusiasm—among political elites across the political spectrum but also with the public at large—for nuclear energy as an innovative technology epitomizing progress and prosperity. This enthusiasm paved the way for highly ambitious nuclear research programs. By the early 1960s, demonstration power reactors were in operation in all the leading industrial countries, and from the second half of the decade commercial nuclear plants were being built in growing numbers in some thirty countries. During the peak year of 1975, construction started on forty-four plants worldwide.[5] In fact, over 300 of over 400 reactors in operation throughout the world today began their construction between 1965 and 1979.[6]

In the mid-1970s, a new kind of public engagement with the atom emerged from below, when antinuclear movements were born in a number of West European countries, largely inspired by similar movements in the United States. These movements were rooted in a critique of nuclear technology and concerns about its potentially detrimental consequences. Their spokespersons argued that accidents in nuclear power plants would lead to the release of radioactive isotopes harmful to human health and genes, that there were no plans for the management of the spent fuel, and that there was a risk of proliferation—namely, that nuclear material from reactors could end up being used for constructing nuclear weapons. In many West European countries, movements opposing nuclear power grew quickly and organized large demonstrations and other actions to try and stop the construction of nuclear plants, as chapter 3, "One Movement or Many? The Diversity of Antinuclear Movements in Europe," illustrates in greater detail.

In February 1975, local citizens, among them farmers and winegrowers, occupied the building site of a reactor at Wyhl in southern West Germany and were forcibly removed by police two days later. This spurred support from the nearby university town of Freiburg, and a few days later about 30,000 people reoccupied the site, following the example of similar protests in neighboring France. The confrontation at Wyhl kick-started a series of protests in Germany and across the border in Switzerland. A long series of legal proceedings instigated by the opposition led to the reactor never being built.[7]

In the summer of 1976, antinuclear activists from France, Germany, and Switzerland organized a similar occupation at the building site of the Superphénix fast-breeder nuclear reactor, 60 kilometers east of Lyon. The antinuclear movement considered this plant to be particularly dangerous. The reactor was intended to "breed" plutonium, and as such provoked concern for proliferation risks. Moreover, the breeder was cooled by liquid sodium, which burns when in contact with air and explodes when placed in water. The French police terminated the site occupation, using considerable force to expel the occupants. The antinuclear activists returned in even larger numbers the following summer. On July 31, around 50,000 demonstrators gathered on a field a few kilometers from the building site. Among them were militant German activists wearing motorcycle helmets. This kind of gear was unusual among French protesters and was one of the reasons for the French police's use of exceptional force. In the ensuing confrontation, as the police blocked the demonstrators' access to the building site, one activist died, and three were seriously injured by police gas grenades. This tragic event became a turning point for the antinuclear movement in France, which gave up on large-scale demonstrations as a strategy.[8]

In most countries, the antinuclear movement continued to gain momentum in the following years, not least due to the accidents at the Three Mile Island (US, March 1979) and Chernobyl (USSR, April 1986) nuclear power plants. For example, in Austria, Italy, Denmark, and Portugal, citizen opposition led to formal decisions—sometimes preceded by referendums—to abandon nuclear power as an electricity source. Other countries decided to phase out nuclear power within a few decades (Sweden) or to temporarily halt their nuclear programs (Spain). In a number of Eastern and Central European countries, antinuclear movements emerged in the wake of Chernobyl as part of the struggle against the Soviet system and for independence from the Soviet Union. The two accidents also accelerated the measures, already under way in many countries, to clearly separate nuclear regulation and promotion duties within their administrative structures. Additional safety measures were introduced, and the concept of "safety culture" was coined to increase the awareness and understanding of risk management within nuclear organizations.[9] A number of countries began searching for possible sites for the final disposal of nuclear waste, but the site investigations often met strong local opposition.[10] In parallel, more advanced engagement approaches and strategies emerged slowly, via a complex and volatile learning process. An obstacle to a more dialogical approach was the persistent assumption among many nuclear-sector actors that antinuclear views either resulted from the negative tone set in public debates on any topic related to nuclear or reflected irrational, "psychological" fears and emotional reactions of a poorly informed public.[11] In many countries, antinuclear movements' growing mistrust of authorities and utilities also made dialogue difficult.

Sweden provides an example of a shift toward a more dialogical approach, and of the manifold consequences of such change.[12] In the early 1980s, the Swedish nuclear industry made a number of attempts to find a suitable site for a nuclear spent-fuel repository. Yet each time a drilling team arrived at a new location, it met fierce opposition from local groups. SKB, the Swedish Nuclear Fuel and Waste Management Company, established jointly by the country's nuclear operators to handle the spent fuel, finally concluded that it would be impossible to establish a repository at a site in which the local population was strongly against the project. The company therefore abandoned all the drillings and developed a new strategy. In 1992, SKB stressed that the subsequent process would be based on voluntariness and that no municipality would be forced to accept a repository. This in turn led SKB to focus on two municipalities that already had nuclear plants and whose inhabitants therefore appeared as less critical toward new nuclear facilities: Östhammar and Oskarshamn.[13]

To gain the trust of the local citizens, SKB arranged a number of meetings and consultations in both municipalities to provide information and to discuss the repository project—its design, safety, and consequences for the local communities. After long investigations, SKB proposed Östhammar as the most suitable host, mainly referring to Östhammar's favorable geological conditions. In fall 2017, the Swedish Land and Environment Court organized a seven-week hearing concerning SKB's proposal, with all relevant parties participating. During this long process, the character of the public engagement changed fundamentally. It had begun with angry local opposition to what were seen as hostile and unannounced attempts to start proof drillings for a hazardous plant but ended with an open and transparent court hearing at which a number of engaged and well-informed actors scrutinized SKB's application.

Finland has experienced a similar development. There, too, opposition to drillings led the utilities' joint waste-management company, Posiva, to shift its strategy away from searching for the geologically most suitable site toward a more pragmatic approach of looking for a site with "sufficiently good" geology. As a result, Posiva opted for a municipality that already hosted a nuclear power plant and, as a "nuclear community," had its fate already closely tied with the nuclear industry. Also in France and the UK, the government and the nuclear sector have made efforts to earn trust from local municipalities for their nuclear waste plans, albeit with little success so far. In Germany, the process of identifying a host community for a repository was recently relaunched and new institutions established. However, the project is still subject to widespread public suspicion, in particular because the Gorleben salt dome—selected as a site in 1977, and a venue for protest ever since—was only excluded from the candidate list in 2020. While this was welcomed locally, other observers criticized this decision as an overtly political move, and against the principles of a supposedly science-based search.[14]

Nuclear technology is an exceptional technology when it comes to engaging the public, especially in the early years in generating enthusiasm and over time increasingly in spurring opposition. As such, nuclear power is a prime example of a "public technology" (see chapter 8, "Nuclear Energy in Europe: A Public

Technology"). In many European countries, various forms of public engagement decisively shaped the development of nuclear programs—sometimes speeding it up but more often slowing down, halting, or preventing the development of these programs. Focusing our book on public engagement inevitably implies giving less attention to other significant aspects. These notably include the importance of military considerations in the development of civilian nuclear power, the support to nuclear programs by the US and the USSR as part of their Cold War strategies, and the geostrategic energy policy considerations relating to the trade-offs between the development of nuclear power as an expensive new electricity generation option and the massive recourse to the increasingly cheap oil and gas made possible by the opening of the Middle East. In chapter 1, "Nuclear-Society Relations from the Dawn of the Nuclear Age," we briefly outline this wider military, strategic, and political context, but the subsequent chapters center on the key actors and various perspectives toward public engagement with nuclear technology.

A Brief Overview of Existing Literature on the History of Nuclear Energy and Society in Europe

This is certainly not the first book to address the interaction between nuclear energy and society in Europe. At least since the 1970s, with the rise of critique and protest, contemporary social science has studied—often in a comparative manner—not only the emerging movements but also the government policies and the attitudes of political parties toward this new source of energy, including how its proponents sought to "market" nuclear energy to the public.[15] Such analyses were often substantially informed by the controversies and disciplinary debates of the day, including those concerning the "new social movements."[16] Nevertheless, some of these studies amass significant comparative evidence and demonstrate impressive analytical depth.[17] Many issues of ongoing relevance—such as the assessment of risk, the various dimensions of fairness and justice, and the role of varying opportunity structures or the mobilization of resources for antinuclear movements—have been covered by contemporary social science.[18] Different stages of the fuel cycle, ranging from the uranium mining to waste disposal and its governance, have also attracted growing interest amongst social scientists, especially since the 1990s.[19] For the interdisciplinary research underlying this book, these studies provided an important empirical, conceptual, and theoretical resource.

In recent years, there has been a veritable boom in historical research on certain issues regarding nuclear energy and society in Europe. While the literature on the history of the promotion of nuclear power and the controversy in the media is more limited,[20] a considerable number of social and environmental historians have explored the history of antinuclear protest—often focusing on local or regional cases and sites of protest, including waste repositories and reprocessing plants.[21] Some studies have examined national movements[22] and the role of these movements in widening the scope for public participation and shaping attitudes toward democracy. Others have explored the role of specific groups, such as Protestant pastors, within the protest movements.[23] Most recently, research has highlighted the transnational, and occasionally also the transboundary, dimension of protest.[24]

Scholars in the field of the history of technology have explored the history of nuclear research centers.[25] These studies also include—albeit often not very systematically—information about the role of nuclear researchers as experts and counter-experts in their countries' respective debates, such as those concerning nuclear waste disposal at sea.[26] The issue of geological disposal of nuclear waste has only recently attracted noteworthy attention among historians.[27]

At the intersection between political history, diplomatic history, and the history of technology, some studies also try to bridge the gap between military and civil uses and cover the international dimensions of nuclear affairs.[28] A more systematic analysis of the role of nuclear-sector intergovernmental organizations and their role in dealing with the public still remains to be carried out.[29]

Finally, nuclear accidents have attracted ample attention from historians. Recently, some research has analyzed the reactions to, debates on, and other consequences of Three Mile Island,[30] while Chernobyl has given rise to a considerable body of publications. These include factual histories concerning topics such as the events and local political consequences in the late Soviet Union, Chernobyl's health consequences, and the "production of ignorance" by intergovernmental organizations like the International Atomic Energy Agency (IAEA).[31] A number of studies explore how Chernobyl linked West and East, for instance, via the Chernobyl aid programs for children,[32] how the disaster impacted politics and society at the time,[33] and how it has become a point of reference in public memory across Europe.[34]

Much of the history of nuclear energy and society, however, remains to be written.[35] Our book is a contribution toward this objective.

The Core Argument

Why is nuclear energy more contentious in some countries than in others? Why have some countries never gone nuclear, or decided to abandon it, while their neighbors have pursued new-build programs and repeatedly placed their faith in a "nuclear renaissance"? How has the character of public engagement varied across countries and changed over time? These are some of the key questions that this book seeks to answer.

Although scholars from a wide variety of fields have studied the past, present, and future of the nuclear sector, including—to a greater or lesser extent—its relations with society, the different strands of this literature hardly engage with each other. Much of the existing literature consists of specific, often national and subnational case studies that frequently make only superficial references to other (inter)national contexts and experiences. This volume tackles the twin challenge of (1) assembling new knowledge acquired through historical research, and (2) overcoming the disciplinary and national biases that have thus far characterized research on nuclear energy. It combines historical research approaches—social, political, economic, and environmental history, as well as the history of technology—with those from the multiple strands of social science research, including political science, sociology, psychology, science and technology studies (STS), human geography, and environmental studies. The book addresses topics such as nuclear policy and politics, economic and industrial policy, social movements and civil society, and public perceptions of nuclear power. However, the volume does not delve into the specifics of the political dynamics and battles in the twenty countries studied in the HoNESt project, such as party politics or the shifting positions held by particular political actors in these countries. None of the chapters presents a single national case study; each draws on insights from a number of countries. In our experience, this approach has proven very effective in helping to overcome the traditional methodological nationalism.

The concept of public engagement—broadly defined as relations between the nuclear sector and society at large—is the unifying element in the interdisciplinary approach shared by all contributors to this book. Even if not all chapters explicitly mention the term "engagement," each explores the different ways in which actors in the nuclear sector engaged with society, whether via communicating, consulting, and interacting with citizens to win their hearts and minds; responding to people's concerns and seeking to earn trust; exercising technocratic, economic, and/or political power; or combining these elements in multiple ways. The chapters show how the diverse modes of interaction with societies not only contributed to solving or aggravating nuclear conflict at a given point in time but also had long-lasting effects on how the relations concerning nuclear projects would evolve in the future.

Crucially, engagement also has its bottom-up dimension. Civil society actors engaged with nuclear power on their own terms and initiative, within and—in many cases—beyond political, economic, cultural, or religious institutions. A variety of individuals and groups within society not only expressed concern about the environmental and health effects of nuclear power but associated nuclear with diverse visions of a good society, modernity, prosperity, and democracy. Antinuclear activists variously criticized the nuclear power sector for reinforcing and relying on state-dominated systems and visions (whether capitalist or communist), for undermining democracy, for serving and collaborating with undemocratic regimes (e.g., in Portugal, Spain, and countries of the former Soviet bloc), or for aggravating the risk of proliferation of fissile material, notably via nuclear technology exports. Active engagement of society with the nuclear sector—whether through citizen initiatives or invited participation by incumbent players—has influenced the development of nuclear technology, which we argue can be understood as a "public technology": nuclear technology has coevolved with society. Not only did nuclear technology provoke various forms of public engagement, but it shaped public perception, societal discourse, action, and controversy.[36]

The central hypothesis of the book springs from the observation that while the issue of engagement is key to understanding the nuclear-society nexus, it is necessary to examine engagement in its diverse and context-specific manifestations. Only by scrutinizing how societies have engaged with nuclear energy, how the nuclear energy sector has engaged with societies, and how this interaction has changed over time, can we make sense of the multiple and multifaceted nuclear-societal relations in Europe and elsewhere. It may also help us better understand the future challenges of nuclear-society interactions.

The Contribution of the Book

This book's approach is distinctive in three key respects. First, our coverage is broad in space and time, as we present historical accounts of nuclear developments in twenty countries (nineteen Western and Eastern European countries, including the Soviet Union, plus the United States) since the 1950s. Second, our comparative approach, based on primary research, juxtaposes observations and singles out transnational connections within and beyond Europe, notably the United States as the technological and political leader of the Western world. Third, we examine the topic from an interdisciplinary perspective, based on a research and writing process involving intense exchange and cooperation across disciplinary boundaries. The new findings thus contribute to the history of technology, social, economic, and environmental history, as well as to social science research on public engagement, governance and perceptions of risk, and social movements.

The volume presents a unique examination of past experiences of nuclear-societal relations in Europe—East and West—thus overcoming the Cold War division that is frequently reproduced in historiography. This history of the nuclear sector's engagement with society, and society's engagement with nuclear energy, encompasses the Atoms for Peace campaigns of the 1950s; the violent and nonviolent protests, societal debates, public hearings, and referendums of the 1970s and 1980s; the diverse responses to the Three Mile Island, Chernobyl, and Fukushima accidents in the East and West; and, more recently, the increasingly professionalized public engagement processes designed to facilitate siting of facilities for interim storage and permanent disposal of nuclear waste. Moreover, the book aims to foster a more reflexive societal debate on future energy policies and on transitions toward more sustainable systems of energy supply and consumption.[37] Many of the patterns and elements of conflict and societal engagement observed throughout the history of nuclear energy indeed seem to repeat themselves, as demonstrated by the recent conflicts over other energy technologies, such as those concerning wind power or unconventional hydrocarbon exploitation via fracking.[38]

Addressing in a single volume the entire history of interactions between the nuclear sector and society in all of its richness would obviously be impossible. A number of issues were excluded, and left instead for future research. Our primary interest is in civil applications of nuclear technology. The development of nuclear weapons indeed laid the foundations for these civil applications, given that much of the early research and development in the nuclear field focused on weapons.[39] Later on, critique of the links between military and civilian applications, and the risk of proliferation of fissionable materials, greatly influenced the development of the antinuclear movement, as well as diplomacy and perceptions around nuclear power. However, this book only occasionally addresses the military applications (e.g., chapters 1 and 7).

The book does engage with the issue of culture, which has gained prominence in historical research in recent years.[40] In particular, chapter 8, by Stathis Arapostathis, Robert Bud, and Helmuth Trischler, considers nuclear as a "public technology," exploring how imaginaries and perceptions were negotiated in the public sphere. However, we do not deal at length with *atomic culture*—that is, how any given national or regional culture interprets, understands, and presents the topic of nuclear power; nor do we include systematic research on the public display of nuclear technologies and their "banalization."[41] Furthermore, we do not address the highly topical issue of nuclear memory, associated in particular with accidents and the legacies left by disused nuclear installations and uranium mines.[42]

Finally, a state-of-the-art history of nuclear energy and society should also address gender perspectives. Some of the country studies on which the chapters in this book are based indeed address the role of women, for example in nuclear research in Austria, in antinuclear movements in West Germany, or among leading politicians opposing nuclear, such as Birgitta Dahl in Sweden and Ritt Bjerregaard in Denmark.[43] However, plenty of topics still await study: for example, gendered discourses on nuclear technology, rife with assumptions about male rationalism and irrational female fears; or gender roles and inequalities in nuclear research and industry.[44] While a number of case studies exist on women in antinuclear movements,[45] the various roles that women play as nuclear advocates—via pronuclear nongovernmental organizations (NGOs) and networks,[46] as industry leaders, communication specialists, or experts at governments and international organizations—would be worth examining from a gender perspective.

How This Book Came About: Crossing Disciplinary Boundaries

This book emerges from the findings of a large collaborative research project, *The History of Nuclear Energy and Society (HoNESt)*, funded by the European Atomic Energy Community (Euratom) under the European Union's Horizon 2020 research framework program,[47] which over a period of three and a half years brought together the expertise of academics from twenty-three high-profile research institutions.[48] This unprecedented large-scale funding enabled us—together with a large number of partners whose expertise has informed this book but who have not contributed directly by authoring chapters—to conduct empirical research in twenty countries, and to share and discuss findings across a range of highly diverse national contexts. As a multidisciplinary consortium involving both historians and social scientists, we employed numerous methodologies, utilizing various task-specific methods, producing a rich and unique body of findings on international nuclear-societal interactions over the approximately seven decades of existence of civilian nuclear power.

The research process was guided by an interdisciplinary framework combining historical accounts of nuclear developments with social science analyses of public perceptions and stakeholder engagement.[49] At the outset of the project, the social scientists formulated a guiding framework with a number of research questions for the historians to explore. Based on this framework, the historians conducted detailed research into the country-specific experiences with nuclear energy development and societal interactions, focusing on how the changes in public engagement over time were placed within the broader political and societal context of the country they were studying. This historical analysis, in turn, informed the social scientists of the historical dynamics underpinning key events and the rationales that lay behind the decisions taken.[50] The historians participating in the project thus considered nuclear events from both national and transnational perspectives to account for their cross-border nature and the potential implications of the international circumstances.

The choice to conduct cross-disciplinary analysis involving history and the social sciences implied significant challenges. Generating an integrated approach required linking often-separate knowledge pools, translating concepts across disciplines but also between multiple national languages, integrating approaches, and jointly identifying key issues. This helped to ensure that the historical research was comprehensive while meeting the needs of social scientists, and that the social science analyses took the broader historical context into account. The concept of "public technology," coined by three of the historians in the project, was a direct outcome of this stimulating interdisciplinary exchange (see chapter 8).

The result of the systematic process of collecting, analyzing, and interpreting historical evidence is summarized in twenty short country reports (SCRs). These constitute a rich body of over 1,500 pages of international historical narratives and data on nuclear-society interactions. The SCRs describe chronologically and in a concise manner the nuclear energy development and its relations with society in each country, and identify significant country-specific events reflecting the various kinds of societal engagement with nuclear energy and nuclear energy's engagement with society. The reports also single out relevant actors in the nuclear field and society involved in these interactions. Thus, the SCRs constitute the core empirical basis for this volume but also provide a complementary source of material, allowing the interested reader to delve deeper into the developments of any specific country. All SCRs are freely available for download.[51]

The Book's Contents

This book's nine chapters are distributed into three sections, "Contexts," "Actors," and "Perspectives." "Contexts" contains two chapters focusing, respectively, on the sociopolitical and the economic context of nuclear developments. In chapter 1, "Nuclear-Society Relations from the Dawn of the Nuclear

Age," Paul Josephson, Jan-Henrik Meyer, and Arne Kaijser provide a historical overview based on the comparative research of twenty countries conducted in the HoNESt project. It introduces a periodization that provides a framework for the subsequent chapters and discusses the role of nuclear energy as both a national and an international enterprise. The chapter describes how public attitudes evolved from the dawn of the nuclear age, from largely enthusiastic embrace of the so-called "peaceful uses" of nuclear technology toward greater concern and awareness of environmental risks. It further explores the role that transitions to democracy in Southern and Eastern Europe played in shaping the relations between the nuclear sector and society, as well as the impact of nuclear accidents on public debates and decision-making. It also highlights the growing concern over nuclear waste in recent decades and closes with a brief discussion on possible nuclear futures in Europe.

The economics of nuclear energy constitute the focus of chapter 2, "The Changing Economic Context Influencing Nuclear Decisions," by Mar Rubio-Varas. Nuclear power plants are large, capital-intensive, and risky investments that take a long time to complete. The scale of these endeavors poses particular challenges—not least of economic nature. First, a country's electricity grid and market have to be sufficiently large and developed to absorb and manage the electricity flows from large and centralized plants. Second, a minimum industrial base is required to develop domestic nuclear technology or to successfully implement technology imported from abroad. Not surprisingly, many of the countries that have refrained from using nuclear power are small. Because nuclear power plants take a long time to build, they are highly dependent on various kinds of government support as well as large loans for their financing. The real rate of interest is therefore of crucial importance for the final cost of nuclear electricity. After a period of low rates in the 1960s, real interest rates grew significantly in the 1970s and in the 1980s. Moreover, with the advent of neoliberalism from the 1980s onwards, utilities in many countries were privatized and government financial support for nuclear investments cut down. The chapter demonstrates that the changing economic context has been decisive in shaping the development of nuclear power.

The second section, "Actors," focuses on two sets of key actors that shaped nuclear-societal relations; on the one hand, antinuclear movements challenging the nuclear sector in European countries from below, and on the other hand, intergovernmental organizations supporting them from above. In chapter 3, "One Movement or Many? The Diversity of Antinuclear Movements in Europe," Albert Presas i Puig and Jan-Henrik Meyer critically revisit the nature of antinuclear movements. They argue that social science research has frequently brushed over the considerable diversity in terms of origins, social composition, and strategies of antinuclear groups across countries. When antinuclear activism emerged in Western Europe in the 1970s, it came in different forms. Not all were motivated by a desire to oppose nuclear technology as such; there were highly diverse reasons for opposing a specific nuclear installation or planning process. Thus, rather than essentializing the antinuclear movement as a single homogeneous phenomenon, we should speak of antinuclear movements in the plural. Chapter 3 focuses on these movements in Austria, Denmark, Spain, Sweden, and West Germany, highlighting their distinct origins and members; perceptions and ideological orientations; ways of framing the conflicts; strategies and protest behavior; and the varying geographical scope of their activities. The chapter equally highlights the transnational connections between the movements, which were vital for the exchange of scientific and technological knowledge and the elaboration of protest strategies. Because of the differences between the movements, however, transnational cooperation was often fragile, and involved only a small and select number of members of the various movements.

In chapter 4, "International Organizations and the Atom: How Comecon, Euratom, and the OECD Nuclear Energy Agency Developed Societal Engagement," Paul Josephson and Markku Lehtonen describe how international organizations were established during the Cold War with the double aim of regulating and promoting nuclear energy. International organizations coordinated joint research and development efforts in the nuclear field on both sides of the Iron Curtain. They also took on an active promotional role, with public communication designed to convince the public of the virtues of nuclear power. In the 1970s and '80s, the increasing public skepticism about the benefits of nuclear power, as well as the Three Mile Island and Chernobyl accidents, challenged the promotional role of these organizations. The collapse of the Soviet Union and the dissolution of Comecon led to the gradual integration of the former Eastern bloc countries into the Western nuclear community. In this process, Euratom and the OECD Nuclear Energy Agency (NEA) undertook efforts to westernize Eastern European nuclear facilities and regulating practices. Moreover, these organizations gradually changed their approach to public information, communication, and participation. The chapter exposes some of the key differences between the Western and Eastern traditions that shaped the nuclear-society relationship and also describes how some international cooperation efforts were constrained by member countries' desires to retain their sovereignty in a strategically sensitive policy area.

The chapters in the third section approach the issue of nuclear energy and society from a variety of different perspectives. In chapter 5, "Risky or Beneficial? Exploring Perceptions of Nuclear Energy over Time in a Cross-Country Perspective," Josep Espluga, Wilfried Konrad, Ann Enander, Beatriz Medina, Ana Prades, and Pieter Cools compare public perceptions of nuclear power in Bulgaria, Finland, the Netherlands, Spain, Sweden, the UK, the US, and West Germany. They show that the types of perceived risks and benefits were very similar across countries, and that the evolution of these perceptions over time followed relatively similar patterns. In brief, the perceived risks were relatively few in the first period (1950–1970), increased during the second period (1970–1990), and then decreased somewhat in the third period (1990–2015). The expectations of economic benefits were quite constant but decreased over time, while arguments about the environmental benefits of nuclear energy, especially in helping to combat climate change, gained greater attention as time went by. The chapter highlights two main findings: first, public perceptions were fundamentally shaped by many factors other than the health and environmental risks and economic benefits, and second, public reactions were highly dependent on the context in which the nuclear technology was deployed. The key to understanding consensus and conflict on nuclear energy lies in the sphere of trust in institutions, on the one hand, and sociocultural contextual factors, on the other. These contextual factors concern issues such as responsibility, moral judgments, principles of justice and equity, as well as accountability and legitimacy.

In chapter 6, "Trust and Mistrust in Radioactive Waste Management: Historical Experience from High- and Low-Trust Contexts," Markku Lehtonen, Matthew Cotton, and Tatiana Kasperski compare radioactive waste-management policies in Finland, France, Sweden, and the UK, and contrast the findings from these Western countries with those from Russia. In the 1970s, various Western European countries introduced legislation obliging the utilities to ensure safe disposal of nuclear waste or established public agencies and committees to deal with the issue. During the same period, the international nuclear community endorsed deep geological disposal as the best solution for the problem of high-level nuclear waste. Since then, the four Western countries addressed faced the delicate task of identifying local communities with not only adequate geological conditions but also a population willing to host a repository. Often, local populations vehemently opposed such facilities and mistrusted the industry and government actors promoting and implementing repository projects. Gaining the trust of local populations has therefore been a major priority for the project proponents and implementers. The chapter analyzes how trust and mistrust have developed over the years, how different measures were undertaken to build trust, as well as how and why these measures have worked or not. The case of Russia provides a highly illuminating contrasting example that underlines some of the key features that shape public engagement in nuclear waste management.

In chapter 7, "Nuclear Power and Environmental Justice: The Case for Political Equality," Matthew Cotton analyzes nuclear-sector policymaking by combining concepts from the fields of environmental justice and political philosophy. Environmental justice is concerned with fairness in the distribution of environmental risks and benefits, opportunities for citizen participation, and recognition of local cultures when siting environmentally risky activities and installations. Political philosophy, in turn, focuses on conditions for political decision-making. Cotton analyzes the evolution of nuclear policies in the postwar UK, starting from the normative assumption that the potentially affected local community must have the opportunity to participate in planning and decision-making concerning any given nuclear facility. Cotton distinguishes three different periods: a first long phase, until 1997, when military interests and national security motivated very centralized and secretive decision-making, with practically no public involvement. The second phase began as local opposition to planned nuclear waste facilities reached its culmination point, nuclear expansion was off the political agenda, and Tony Blair's New Labour government entered into power. New Labour policy fostered greater citizen engagement and helped to trigger a deliberative turn in nuclear policymaking. The third phase marked a return to technocracy. The government radically changed its position again, now advocating nuclear as necessary for combating climate change. To facilitate new build, it introduced a planning reform that curtailed the possibilities for local citizen participation. The chapter ends with a brief comparison between the United Kingdom and other European countries regarding considerations to environmental justice in nuclear policymaking.

In chapter 8, "Nuclear Energy in Europe: A Public Technology," Stathis Arapostathis, Robert Bud, and Helmuth Trischler introduce the concept of "public technology" for analyzing the specifics of the history of nuclear power in Europe. This approach draws on the literature on "public science," which stresses the role of public engagement as essential to our understanding of science. In particular, they highlight the vital role of the various and heterogeneous publics in shaping the evolution of nuclear technology, both in East and West. Discourses about modernity, industrialization, and progress shaped the public representations of nuclear energy in its early period on both sides of the Iron

Curtain. Gradually, views on nuclear technology became increasingly polarized between supporters and opponents among the general public in various societies. The authors argue that social movements were active in assessing nuclear technologies and their institutional settings, and at times were able to compel the nuclear regimes to change their strategies, framings, and even technical designs.

Transboundary relations constitute the key concept in chapter 9, "Nuclear Installations at European Borders: Transboundary Collaboration and Conflict," by Arne Kaijser and Jan-Henrik Meyer. The authors take as a point of departure a salient characteristic of the nuclear industry in Europe—namely, the fact that a large number of nuclear facilities are located close to national borders. They discuss the implications of such a location and in particular the role of actors on the other side of the border that are potentially affected. Based primarily on four cases of nuclear installations near borders between Sweden and Denmark; Spain and Portugal; France, West Germany, Italy, and Switzerland; and at the intra-German border, they analyze transboundary relations between the main types of actors—power companies, nuclear authorities, antinuclear movements, and municipal, regional, and national governments. They show that close cooperation between peers across borders often emerged—power companies exchanged electricity, antinuclear movements coordinated demonstrations, and so on. Governments were usually careful not to openly criticize nuclear plants in a neighboring country. However, occasionally, citizen mobilization against foreign plants on the other side of the border persuaded regional and national governments to undertake legal and diplomatic steps and demand the closure of such plants.

In the conclusion, "Future Challenges for Nuclear Energy and Society in Historical Perspective," we first summarize the main lessons from the analysis concerning the interaction between the nuclear sector and society. In particular, we highlight the diversity of the practices, functions, and outcomes of public engagement across time and space. These diverse manifestations of public engagement have had lasting impacts on not only the nuclear sector but also the development of European societies more generally. This experience furthermore confirms and illustrates the findings from earlier research on public engagement in various sectors of policymaking: although vital and indispensable, public engagement as such is no guarantee for smooth and problem-free policymaking. The second part of the conclusion builds on lessons from history to discuss the challenges facing the nuclear sector today and in the future, as the center of gravity in the development of nuclear continues to shift from the Western countries toward Asia. The main potential markets for the international nuclear industry seem to lie in countries such as China, Russia, India, and perhaps even some countries in Africa. However, although European countries seem unlikely to launch extensive new-build programs, they will also need to grapple with an enduring legacy of their past nuclear programs. Upgrading aging reactors to current safety standards, decommissioning old reactors, as well as safely managing nuclear waste will require considerable human, technical, and financial resources. However, attracting prospective students to an industry that may seem to be dying, rather than promising a glorious future as it did in the 1950s and 1960s, is likely to be a formidable challenge. Successfully addressing these challenges will likely require a significant contribution from the humanities and social sciences.

Notes

1. Dwight D. Eisenhower, "Text delivered by the President of the United States before the General Assembly of the United Nations in New York City Tuesday Afternoon, Dec. 8, 1953. Press Release," Eisenhower Library, Abilene, KS, DDE's Papers as President, Speech Series, Box 5, United Nations Speech 12/8/53, https://www.eisenhowerlibrary.gov/sites/default/files/file/atoms_Binder13.pdf. Eisenhower's speech highlighted this issue. Listen to HoNESt podcast, "Episode 3: How Events Shaped the Relationship Between Societies and Nuclear Energy," 2017, https://soundcloud.com/honest-podcast/episode3.
2. Eric Berkers, "The Netherlands Short Country Report (version 2018)," in *History of Nuclear Energy and Society (HoNESt) Consortium Deliverable no. 3.6* (2019), 34–39, https://hdl.handle.net/2454/38269.
3. For instance, in Austria and Denmark: Christian Forstner, "The Failure of Nuclear Energy in Austria: Austria's Nuclear Energy Programmes in Historical Perspective," in *Pathways into and out of Nuclear Power in Western Europe: Austria, Denmark, Federal Republic of Germany, Italy, and Sweden*, ed. Astrid Mignon Kirchhof (Munich: Deutsches Museum, 2019), 36–73; Jan-Henrik Meyer, " 'Atomkraft—Nej tak': How Denmark Did Not Introduce Commercial Nuclear Power Plants," in Kirchhof, *Pathways into and out of Nuclear Power in Western Europe*, 74–123.
4. Stuart W. Leslie and Joris Mercelis, "Expo 1958: Nucleus for a New Europe," in *World's Fairs in the Cold War: Science, Technology, and the Culture of Progress*, ed. Arthur Molella and Scott Gabriel Knowles (Pittsburgh: University of Pittsburgh Press, 2019), 11–26.
5. Mycle Schneider et al., *The World Nuclear Industry Status Report 2019* (Paris and London: Mycle Schneider Consulting, 2019).
6. International Atomic Energy Agency, PRIS database, https://pris.iaea.org/pris/.
7. Natalie Pohl, *Atomprotest am Oberrhein: Die Auseinandersetzung um den Bau von Atomkraftwerken in Baden und im Elsass (1970–1985)* (Stuttgart: Steiner, 2019); Astrid Mignon Kirchhof and Helmuth Trischler, "The History behind West Germany's Nuclear Phase-Out," in Kirchhof, *Pathways into and out of Nuclear Power in Western Europe*, 124–69.
8. Claire Le Renard, "The Superphénix Fast Breeder Nuclear Reactor—Cross-Border

Cooperation and Controversies," *Journal for the History of Environment and Society* 3 (2018): 107–44, https://doi.org/10.1484/J.JHES.5.116796.

9. Arne Kaijser et al., *The Past and the Present of Nuclear Safety Regulation in Europe: Transcript of the Discussions of the Witness Seminar in Barcelona, 16 October 2018* (Mimeo, Barcelona: Pompeu Fabra University, 2019).
10. Ulf Roßegger, *Entsorgung radioaktiver Abfälle in Deutschland in der 16. Legislaturperiode von 2005–2009: Eine Untersuchung der politischen Prozesse im Politikfeld und der Wirkungsmächtigkeit von Akteursinteressen* (Berlin: Berliner Wissenschafts-Verlag, 2019); Achim Brunnengräber and Maria Rosaria Di Nucci, eds., *Conflicts, Participation and Acceptability in Nuclear Waste Governance: An International Comparison Volume III* (Wiesbaden: Springer VS, 2019); Achim Brunnengräber et al., eds., *Challenges of Nuclear Waste Governance: An International Comparison Volume II* (Wiesbaden: Springer VS, 2018); Achim Brunnengräber et al., eds., *Nuclear Waste Governance: An International Comparison Volume I* (Wiesbaden: Springer VS, 2015).
11. Dieter Rucht, *Von Wyhl nach Gorleben: Bürger gegen Atomprogramm und nukleare Entsorgung* (Munich: C.H. Beck, 1980), 79; Martin Schlak, "Wir brauchen mehr Atomkraft: Harvard-Professor Steven Pinker hält die Angst vieler Menschen vor Kernenergie für irrational. Kohlestrom sei viel gefährlicher," *Der Spiegel*, December 13, 2019, https://www.spiegel.de/wissenschaft/harvard-professor-ueber-risiko-mythen-wir-brauchen-mehr-atomkraft-a-00000000-0002-0001-0000-000167507159.
12. Arne Kaijser, "The Referendum that Preserved Nuclear Power and Five Other Critical Events in the History of Nuclear Power in Sweden," in Kirchhof, *Pathways into and out of Nuclear Power in Western Europe*, 238–93; Jonas Anshelm, *Mellan frälsning och domedag: Om kärnkraftens politiska idéhistoria i Sverige 1945–1999* (Stockholm: Symposion, 2000).
13. Such municipalities, whose economic fate was closely tied to the nuclear industry and its permanent legacies, have been characterized as "nuclear communities," whose inhabitants tended to be more positive and welcoming toward nuclear installations. Astrid M. Eckert, *West Germany and the Iron Curtain: Environment, Economy and Culture in the Borderlands* (Oxford: Oxford University Press, 2019), 223. See also Kate Brown, *Plutopia: Nuclear Families, Atomic Cities, and the Great Soviet and American Plutonium Disasters* (Oxford: Oxford University Press, 2013).
14. Eckert, *West Germany and the Iron Curtain*, 201–43; Roßegger, *Entsorgung radioaktiver Abfälle*; Astrid Mignon Kirchhof, "East-West German Transborder Entanglements through the Nuclear Waste Sites in Gorleben and Morsleben," *Journal for the History of Environment and Society* 3 (2018): 145–78, https://doi.org/10.1484/J.JHES.5.116797; "Heftige Proteste: Suche nach Atomendlager: Gorleben raus–aber Ärger bleibt," *Die Zeit*, October 15, 2020, https://www.zeit.de/news/2020-10/15/suche-nach-atomendlager-gorleben-raus-aber-aerger-bleibt.
15. For instance: Lutz Mez, *Der Atomkonflikt: Atomindustrie, Atompolitik und Anti-Atom-Bewegung im internationalen Vergleich* (Berlin: Olle & Wolter, 1979); Lutz Mez and Birger Ollrogge, *Energiediskussion in Europa: Berichte und Dokumente über die Haltung der Regierungen und Parteien in der Europäischen Gemeinschaft zur Kernenergie* (Villingen: Neckar-Verlag, 1979); Herbert Kitschelt, *Kernenergiepolitik: Arena eines gesellschaftlichen Konflikts* (Frankfurt: Campus, 1980); Herbert Kitschelt, *Politik und Energie: Eine vergleichende Untersuchung zur*

Energie-Technologiepolitik in den U.S.A. der Bundesrepublik, Frankreich und Schweden (Frankfurt: Campus, 1983); Herbert Kitschelt, *Der ökologische Diskurs: Zur Wissenssoziologie der Energiekontroverse* (Frankfurt: Campus, 1984).

16. Alain Touraine, *La prophétie anti-nucléaire* (Paris: Seuil, 1980); Claus Offe, "New Social Movements: Challenging the Boundaries of Institutional Politics," *Social Research* 52, no. 4 (1985): 817–68.
17. Wolfgang Rüdig, *Anti-Nuclear Movements: A World Survey of Opposition to Nuclear Energy* (London: Longman, 1990); Helena Flam, ed., *States and Anti-Nuclear Movements* (Edinburgh: Edinburgh University Press, 1994); Dorothy Nelkin and Michael Pollak, *The Atom Besieged: Extraparliamentary Dissent in France and Germany* (Cambridge, MA: MIT Press, 1981). Most recently: Wolfgang C. Müller and Paul W. Thurner, eds., *The Politics of Nuclear Energy in Western Europe* (Oxford: Oxford University Press, 2017).
18. Tony Chafer, "Politics and the Perception of Risk: A Study of the Anti-Nuclear Movements in Britain and France," *West European Politics* 8, no. 1 (1985): 5–32; Herbert P. Kitschelt, "Political Opportunity Structures and Political Protest: Anti-Nuclear Movements in Four Democracies," *British Journal of Political Science* 16, no. 1 (1986): 57–85; Ragnar E. Löfstedt, "Risk Communication: The Barsebäck Nuclear Plant Case," *Energy Policy* 24, no. 8 (1996): 689–96, https://doi.org/10.1016/0301-4215(96)00042-0; Ragnar E. Löfstedt, "Fairness across Borders: The Barsebäck Nuclear Power Plant," *Risk: Health Safety and Environment* 7 (Spring 1996): 135–44.
19. E.g., Herbert Kitschelt, "Four Theories of Public Policy Making and Fast Breeder Reactor Development," *International Organization* 40, no. 1 (1986): 65–104; Mark Elam and Göran Sundqvist, "Meddling in Swedish Success in Nuclear Waste Management," *Environmental Politics* 20, no. 2 (2011): 246–63; Brunnengräber et al., *Nuclear Waste Governance*; Andrew Blowers, *The Legacy of Nuclear Power* (London: Routledge, 2016); Gabrielle Hecht, *Being Nuclear: Africans and the Global Uranium Trade* (Cambridge, MA: MIT Press, 2012).
20. Dick van Lente, ed., *The Nuclear Age in Popular Media: A Transnational History 1945–1965* (Basingstoke: Palgrave, 2012); Michael Schüring, "Advertising the Nuclear Venture: The Rhetorical and Visual Public Relation Strategies of the German Nuclear Industry in the 1970s and 1980s," *History and Technology* 29, no. 4 (2013): 369–98; Christoph Laucht, "Atoms for the People: The Atomic Scientists' Association, the British State and Nuclear Education in the Atom Train Exhibition, 1947–1948," *British Journal for the History of Science* 45, no. 4 (2013): 591–608, https://doi.org/10.1017/S0007087412001070.
21. E.g., on Germany: Janine Gaumer, *Wackersdorf: Atomkraft und Demokratie in der Bundesrepublik 1980–1989* (Munich: Oekom, 2018); Eckert, *West Germany and the Iron Curtain*, 201–43; Ute Hasenöhrl, *Zivilgesellschaft und Protest: Eine Geschichte der Naturschutz- und Umweltbewegung in Bayern 1945–1980* (Göttingen: Vandenhoeck & Ruprecht, 2011), 200–34.
22. Dolores L. Augustine, *Taking on Technocracy: Nuclear Power in Germany, 1945 to the Present* (New York: Berghahn, 2018); Stefania Barca and Ana Delicado, "Anti-Nuclear Mobilisation and Environmentalism in Europe. A View from Portugal," *Environment and History* 22, no. 4 (2016): 497–520.
23. Michael Schüring, "West German Protestants and the Campaign against Nuclear Technology," *Central European History* 45, no. 4 (2012): 744–62, https://doi

.org/10.1017/S0008938912000672; Michael Schüring, *"Bekennen gegen den Atomstaat": Die evangelischen Kirchen in der Bundesrepublik und die Konflikte um die Atomenergie 1970–1990* (Göttingen: Wallstein, 2015); Luise Schramm, *Evangelische Kirche und Anti-AKW-Bewegung. Das Beispiel der Hamburger Initiative kirchlicher Mitarbeiter und Gewaltfreie Aktion im Konflikt um das AKW Brokdorf 1976–1981* (Göttingen: V&R, 2018).

24. Pohl, *Atomprotest am Oberrhein*; Stephen Milder, *Greening Democracy: The Anti-Nuclear Movement and Political Environmentalism in West Germany and Beyond, 1968–1983* (Cambridge: Cambridge University Press, 2017); Andrew Tompkins, *Better Active than Radioactive! Antinuclear Protests in 1970s France and West Germany* (Oxford: Oxford University Press, 2016); Astrid Mignon Kirchhof and Jan-Henrik Meyer, "Global Protest Against Nuclear Power: Transfer and Transnational Exchange in the 1970s and 1980s," *Historical Social Research* 39, no. 1 (2014): 163–273, https://doi.org/10.12759/hsr.39.2014.1.165-190 (and the contributions to this special issue).
25. Bernd A. Rusinek, *Das Forschungszentrum: Eine Geschichte der KFA Jülich von ihrer Gründung bis 1980* (Frankfurt: Campus, 1996); Henry Nielsen et al., *Til samfundets tarv—Forskningscenter Risøs historie* (Risø: Forskningscenter Risø, 1998).
26. Christian Forstner, *Kernphysik, Forschungsreaktoren und Atomenergie: Transnationale Wissensströme und das Scheitern einer Innovation in Österreich* (Wiesbaden: Springer Spektrum, 2019); Davide Orsini, "Experts at Risk: Military Secrets and Italian Radioecology around the US Naval Installations on La Maddalena," in *Nuclear Portraits: Communities, the Environment, and Public Policy*, ed. Laurel Sefton MacDowell (Toronto: University of Toronto Press, 2017), 94–120; Jacob Darwin Hamblin, *Poison in the Well: Radioactive Waste in the Oceans at the Dawn of the Nuclear Age* (New Brunswick, NJ: Rutgers University Press, 2008); Christoph Laucht, "Scientists, the Public, the State and the Debate over the Environmental and Human Health Effects of Nuclear Testing in Britain, 1950–1958," *Historical Journal* 59, no. 1 (2016): 221–51, https://doi.org/10.1017/S0018246X15000096.
27. E.g., Eckert, *West Germany and the Iron Curtain*, 201–43; Kirchhof, "East-West German Transborder Entanglements"; Maria del Mar Rubio-Varas, António Carvalho, and Joseba de la Torre, "Siting (and Mining) at the Border: Spain-Portugal Nuclear Transboundary Issues," *Journal for the History of Environment and Society* 3 (2018): 33–69, https://doi.org/https://doi.org/10.1484/J.JHES.5.116794; James W. Feldman, "Permanence, Justice and Nuclear Waste at Prairie Island," in MacDowell, *Nuclear Portraits*, 190–216; Michael Greenberg et al., "Nuclear Waste Management and Nuclear Power: A Tale of Two Essential US Department of Energy Sites in Idaho and New Mexico," in MacDowell, *Nuclear Portraits*, 217–37; Detlev Möller, *Endlagerung radioaktiver Abfälle in der Bundesrepublik Deutschland: Administrativ-politische Entscheidungsprozesse zwischen Wirtschaftlichkeit und Sicherheit, zwischen nationaler und internationaler Lösung* (Frankfurt: Peter Lang, 2009); Anselm Tiggemann, *Die "Achillesferse" der Kernenergie in der Bundesrepublik Deutschland: zur Kernenergiekontroverse und Geschichte der nuklearen Entsorgung von den Anfängen bis Gorleben, 1955 bis 1985* (Lauf an der Pegnitz: Europaforum Verlag, 2004).
28. E.g., Elisabetta Bini and Igor Londero, eds., *Nuclear Italy: An International History of Italian Nuclear Policies during the Cold War* (Trieste: Edizioni Università di Trieste,

2017); John Krige, *Sharing Knowledge, Shaping Europe: US Technological Collaboration and Nonproliferation* (Cambridge, MA: MIT Press, 2016); John Krige and Jessica Wang, "Nation, Knowledge, and Imagined Futures: Science, Technology, and Nation-Building, Post-1945," *History and Technology* 31, no. 3 (2015): 171–79, https://doi.org/10.1080/07341512.2015.1126022; Jacob Darwin Hamblin, "Quickening Nature's Pulse: Atomic Agriculture at the International Atomic Energy Agency," *Dynamis* 35, no. 2 (2015): 389–408; Jacob Darwin Hamblin, "Gods and Devils in the Details: Marine Pollution, Radioactive Waste, and an Environmental Regime circa 1972," *Diplomatic History* 32, no. 4 (2008): 539–60, https://doi.org/10.1111/j.1467-7709.2008.00712.x.

29. Elisabeth Röhrlich, *Dual Mandate: A History of the IAEA* (Baltimore: Johns Hopkins University Press, forthcoming 2021); John Krige, "Euratom and the IAEA: The Problem of Self-Inspection," *Cold War History* 15, no. 3 (2015): 341–52, https://doi.org/10.1080/14682745.2014.999046; Ilina Cenevska, *The European Atomic Energy Community in the European Union Context: The "Outsider" Within* (Leiden: Nijhoff, 2016).
30. Frank Bösch, "Taming Nuclear Power: The Accident near Harrisburg and the Change in West German and International Nuclear Policy in the 1970s and early 1980s," *German History* 35, no. 1 (2017): 71–95, https://doi.org/10.1093/gerhis/ghw143; Natasha Zaretsky, *Radiation Nation: Three Mile Island and the Political Transformation of the 1970s* (New York: Columbia University Press, 2018).
31. Kate Brown, *Manual for Survival: A Chernobyl Guide to the Future* (New York: W.W. Norton, 2019).
32. Melanie Arndt, *Tschernobylkinder: Die transnationale Geschichte einer nuklearen Katastrophe* (Göttingen: Vandenhoeck & Ruprecht, 2020).
33. Melanie Arndt, *Tschernobyl: Auswirkungen des Reaktorunfalls auf die Bundesrepublik und die DDR* (Erfurt: Landeszentrale für politische Bildung Thüringen, 2011); Melanie Arndt, ed., *Politik und Gesellschaft nach Tschernobyl: (Ost-)Europäische Perspektiven* (Berlin: Chr. Links, 2016).
34. Susanne Bauer, Karena Kalmbach, and Tatiana Kasperski, "From Pripyat to Paris, from Grassroots Memories to Globalized Knowledge Production: The Politics of Chernobyl Fallout," in MacDowell, *Nuclear Portraits*, 149–89; Karena Kalmbach, *Tschernobyl und Frankreich: Die Debatte um die Auswirkungen des Reaktorunfalls im Kontext der französischen Atompolitik und Elitenkultur* (Frankfurt: Peter Lang, 2011); Melanie Arndt, "Memories, Commemorations, and Representations of Chernobyl," *Anthropology of East Europe Review* 30, no. 1 (2012): 1–140.
35. Karena Kalmbach, "Revisiting the Nuclear Age: State of the Art Research in Nuclear History," *Neue Politische Literatur* 62, no. 1 (2017): 49–69; Karena Kalmbach, *The Meanings of a Disaster: Chernobyl and Its Afterlives in Britain and France* (New York: Berghahn, 2020).
36. See chapter 8 in this volume and Helmuth Trischler and Robert Bud, "Public Technology: Nuclear Energy in Europe," *History and Technology* 34, no. 3–4 (2018): 187–212, https://doi.org/10.1080/07341512.2018.1570674.
37. Ute Hasenöhrl and Jan-Henrik Meyer, "The Energy Challenge in Historical Perspective," *Technology and Culture* 61, no. 1 (2020): 295–306.
38. Ute Hasenöhrl, "Just a Matter of Habituation? The Contentious Perception of (Post)energy Landscapes in Germany, 1945–2016," *Environment, Place, Space* 10, no. 1 (2018): 63–88.

39. Robin Cowan, "Nuclear Power Reactors: A Study in Technological Lock-in," *Journal of Economic History* 50, no. 3 (1990): 560–61, doi:10.2307/2122817.
40. On issue of culture, see Sheila Jasanoff and Sang-Hyun Kim, eds., *Dreamscapes of Modernity: Sociotechnical Imaginaries and the Fabrication of Power* (Chicago: University of Chicago Press, 2015); Lente, *The Nuclear Age in Popular Media*; Gabrielle Hecht, "Nuclear Ontologies," *Constellations* 13, no. 3 (2006): 320–31, https://doi.org/10.1111/j.1467-8675.2006.00404.x.
41. Gabrielle Hecht, *The Radiance of France: Nuclear Power and National Identity after World War II* (Cambridge, MA: MIT Press, 1998); Jaume Sastre-Juan and Jaume Valentines-Álvarez, "Fun and Fear: The Banalization of Nuclear Technologies through Display" *Centaurus* 61, no. 1–2 (2019): 2–13 (and the contributions to this special issue).
42. E.g., Anna Storm, *Post-Industrial Landscape Scars* (Basingstoke: Palgrave, 2014); Anna Storm, "Kärnkraftverk som minnesplatser: Barsebäck," *Nordisk museologi*, no. 2 (2010): 96–102; Arndt, "Memories, Commemorations, and Representations of Chernobyl."
43. Forstner, "The Failure of Nuclear Energy in Austria"; Kirchhof and Trischler, "The History behind West Germany's Nuclear Phase-Out"; Kaijser, "The Referendum that Preserved Nuclear Power"; Meyer, " 'Atomkraft—Nej tak.' " See for Greece: Maria Rentetzi, "Gender, Science and Politics: Queen Frederika and Nuclear Research in Post-War Greece," *Centaurus* 51, no. 1 (2009): 63–87.
44. Brown, *Plutopia*.
45. Jens Ivo Engels, "Gender Roles and German Anti-Nuclear Protest. The Women of Wyhl," in *The Modern Demon: Pollution in Urban and Industrial European Societies*, eds. Christoph Bernhardt and Geneviève Massard-Guilbaud (Clermont-Ferrand: Presses Univ. Blaise Pascal, 2002), 407–23; Astrid Mignon Kirchhof, "Frauen in der Antiatomkraftbewegung. Das Beispiel der Mütter gegen Atomkraft," *Ariadne* 64 (2013): 48–57; Sarah E. Summers, " 'Thinking Green!' (and Feminist): Female Activism and the Greens from Wyhl to Bonn," *German Politics and Society* 33, no. 4 (2015): 40–52, https://doi.org/10.3167/gps.2015.330404; Pohl, *Atomprotest am Oberrhein*, 234–41.
46. See for instance the international networks Women in Nuclear (https://www.win-global.org) and Mothers for Nuclear (https://www.mothersfornuclear.org).
47. This project received funding from the Euratom research and training program 2014–2018 under grant agreement no. 662268. Despite this, the views expressed here are entirely those of the authors, who are also responsible for any errors and/or omissions.
48. All public deliverables can be found at http://www.HoNESt2020.eu.
49. This section draws substantially on Ioan Charnley-Parry, John Whitton, and Mar Rubio-Varas, "Final Project Summary Report," in *History of Nuclear Energy and Society (HoNESt) Consortium Deliverable no. 3.7* (2019).
50. John Whitton and Mar Rubio-Varas, "Social Science Research Question and Guiding Framework," in *History of Nuclear Energy and Society (HoNESt) Consortium Deliverable no. 3.1* (2016).
51. The complete collection of short country reports (SCRs) produced by the History of Nuclear Energy and Society (HoNESt) consortium is freely available online at https://hdl.handle.net/2454/38269. The preferred citation is to individual SCRs,

acknowledging the authorship to the responsible partners but recognizing the nature of the document as an HoNESt consortium deliverable output. In addition, five of the SCRs were rewritten and edited in Astrid Mignon Kirchhof, ed., *Pathways into and out of Nuclear Power in Western Europe: Austria, Denmark, Federal Republic of Germany, Italy, and Sweden* (Munich: Deutsches Museum Verlag, 2019). Available at: https://www.deutsches-museum.de/fileadmin/Content/010_DM/060_Verlag/Studies-4-online-2.pdf.

PART 1

Context

CHAPTER 1

Nuclear-Society Relations from the Dawn of the Nuclear Age

Paul Josephson, Jan-Henrik Meyer, and Arne Kaijser

Introduction

The relationship between industry, government, and the public has evolved since the dawn of the nuclear age and the Atoms for Peace programs in the 1950s. Throughout most European countries, public support for nuclear power has remained relatively constant. Initially, however, such support was based on limited public knowledge of nuclear technologies, restricted by military secrecy or a paucity of information. Specialists and representatives of industry, utilities, and other owners, as well as those in incipient and usually non-independent regulatory bodies, largely shaped public knowledge of the nuclear age and promoted its peaceful futuristic applications. If worries about safety or other risks existed, they mostly concerned nuclear weapons and the spread of fallout from bomb tests.[1]

The late 1960s and early 1970s witnessed increasing public participation and shifting citizen concerns over siting, safety, cost, environmental impacts, and the perceived authoritarianism of nuclear power.[2] As part of rising environmentalism, antinuclear protest broke out by the early 1970s, directed against the technocratic, top-down decision-making by experts that was standard practice well into the 1980s in practically all countries, regardless of their political systems. Societies with well-developed civic cultures—practicing and recognizing public participation—were much more likely to see protest as nuclear power expanded across the globe.[3] On occasion, protests were violent (e.g., in Germany, France, Spain) and led to intensive debates about the nature and legitimacy of "resistance."[4] Nuclear governance in authoritarian regimes—in Greece, Portugal, and Spain in the 1950s and 1960s and in the socialist world until 1989—essentially involved only specialists, officials, and operators, while free and open public debate, let alone protest, were not tolerated.

Major accidents had a significant negative impact on public perceptions of nuclear power and gave great impetus to further protest. In several countries they led to a scaling back of plans to expand nuclear power, although cost overruns in the industry had already slowed development. The Three Mile Island and Chernobyl disasters dampened public support in the US and many places in Europe, though responses varied.[5] After Fukushima, Chancellor Angela Merkel announced and implemented plans to shut all nuclear power stations in Germany by 2022.[6]

This brief overview suggests that a historical perspective on nuclear energy and society is necessary to better understand societal perceptions of and societal engagement with nuclear energy in Europe. Nuclear-societal relations have been in flux since the dawn of the nuclear age, from the first efforts of countries around the world to commercialize nuclear energy and employ other nuclear applications in medicine, agriculture, and industry. Furthermore, there is great diversity of attitudes and experiences in the different European countries. In countries with Cold War military ambitions, military interests greatly influenced how the nascent nuclear sector evaluated the costs, scope, and scale of programs and assessed secrecy requirements. Indeed, it has been suggested that this initial predominance of military concerns put the sector on a path of secrecy, unaccountability, and recurrent cost overruns. This lasting legacy of the technology's original attachment to the military may also explain some of the subsequent problems in nuclear-societal relations.[7]

This chapter analyses evolving public views toward—and growing conflict about—nuclear power in Europe, with occasional reference to the US experience. It is based on a comparative analysis of twenty short country reports (SCRs), produced in the History of Nuclear Energy and Society (HoNESt) research project, which analyzed the history of relations between the nuclear sector and society since the 1940s.[8] Drawing on primary and secondary sources, each SCR included a general history of nuclear power in the given country; a brief discussion of institutional actors from government, industry, and the public; and five case studies characterizing the industry-society relations at different points in time in greater detail. An annex provides data and bibliographical references.

The unique empirical basis provided by the SCRs is intended to advance the state of art in nuclear history, in particular our understanding of nuclear industry-society relations. It allows a fresh, much broader, and much more thorough comparative view of the history of what we call engaging the atom. To date, studies in sociology, political science, and anthropology have largely focused on protest movements, whereas industry- and country-specific analysis has dominated historical research.[9] Furthermore, countries with small programs, as well as those whose programs failed to materialize (e.g., Denmark, Greece, and Portugal), have frequently been overlooked in the literature.[10] The research conducted within HoNESt draws attention to a range of contexts, explanatory factors, and perspectives beyond mere protest or its absence.[11]

Nuclear power is at once national, international, and in many ways transnational.[12] Ensuring security of energy supply has traditionally constituted a key element of national security.[13] Nuclear energy has therefore been fundamentally national in its organization, underlying justifications, shape of research and of development programs, reactor design, regulation and legal issues, and relationship with the public. As the SCRs make clear, a wide range of actors made this history: government ministries and bureaucracies; scientific organizations, universities, laboratories, and professional societies; producers, operators, utilities, and construction firms; citizens' groups; and individual citizens themselves. Yet in spite of the large number of organizations, there has been a significant overlap between government and industry. Until the 1990s, most utilities in Europe were publicly owned, while almost all governments funded research and sought to promote nuclear-technology exports.[14] Until the mid-1970s, on both sides of the Atlantic, the boundaries between the promotional and regulatory functions of government bodies responsible for nuclear power remained blurred.

Public reactions to nuclear power and industry responses varied widely between nation-states and in conjunction with national discursive and institutional contexts. National political systems varied significantly in terms of the opportunities they provided for critics of nuclear power to make their voices heard, be taken seriously, and have an impact.[15] In many European countries, most mainstream political parties remained staunchly pronuclear until the 1980s, that is, until the arrival of Green parties in France and West Germany. By contrast, in Denmark and Sweden, protesters found support in parliament already in the 1970s. In some countries, political leaders in the representative systems in Western Europe felt compelled to force a decision via direct democracy in a referendum. In Austria and Italy, this put an end to the pursuit of nuclear power, while in Sweden, the result of the referendum effectively legitimated the expansion of nuclear power in the short term.[16]

However, nuclear power also has crucial international and transnational elements. The so-called fuel cycle—including mining, milling, conversion, enrichment, fuel fabrication, reactor operation, reprocessing, interim storage and waste disposal—is highly transnational in character.[17] Moreover, the generation and transfer of technical know-how is highly international and has been facilitated by international organizations (see chapter 4). It has proven to be transnational regarding accidents, and fear of such accidents has created "transboundary issues,"[18] notably where the construction of a nuclear power station close to a national boundary worried people and governments on the other side of the border (see chapter 9).[19]

Antinuclear protests as a kind of public engagement with nuclear power from below have been transnational in character. But even though critical views about nuclear technology traveled across borders and protesters felt they were part of one transnational movement, effective cross-border antinuclear protests often proved difficult to organize.[20] For example, nuclear accidents with transnational impacts, like Chernobyl, hit home very differently. In some countries, publics responded with a kind of acquiescence (in the United Kingdom), with quiet appreciation of the work of experts (in the socialist world), a rapid decline of public trust in government believed to conceal the true impacts of the accident (in France), large demonstrations (in Sweden and Denmark), and with violent protests and/or civil disobedience (in the Federal Republic of Germany and France).

This chapter presents and analyzes how public attitudes evolved from the dawn of the nuclear age, from largely enthusiastic embrace to greater concern owing to the intractable problem of high-level radioactive waste, and hence environmental uncertainties. Organized in a roughly chronological fashion, the narrative starts out with the early years of nuclear research and Atoms for Peace. The early 1970s saw the emergence of the first large-scale nuclear programs and the rise of protest involving violence but also attempts to solve the issue by direct democracy. We explore the role that transitions to democracy in Southern and Eastern Europe played in shaping the relations between nuclear energy and society, as well as the impact of nuclear accidents on public debate and civil society action. We close with considerations concerning nuclear waste—an issue that raises increasing concern and is set to continue doing so long into the twenty-first century—and a brief discussion on possible nuclear futures on the European continent.

A few limitations deserve mentioning. The first concerns the *geographical scope* and the European focus. It is impossible in a chapter of this length to systematically discuss all twenty countries considered by the HoNESt research project. Instead, we identify themes common to many countries and evoke several crucial events in specific settings. For the purposes of this chapter—and the book more generally—"nuclear Europe" stretches from the Atlantic Ocean to the foothills of the Ural Mountains, encompassing 135, or one-third, of the world's nuclear power stations, many dating back to the earliest years of commercialization. However, a number of nations that fall within this geography were not covered in HoNESt and are thus not considered here. The focus on the entire European continent, broadly defined, enables us to overcome the historiographical perpetuation of the East-West divide, which has prevailed in the nuclear sphere as well as in many other spheres of Cold War history.[21] We thus include many of the former socialist nations of Central and Eastern Europe, and also former republics of the Soviet Union, recognizing the specific nature of interaction between public and industry in formerly closed political systems. In these countries, public input in nuclear policy was limited, and civic culture—the notion that citizens ought to be active in civic duties—was generally either not recognized or suppressed until the fall of the Berlin Wall and the collapse of the Soviet Union.[22]

A second limitation concerns several key concepts. While the nuclear sector seems like a relatively tangible object, notions of the public, civil society, and society are much more difficult to pin down. It is hard to adequately classify groups and actors, given that different kinds of actors behaved differently in different countries and at different points in time. For instance, students violently opposed nuclear power in Germany but were enthusiastic supporters in Bulgaria and the Czech Republic during the same time period. Nuclear issues seemed to also have cut across traditional political divides. While leftist students became increasingly antinuclear, working-class labor unionists staunchly defended nuclear in Finland, France, Sweden, and the UK. However, not only young people protested—hence, the divide was not only generational. It is perhaps tautological to suggest that industry representatives and nuclear engineers were pronuclear—but they did not always agree on the desirable scope of nuclear power programs or on the choices between different reactor technologies, for instance. Moreover, some of the most knowledgeable and influential critics of nuclear power emerged from their ranks.[23] We consider as "the public" citizens who do not work in industry, trade associations, international organizations tethered to government agencies in the area of nuclear power, or financial institutions with a stake in nuclear power.

The Cold War and Atoms for Peace

If the profusion of nuclear military research institutions preceded the rise of nuclear energy programs, then commercialization proceeded rapidly after US President Dwight D. Eisenhower's "Atoms for Peace" speech at the United Nations in 1953. Governments offered financial and regulatory support and set up promotional and supervisory institutions. Nation-states saw economic potential and political legitimacy in the atom, engaging in "technological nationalism."[24] At first, many governments had difficulty convincing manufacturers and utilities to adopt and develop reactors of unproven efficiency and uncertain operating parameters, involving significant capital costs. Direct and indirect government support was frequently in place to mitigate utilities' risks. As explored in chapter 2, it was difficult to obtain a precise understanding of nuclear economics at the early stages of the commercialization of nuclear power. Many supporters, to be sure, contended that nuclear power would be competitive with coal, gas, and oil and would become more efficient and less expensive as operating experience accumulated and technological learning took place. Most often, the public did not have direct access to technical and other reports, in part because nuclear technologies were usually hatched in military secrecy. Across the world, the public interface with nuclear—in film, magazines and journals, exhibitions, and displays—involved almost utopian expectations of the new technology as a source of inexpensive and safe energy and as a replacement for rapidly diminishing coal and oil reserves (see chapter 8).[25]

The context of the Cold War was prominent in the promotion of nuclear power. The Soviet and American programs for the peaceful atom—the spread of nuclear technology, the choice of reactors, and so on—reflected their geopolitical concerns.[26] Enthusiasm for Atoms for Peace led essentially all of the nations of Europe to embark on programs to establish at the very least nuclear physics programs and in many cases also to build research reactors. However, these programs did not mean that a nation would rapidly commercialize nuclear power and there was no common European response to Atoms for Peace. Domestic considerations such as national imaginaries and political constraints were crucial in determining how far a nation pursued nuclear power.

In Western Europe, the US supported nuclear programs and also encouraged the establishment of the European Atomic Energy Community (Euratom) and European Nuclear Energy Agency (ENEA) in 1957 (see chapter 4). The US promoted nuclear programs not least in countries that played an important role in the Cold War. One was West Germany, which fully embraced Atoms for Peace and built its first research reactor in 1957. But—as in many other countries—commercial nuclear power did not initially attract utility interest because the technology remained unproven and costly (see chapter 2).[27] These problems overcome, concerns about nuclear power were publicly expressed in West Germany for the first time in the 1950s and 1960s and focused on costs and uncertainties associated with nuclear waste disposal.[28]

Although at the time one of the less developed among the Western European countries, and an authoritarian regime, Spain pioneered in nuclear power and connected its first nuclear reactor to the grid in 1968. The Cold War and internal demands to show "progress" to the citizenry underlay the government's determination to pursue nuclear power. The US supported Spain's atomic program, more worried about the risk of communism's spread to Western Europe than about supporting a fascist regime. Nuclear research commenced in 1948. By the mid-1970s, Spain had become the largest nuclear client of the US, and it advanced an ambitious program for up to forty reactors with American (Westinghouse, General Electric), West German (Kraftwerk Union), and French (EDF) technology.[29]

Another politically important country in the Cold War was Greece. In the 1950s, Greece's leaders tried to fuse military thinking, nuclear hopes, and national independence through a nuclear energy program. Like most of those European countries to the west of the Iron Curtain, Greece established an atomic energy commission and a nuclear research center (Demokritos) in the early 1950s, also taking advantage of US Cold War support. If at first the scientific community supported nuclear efforts, other specialists pointed out that cost, siting, and other issues were problematic. But during the junta (1967–1974) the ruling military pushed for nuclear energy, entertaining a discourse of political autonomy, sovereignty, and energy autarky.

Straddling the blocs, Austria, too, embraced Atoms for Peace. In March 1955, when Austria had its national sovereignty restored on the condition of a commitment to eternal neutrality, the government and scientists decided to bring a research reactor on line with US help. Industry and government created the Austrian Research Center for the Peaceful Use of Nuclear Energy in Seibersdorf, and three research reactors were built between 1959 and 1965. During the 1960s the publicly owned electricity companies became increasingly interested in nuclear power and started to incorporate it in their future planning. In 1971 a decision was made to build a nuclear power plant in Zwentendorf.[30]

In Eastern and Central Europe, the USSR pursued bilateral and international Warsaw Pact cooperation to secure the region economically, militarily, and politically, including through cooperation on the peaceful atom. Within the Soviet-sponsored Council for Mutual Economic Assistance (known as CMEA or Comecon), the socialist states agreed to a permanent commission on nuclear energy in 1955 (see chapter 4). Under Moscow's control, the commission created an opportunity for nuclear scientists and research institutions in Bulgaria, Poland, Czechoslovakia, Romania, Yugoslavia, the German Democratic Republic (GDR), and Hungary to access recent Soviet research results and work in such Soviet research centers as the Joint Institute for Nuclear Research.[31] Soviet reactor design reflected engineering considerations without the benefit of public scrutiny. This may explain why the first-generation Soviet VVER (a pressurized water reactor, or PWR) was built without containment vessels, and the Chernobyl-type RBMK (channel graphite) reactor was in many senses like any other factory building, lacking safety structures specific to nuclear. It also had a positive void coefficient that made it unstable at low power.

The experience of the GDR reveals how the socialist planning system handicapped the effort to commercialize nuclear power. In response to the first Geneva conference on the peaceful uses of atomic energy, in 1955 East Germany created an Office for Nuclear Research and Technology with plans to build up to twenty nuclear power plants. This forecast was an appeal to modernity and a wishful dream of a GDR-style Atoms for Peace. However, the industry fell far short of its targets. Every few years, a new forecast set out a target of a smaller number of reactors. Serial production of pressure vessels from the Atommash plant in Volgodonsk, Russia, never materialized, and defective equipment was delivered to the GDR. Five PWRs were eventually brought into operation by the late 1970s. After the unification of Germany, all of these were deemed unsafe and shut down.

The Rise of Public Protest

Antinuclear protest emerged in the late 1960s, first in the United States and then in Western Europe, reaching its peak in the 1970s (see also chapter 3).[32] Protest was connected, first, with environmental and other movements, and with subsequent governmental reforms that permitted greater access to the policy process in the scientific, technical, and environmental domains. A number of nations created environmental protection agencies or ministries, which increasingly included a nuclear regulatory function, such as in Denmark. Second, protest emerged because in Europe, it was only in the early 1970s that large-scale nuclear power programs turned into concrete local realities that saw the construction of numerous nuclear power plants. Nuclear power was now a reality and no longer only a vision.[33] Opposition reflected a variety of concerns: safety, siting, cost, violation of nature, and the perceived authoritarianism of nuclear power. As reactors became larger in size and more complex, safety became an issue of great concern as the potential risk of catastrophic accidents grew. While promoters of nuclear power had operated with the understanding that they would find solutions to safety problems as they arose, increasing numbers of experts and citizens worried about the seemingly *ad hoc* ways in which designers and engineers found solutions to potential catastrophic accidents.[34]

Safety and siting questions went hand in hand. Lacking the requisite specialists in a variety of fields—notably in hydrology and seismology—to carry out complete studies before awarding licenses, early atomic energy commissions relied largely on industry recommendations. For example, in Greece, the Karystos nuclear power plant was planned for the Gulf of Corinth, but in 1981, before construction commenced, a major earthquake hit the area. It boosted public opposition against the plant, and the socialist government decided to abandon nuclear power. The government instead turned to domestic lignite, imported natural gas, and became indirectly nuclear via electricity imports from the Bulgarian Kozloduy plant. This latter plant had in fact experienced a fairly severe earthquake in 1977 but luckily without suffering major damage to the reactors.

In 1970, the Swedish government approved the building of a nuclear power plant in Barsebäck, only twenty kilometers from the Danish capital, Copenhagen, and also very close to three Swedish cities. Danish authorities had been informed about the plans and had no objections. Denmark at the time had its own ambitious nuclear program, partly due to its long history of nuclear physics research. However, it had not yet taken any definite steps toward commercialization. In 1974 an emerging antinuclear movement led by the Organization for Nuclear Information (Organisationen til Oplysning om Atomkraft, OOA), started questioning the country's own nuclear program as well as the building of the Barsebäck plant. After the first reactor at Barsebäck was commissioned in 1975, the closing of this plant became OOA's major goal. In the following year a big demonstration against the plant was organized together with Swedish antinuclear organizations, and similar demonstrations were organized almost annually in the following years. After the Three Mile Island accident, more than 300,000 signatures were collected in Denmark calling upon the prime minister to demand the Swedish government close the Barsebäck plant. In 1985, the Danish government excluded nuclear power from future energy planning, and from then on it started demanding that the Swedish government close down the Barsebäck plant. Eventually, in 1999 and 2005 the two reactors were indeed closed.[35]

Protest against nuclear power plants in France started in the early 1970s, a few years earlier than in West Germany. In fact, French protest tactics crossed the Upper Rhine and provided an example for the incipient West German antinuclear movement.[36] French protests frequently involved violent clashes, culminating in the deadly conflict in 1977 at the Superphénix fast breeder site in Creys-Malville.[37] In West Germany, the planning and building of nuclear power plants as well as radioactive waste disposal and reprocessing facilities in the federal states of Baden-Württemberg (Wyhl), Schleswig-Holstein (Brokdorf),[38]

Lower Saxony (Gorleben),[39] North Rhine-Westphalia (Kalkar),[40] and Bavaria (Wackersdorf) provoked massive and long-drawn protests throughout the 1970s and 1980s. Demonstrations at the village of Wyhl are widely recognized as the starting point of West Germany's antinuclear movement.[41] In 1985, the Deutsche Gesellschaft zur Wiederaufarbeitung von Kernbrennstoffen mbH (DWK) decided to build a reprocessing plant—originally planned for Gorleben—in Wackersdorf in northern Bavaria, not far from the Czech border. When DWK started clearing woodland, 30,000 people occupied the building site and erected a tent village. After Chernobyl, a violent dispute between police and antinuclear activists reached its peak. Eventually, in 1988, the protest—and ensuing uncertainty about the project's political and economic viability—convinced industry managers to abandon the plans and instead to cooperate with French industrial partners to arrange for reprocessing in La Hague.[42]

What brought about the massive and violent protests? Many German antinuclear activists associated nuclear energy with authoritarianism, seeing in it an effective continuation of Germany's National Socialist past. They argued that the nuclear lobby lacked transparency and honesty and thus trustworthiness.[43] Many critics saw industry as traitors and themselves as victims. They believed that the lack of truthfulness of the state and nuclear industry justified their actions. Their "faith" motivated them to stand up to authority as many of them believed that their parents had failed to stand up against Nazism. The police's brutal response and the intention to prosecute protestors further increased suspicion of authorities and utilities.[44]

National Referenda

One of the ways to resolve the political conflict over nuclear power has been to let the public vote on its use. National referenda enabled the public to participate in decisions over nuclear power, although they left policymakers substantial leeway to phrase the questions and the alternatives. Wherever they were held, they produced close results. In Austria, Chancellor Bruno Kreisky called a referendum in mid-1978, and 50.47 percent of those turning out voted against nuclear power, killing the essentially complete Zwentendorf plant before it went critical and leading Parliament to enact the *Atomsperrgesetz* in December 1978, the law banning the use of nuclear power. Kreisky's decision to put the issue to a public vote was tactically informed by a lesson he had drawn from the September 1976 elections in Sweden. The Swedish Social Democrats under Olof Palme, with whom Kreisky collaborated closely on numerous international issues,[45] were held to have lost the election partly because of their nuclear policy. Following the Three Mile Island accident in 1979 and several attempts to withdraw the *Atomsperrgesetz*, Austria ended its plans to pursue nuclear power.[46] Subsequently, Austria has been a vocal critic of nuclear power plants near its borders in Slovakia and the Czech Republic (see chapter 9).[47]

By the early 1970s, when an antinuclear movement emerged in Sweden, it was facing what can be described as a well-developed nuclear-industrial complex and a number of commercial nuclear plants already under construction. Nevertheless, two of the five political parties in the Parliament took an antinuclear stance around 1975, and the coalition government that took office after the 1976 election was headed by Center Party leader Thorbjörn Fälldin, who opposed nuclear power. In 1978, a broad but internally diverse umbrella organization, Folkkampanjen mot Atomkraft (People's Campaign against Atomic Power), demanded a referendum to halt nuclear power, but this demand was rejected by a majority in Parliament. After the Three Mile Island accident in the US, Parliament reversed its position. The resulting ballot, however, included three options, rather than two clear alternatives. Hundreds of thousands of activists were engaged in the referendum process. The people's campaign lost, however, and Parliament decided to continue nuclear expansion, albeit agreeing to phase out all nuclear power by 2010. At this time, Sweden generated the most nuclear power per capita in the world, and it is still one of the leading nuclear countries in this respect. The phaseout by 2010 did not occur. The two reactors at Barsebäck, however, were closed in 1999 and 2005, and four nuclear plants have been shut down for economic reasons.

The third country where a referendum was organized was Italy. There, the antinuclear movement had grown in the 1970s and 1980s, and the Chernobyl accident had led to enormous demonstrations. By July 1986 a campaign to gather signatures demanding a referendum on the future of nuclear power had commenced, and it eventually led to a referendum in November 1987. The antinuclear side won the election, and Parliament decided to phase out nuclear power. A second referendum took place after Fukushima and confirmed Italy's antinuclear stance.

Transition to Democracy and Antinuclear Protest

The nature of the political system has a significant impact on nuclear-societal interactions. As political scientists studying so-called political opportunity structures have highlighted since the 1980s, an "open" political system—that is, a system of government that is receptive to societal concerns—obviously provides greater opportunities for public concerns to be heard in the discussion on nuclear policy. By contrast, a more closed system, in which political elites are not willing to listen to, take seriously, or even tolerate popular critique—coupled with a weak civil society—reduces the likelihood and effectiveness of protest.[48] This goes a long way toward explaining why public concerns and opposition to nuclear power in Spain, Portugal, East Central Europe, and the former USSR became visible only after regime change.

In Spain in the 1960s and 1960s, supporters portrayed nuclear power as the only plausible alternative allowing the nation to meet its electricity needs, while citizens had little if any knowledge of the technology. Early nuclear projects barely faced opposition, in large measure because of police controls, censorship, and a poorly developed civil society.[49] Yet even before the end of Francoism, a number of informal groups submitted formal complaints against most of the twenty sites being considered for nuclear projects, and several Francoist mayors, provincial governments, religious associations, agricultural unions, and others were also opposed to local siting. After the fall of the dictatorship in the mid-1970s, antinuclear groups opposing the construction of nuclear power plants emerged in several parts of Spain. In the Basque Country, the nuclear industry became the target of the Basque separatist terrorist movement, ETA, from the mid-1970s. ETA perceived nuclear power as an imposition from outside and a potential risk to cultural values and the environment. Entangled with more complex issues and after a number of bombings, kidnappings, and assassinations of engineers in the 1980s, the nuclear projects in the Basque Country were canceled despite being close to completion.

In Portugal, too, greater public access to information and the policy process was possible only after the April 1974 democratic revolution that ended the dictatorship, allowed freedom of the press, and ended police surveillance and control of civic organizations. In March 1976, an attempt to site a nuclear power plant at Ferrel, along the central coast of Portugal, met fierce public opposition. The utility later abandoned the plant. After Chernobyl, the government shelved its nuclear power program, and subsequent attempts to put it back on the agenda have failed.[50]

To varying degrees, the Soviet bloc countries were closed societies, with highly centralized bureaucracies in which planners' preferences for price, mix of product, and directions of economic activities prevailed. These planners' preferences dictated many policy decisions, while party officials used political, cultural, and other institutions to shape and control civil society. Dissent existed in a variety of such forums as underground press, so-called *samizdat* or self-publishing, but focused more on human rights than on environmental concerns, let alone on the future of nuclear power. Were public concerns about nuclear power to be voiced, they would have had to come from within the engineering community, and those voices were rare. Until Chernobyl, the prototypical "socialist citizen" was rather accepting of nuclear power and joined the "socialist engineer" in seeing Atoms for Peace as a sign of modernity. Only after Mikhail Gorbachev instituted glasnost and because of the shock of the Chernobyl disaster was nuclear power openly questioned, as we will discuss below.

Accidents and the Public

Three major accidents have hit the nuclear power industry, Three Mile Island (US), Chernobyl (USSR), and Fukushima (Japan), and they have had major impacts on the public perception of nuclear power.

The Three Mile Island accident was a partial nuclear meltdown on March 28, 1979, in reactor unit 2 of the Three Mile Island power plant near Harrisburg, Pennsylvania.[51] The accident revealed weaknesses in the Nuclear Regulatory Commission's (NRC) regulatory powers and supervision, slow response of federal and state agencies to safety issues, and lack of understanding and trust among the public. A commission under John Kemeny was set up to analyze the causes of the accident. The Kemeny Report was very critical both of NRC and of the owners of the Harrisburg plant, and it led to increased regulatory powers and a renewed safety philosophy among NRC staff and administrators. Three Mile Island dampened the nuclear aspirations of a number of nations, not least the US, where the number of reactor orders came to a halt after having already dropped precipitously during the economic recession following the first oil crisis. The industry only began to see recovery in the twenty-first century.[52] Beyond the US, Three Mile Island unsettled the public in many European countries and gave an impetus to the antinuclear movements in many countries.[53] Moreover, it triggered safety improvements in the nuclear industry. As one example, French utilities responded with safety improvements and early moves toward openness within the highly secretive nuclear technocracy.

The Chernobyl disaster occurred on April 26, 1986, in reactor unit 2 of the nuclear power plant in the Ukrainian Republic in the USSR. A test led to an uncontrolled nuclear chain reaction, resulting in a steam explosion and a major release of radioactive contamination over large parts of Europe. The Chernobyl disaster had a far greater impact than that of Three Mile Island on public perceptions and on the regulation of nuclear power.[54] Industry and government representatives in Western countries producing nuclear energy generally underlined the "Soviet" character of the accident, claiming that a similar disaster could not happen at home because of different, safer reactors and better regulation. In addition, several countries downplayed the seriousness of the accident for public health, and this triggered public distrust.

In Sweden, the Chernobyl disaster provoked extensive discussion; after all, Swedish nuclear experts disclosed the accident before Soviet acknowledgments were made. Parts of northern and eastern Sweden were severely affected by fallout. The Swedish antinuclear movement revived, with demands for an immediate closing of nuclear power plants. Moreover, antinuclear activists in neighboring Denmark collected some 160,000 signatures and demanded the closure not only of Barsebäck but of all Swedish, West German, and East German nuclear plants.[55] The revived activism, however, ebbed fairly quickly in Sweden, and after a few years the postreferendum decision to phase out nuclear power by 2010 was confirmed.

The French state and its elites were strongly committed to nuclear technology as a source of energy, ensuring energy independence, a symbol of modernization and not least of political independence due to the French nuclear bomb. In the face of this commitment, both government and industry downplayed Chernobyl's impact.[56] The French public's discovery of these attempts led to doubt and mistrust, the creation of two citizen-led expert organizations that exist to the present day (ACRO and CRIIRAD), and also to greater industry attention to environmental issues. New European legislation (e.g., Euratom Directive 89/618 on public information during a radiological emergency and the Environmental Impact Assessment Directive) also helped to advance transparency within the French nuclear establishment. In the 1990s, the French government set up safety and public advisory groups, especially in the area of radioactive waste management. The opaque, highly centralized nuclear technocracy had gradually been forced to become more transparent.

Chernobyl had an even greater impact in the socialist world. In the Soviet Union, the disaster was a shock to the industry, to Soviet leadership, and to a public grown accustomed to believing that industrial society was safe. Mikhail Gorbachev was required to address the accident openly in keeping with glasnost, and later he came to see Chernobyl as one of the decisive factors in the breakup of the USSR. For the fifteen years that followed, the media and citizens actively engaged in considering the causes and impact of the accident. Under glasnost, the media reported on Chernobyl and on other past military and civilian accidents and disasters. State and industry officials complained that this led to myths, stereotypes, and "radiophobia." The hundreds of thousands of citizens who had been affected by Chernobyl, including so-called liquidators (soldiers and others brought in to remediate—to liquidate—radiological damage and who exposed themselves to dangerous, in some cases life-threatening, labor), found the media to be an effective ally in the search for objectivity.

On the eve of the breakup of the USSR, environmental concerns became associated with nationalist feelings in the Soviet republics. The Kremlin was accused of having exploited the titular nationalities in Ukraine, Lithuania, and elsewhere. This contributed to rising public protest.[57] In Ukraine, Chernobyl gave impetus to environmental, antinuclear, and nationalist movements in what has been characterized as econationalism,[58] while putting the policies of glasnost and perestroika to the test. After the accident, the public insisted on a moratorium of construction at reactor sites and the closing of Chernobyl's remaining reactors (units 1, 3, and 4). In 1991, after the collapse of the Soviet Union, the Ukrainian parliament (the Rada) passed a law creating such a moratorium. However, two years later, in a period of economic crisis, the Rada reverted its earlier decision, to little public reaction. Citizens wanted electricity and warmth, and industry pushed actively for completion of mothballed reactors, license extensions, and the construction of new reactors. The Rada engaged the public through new or renovated information centers that communicated about issues such as safety and energy independence. With a few exceptions (e.g., antinuclear nongovernmental organizations), it seems that the Ukrainian public has accepted nuclear power as crucial to the nation's energy future in the twenty-first century.[59]

Chernobyl also spurred antinuclear and nationalist movements in Lithuania, which had two Chernobyl-type RBMK-1500 reactors, the largest in the world, at its Ignalina station in Visaginas. Demonstrations against Ignalina were partly a protest against Soviet "colonialism," targeting both an enclave of Slavic nuclear "settlers" in the city and the perception that such a massive and unstable structure had been built in Lithuania to exploit the nation. After independence, the Ignalina station became a site of contestation.[60] The station enabled Lithuania not only to satisfy a lion's share of its electricity demand but also to earn income by exporting electricity to Russia. Local people, especially those living near Ignalina, were understandably against the closure of the plant. Yet Lithuanian admission to the European Union was predicated on its closure. In the late 2000s, after the second unit was finally closed, antinuclear claims reappeared in the context of the debates on whether to revive a national nuclear program to take advantage of human capital—nuclear expertise—in Visaginas and secure the nation's energy future and electricity export earnings.

As in the USSR, in Bulgaria the authorities at first failed to report openly on the Chernobyl disaster even when the radioactive cloud reached the country.

The government eventually set up a committee to consider radiation protection and informational programs. As perestroika and glasnost spread to Eastern Europe, so did public questioning of the regime in Bulgaria, including concern about environmental problems and Chernobyl. As in Ukraine and Russia, the economic crisis that resulted from the fall of socialism and lack of public support after Chernobyl led ongoing projects to be mothballed. And as in those countries, promoters of nuclear power attributed public opposition to "radiophobia," an irrational fear of nuclear power, rather than to the absence of reliable information. However, these promoters could not prevent the shutdown of Soviet-era reactors when the European Union made this a condition for joining.

An earthquake and tsunami on March 11, 2011, triggered a disaster at the Fukushima Daiichi nuclear power plant, where nuclear meltdowns and hydrogen explosions in three of the four reactors led to a major release of radioactive contamination. Like with Chernobyl, it led to a tidal wave of public anguish all over the world, while nuclear proponents were quick to point out that such an accident could not happen in *their* country. This time, the reason was the lack of safety culture in Japan.

Within Europe, Fukushima had the most direct impact in Germany. In 1998, a red-green coalition agreement had already decided to phase out nuclear energy within twenty years.[61] After Fukushima, Chancellor Angela Merkel announced the closure of all German power plants by 2022, with eight of the seventeen operating reactors being shut down immediately.[62] This was an ironic reversal of policy, since in 2010 Merkel's Christian Democratic government had extended the operating licenses of Germany's nuclear power plants.

In most other European countries—except maybe Italy—the response to Fukushima was clear-cut; ongoing reactor construction projects in Finland or France were not halted. However, at the international level, Euratom started a large-scale, open, and self-critical process of stock taking regarding nuclear-societal relations.[63] This in turn informed the European Commission's Directorate-General for Research when drafting Euratom's research priorities.[64] And it led to the call for proposals for "Nuclear Developments and Interaction with Society" in the Euratom Research and Training Programme 2014–2015, to which the HoNESt research project responded.[65]

Nuclear Waste, Spent Fuel, and the Nuclear-Societal Interface

In recent decades, public attention has increasingly been directed toward waste management and remediation of accidents, leaks, abandoned mines, and other nuclear facilities that require cleanup.[66] In many European countries, waste management has been a crucial, and hitherto unresolved, public issue since the late 1970s (see chapter 6). In particular, the Swedish case shows that openness and engagement are far more likely to result in an outcome seen positively by industry and the public alike. At first, local environmentalists strongly opposed efforts to select a repository. In the 1990s, SKB, the organization responsible for nuclear waste, changed strategy, seeking cooperation and emphasizing that a decision about a repository would only be made if a local municipality favored it. After screening potential sites all over Sweden, SKB turned to municipalities in southern Sweden that already had nuclear facilities. Preliminary studies indicated that two of these, Östhammar (where Forsmark is located) and Oskarshamn, had the best conditions, with inhabitants that were not averse to nuclear facilities. In 2009, SKB selected Östhammar for the repository, whereas Oskarshamn would host the waste encapsulation facility.[67] The experience with the Finnish Onkalo spent nuclear fuel repository also demonstrates the virtues of public involvement in technical decision-making.[68]

In the Russian case, a waste dump explosion revealed the tensions between, and paradoxes of, secrecy and openness. On September 29, 1957, a major radiation accident occurred at the Mayak chemical combine (Chelyabinsk-40, today Ozersk). The Kyshtym accident took place when a tank containing seventy to eighty tons of liquid waste exploded, releasing over twenty million curies of radioactivity. All witnesses and the "liquidators" of the accident were made to sign letters of nondisclosure. This secrecy was maintained until the late 1980s. In the 1990s, the general public became actively involved in the events related to Kyshtym. In Ozersk, an environmental advocacy organization called Planet Hope was formed to protect the rights of the liquidators and their descendants. Since then, public openness has eroded. The organization was closed in 2015 because its leaders refused to register as a "foreign agent" according to a new Russian law. The law may push nuclear-societal interactions to other forums, such as the industry information centers, public hearings, and perhaps local-level NGO action. In Ukraine, nuclear plant information centers have been sponsoring children's drawing contests since the 2000s. The drawings reveal sometimes-fascinating juxtapositions of "nature" and nuclear power: smiling cooling towers, children playing peacefully under nuclear power plants, people in national costumes, angels and churches, and other abundant national and religious iconography.[69]

Beyond the reaction to Chernobyl, there was one major case of public opposition to nuclear power in Russia: the industry's decision to import spent nuclear fuel (SNF) for handling and storage to earn billions of dollars. Because of Rosatom's reticence to work with the public on the proposal, environmental activists led several campaigns, beginning in 2000, that brought radioactive waste issues to public attention. They succeeded in mobilizing the public against legislation to allow the import of SNF. In spite of a massive public campaign that led over 90 percent of Russians to oppose the practice—including petitions that attracted over two million signatures and would hence have obliged the government to organize a binding referendum on the proposal—Parliament eventually adopted the law. The government rejected the signatures, and Rosatom has never been forthcoming about the handling and disposition of imported SNF. Major environmental NGOs in Russia with significant financial resources and expertise (World Wildlife Fund, Greenpeace, Bellona) and national environmental associations (the International Social and Ecological Union, the Green League) nevertheless remain active.

Nuclear Futures on the European Continent

In gauging the future of nuclear energy in Europe—and its relation to society—key players are the continent's two major nuclear powers, Russia and France. For both countries, nuclear power has been of tremendous political, cultural, ideological, economic, and technological significance since the dawn of the nuclear age, as part of national self-identity and technoscientific development.[70] It has been publicly celebrated in the press and in postage stamps, and touted by leaders and engineers as a sign of strength.

If the Chernobyl accident had a significant negative impact on the public perception of nuclear power, the industry has become rejuvenated since 2000. After a brief hiatus in construction owing to financial crises of the Yeltsin era (the 1990s), the Soviet nuclear industry—with its glorious and difficult history—transformed itself into a Russian one. This has involved a self-proclaimed renaissance with ambitious plans to greatly expand nuclear new-build at home and to sell reactors abroad. With less and less direct public input, Rosatom has embarked on its own nuclear renaissance with the goal of building dozens of reactors by 2030, and to sell dozens more on world markets.

If France's nuclear energy program also grew out of nuclear enthusiasm, in the twenty-first century it must make a series of difficult and perhaps costly decisions. The industry faces critical choices concerning the future of its substantial yet aging nuclear fleet, and must do so under public scrutiny. By the turn of the century, recovery from periods of doubt led to the promise of a French-led nuclear renaissance, whereby France would have become a major exporter of nuclear technology and know-how. A symbol of this promise was the 2005 choice of the French site Cadarache to host the ITER international experimental nuclear fusion reactor. Yet some of these projects have been in crisis. First, the new "third generation" EPR plant, commissioned in 2003 by the Finnish TVO from the Areva-Siemens consortium, suffered significant cost overruns and delays, and soon turned into a quagmire. The Flamanville (Normandy) EPR project faces not only similar cost and time overruns but also serious technical and supply-chain problems. The British Hinkley Point C plant is also years behind its original schedule, and its cost estimate has been revised upward time and again. How will the public and industry respond to these challenging circumstances?

This chapter has demonstrated that public involvement dates back to the dawn of the nuclear age in the largely enthusiastic response of many citizens worldwide to Atoms for Peace programs. Public involvement in shaping nuclear policy began to expand in the 1970s, often in protest against perceived risks and local land use concerns. Within ten or so years, public concerns also centered on worries of growing costs. Nuclear waste management began to garner significant public attention in the 1980s, with local resistance against the siting of waste disposal facilities. In many countries, in response to public concerns, elected officials debated various nuclear futures, and accepted—willingly, or with reservations—the rights of the public to determine the future of nuclear power through referenda. In the socialist world, industry-societal interactions were constrained by closed political institutions. But these public responses, including protest, became pronounced after perestroika and glasnost, and with the fall of the Berlin Wall.

These examples show that the absence of public involvement in nuclear policymaking runs the risk, at the very least, of permitting poor site selection, untamed costs, and mistrust that was highlighted in the fallout—literal and figurative—after major accidents. All states, including more closed ones, have engaged the public in one way or another, but it is an open question whether the nuclear future will be more promising with greater public openness.

Notes

1. Hazel Gaudet Erskine, "The Polls: Atomic Weapons and Nuclear Energy," *Public Opinion Quarterly*, vol. XXVII (1963): 164. See also Holger Nehring, *Politics of Security: The British and West German Protests against Nuclear Weapons and the Early Cold War, 1945–1970* (Oxford: Oxford University Press, 2013); Christoph Laucht, "Scientists, the Public, the State and the Debate over the Environmental and Human Health Effects of Nuclear Testing in Britain, 1950–1958," *Historical Journal* 59, no. 1 (2016): 221–51.
2. Robert Jungk, *The Nuclear State* (Parchment, MI: Riverrun Press, 1984 [1977]).

3. Gabriel Almond and Sidney Verba, *The Civic Culture: Political Attitudes and Democracy in Five Nations* (Princeton, NJ: Princeton University Press, 1963).
4. Andrew Tompkins, *"Better Active than Radioactive!" Anti-Nuclear Protest in 1970s France and West Germany* (Oxford: Oxford University Press, 2016); Dolores L. Augustine, *Taking on Technocracy: Nuclear Power in Germany, 1945 to the Present* (New York: Berghahn, 2018); Michael Schüring, "West German Protestants and the Campaign against Nuclear Technology," *Central European History* 45, no. 4 (2012): 744–62.
5. Frank Bösch, "Taming Nuclear Power: The Accident near Harrisburg and the Change in West German and International Nuclear Policy in the 1970s and early 1980s," *German History* 35, no. 1 (2017): 71–95; Susanne Bauer, Karena Kalmbach, and Tatiana Kasperski, "From Pripyat to Paris, from Grassroots Memories to Globalized Knowledge Production: The Politics of Chernobyl Fallout," in *Nuclear Portraits: Communities, the Environment, and Public Policy*, ed. Laurel Sefton MacDowell (Toronto: University of Toronto Press, 2017), 149–89.
6. Frank Uekötter, "Fukushima and the Lessons of History: Remarks on the Past and Future of Nuclear Power," in *Europe after Fukushima: German Perspectives on the Future of Nuclear Power*, ed. Salomon Temple (Munich: Rachel Carson Center, 2012), 9–31.
7. Andy Stirling and Phil Johnstone, *Interdependencies Between Civil and Military Nuclear Infrastructures*, SPRU Working Paper Series (SWPS), 2018-13, 1–18, available at: http://www.sussex.ac.uk/spru/documents/2018-13-swps-stirling-and-johnstone.pdf; Robert Jacobs, "Born Violent: The Origins of Nuclear Power," *Asian Journal of Peacebuilding* 7, no. 1 (2019): 9–29.
8. This chapter incorporates by reference all of the twenty short country reports (SCRs) as if fully included herein; all are available at https://hdl.handle.net/2454/38269. Not all of them could be given equal weight, as examples were selected to carry the weight of the argument that political circumstances, especially domestic politics and structures, were crucial in shaping each nation's choice of nuclear program or rejection of nuclear power. As noted in the SCRs, the reports are designed to assemble in an accessible manner information and research results on the history of the relations between nuclear energy and society for all the different country cases, and to document the findings with references. Readers of this book may consult the SCRs for complete references and source materials, although in a few cases some references are listed here. The purpose of the country reports is threefold, addressing three different audiences: (1) to provide basic elements of narrative and analysis for further historical research by HoNESt researchers; (2) to provide information, context, and background for further analysis for HoNESt's social science researchers; and (3) to provide accessible information on nuclear-societal relations in the various countries for the purposes of outreach and communication with stakeholders (civil society, industry, associations, policymakers, journalists). For this purpose, SCR-based chapters on five West European countries were published in Astrid Mignon Kirchhof, ed., *Pathways into and out of Nuclear Power in Western Europe: Austria, Denmark, Federal Republic of Germany, Italy, and Sweden* (Munich: Deutsches Museum Verlag, 2019).
9. E.g., Hein-Anton Van der Heiden, *Social Movements, Public Spheres and the European Politics of the Environment: Green Power Europe?* (Basingstoke: Palgrave, 2010).

10. Resulting from the HoNESt project: Efstathios Arapostathis and Yannis Fotopoulos, "Transnational Energy Flows, Capacity Building and Greece's Quest for Energy Autarky, 1914–2010," *Energy Policy* 127 (2019): 39–50; Jan-Henrik Meyer, " 'Atomkraft—Nej tak': How Denmark Did Not Introduce Commercial Nuclear Power Plants," in *Pathways into and out of Nuclear Power in Western Europe: Austria, Denmark, Federal Republic of Germany, Italy, and Sweden*, ed. Astrid Mignon Kirchhof (Munich: Deutsches Museum, 2019), 74–123.
11. For a large comparative political science study considering a number of mostly quantitative indicators, see Wolfgang C. Müller and Paul W. Thurner, eds., *The Politics of Nuclear Energy in Western Europe* (Oxford: Oxford University Press, 2017).
12. Arne Kaijser and Jan-Henrik Meyer, "Nuclear Installations at the Border: Transnational Connections and International Implications. An Introduction," *Journal for the History of Environment and Society* 3 (2018): 1–32.
13. Rüdiger Graf, *Oil and Sovereignty: Petro-Knowledge and Energy Policy in the United States and Western Europe in the 1970s* (New York: Berghahn, 2018).
14. Henry Nielsen et al., "Risø and the Attempts to Introduce Nuclear Power into Denmark," *Centaurus* 41, no. 1–2 (1999): 64–92; Joachim Radkau and Lothar Hahn, *Aufstieg und Fall der deutschen Atomwirtschaft* (Munich: Oekom, 2013); Dennis Romberg, *Atomgeschäfte: die Nuklearexportpolitik der Bundesrepublik Deutschland 1970–1979* (Paderborn: Ferdinand Schöningh, 2020).
15. Hanspeter Kriesi, "Political Context and Opportunity," in *Blackwell Companion to Social Movements*, ed. David A. Snow, Sarah A. Soule, and Hanspeter Kriesi (Oxford: Oxford University Press, 2007), 67–90.
16. Arne Kaijser, "Redirecting Power: Swedish Nuclear Power Policies in Historical Perspective," *Annual Review of Energy and the Environment* 17 (1992): 437–62.
17. Gabrielle Hecht, *Being Nuclear: Africans and the Global Uranium Trade* (Cambridge, MA: MIT Press, 2012).
18. Kaijser and Meyer, "Nuclear Installations at the Border," 10.
19. See chapter 9, and the contributions to the special issue "Siting Nuclear Installations at the Border," *Journal for the History of Environment and Society* 3 (2018): 1–178.
20. Astrid Mignon Kirchhof and Jan-Henrik Meyer, "Global Protest Against Nuclear Power: Transfer and Transnational Exchange in the 1970s and 1980s. Focus Issue," *Historical Social Research* 39, no. 1 (2014): 163–273; Andrew Tompkins, "Transnationality as a Liability? The Anti-Nuclear Movement at Malville," *Revue Belge de Philologie et d'Histoire/Belgisch Tijdschrift voor Filologie en Geschiedenis* 89, no. 3–4 (2011): 1365–79.
21. For another recent contribution to help overcome this divide, see Astrid Mignon Kirchhof and John R. McNeill, eds., *Nature and the Iron Curtain: Environmental Policy and Social Movements in Communist and Capitalist Countries, 1945–1990* (Pittsburgh: University of Pittsburgh Press, 2019).
22. There may be methodological difficulties in comparing "East" and "West" given the nature, extent, and control of public political participation in closed (socialist) systems. See J. Street, "Political Culture—from Civic Culture to Mass Culture," *British Journal of Political Science* 24, no. 1 (1994): 95–113.
23. Sezin Topçu, "Confronting Nuclear Risks: Counter Expertise as Politics within the French Nuclear Debate," *Nature and Culture* 3, no. 3 (2008): 225–45; Dieter Rucht,

"Gegenöffentlichkeit und Gegenexperten. Zur Institutionalisierung des Widerspruchs in Politik und Recht," *Zeitschrift für Rechtssoziologie* 9, no. 2 (1988): 290–305.

24. Henry Nielsen and Henrik Knudsen, "The Troublesome Life of Peaceful Atoms in Denmark," *History and Technology* 26, no. 2 (2010): 91–118.
25. Dick van Lente, ed., *The Nuclear Age in Popular Media: A Transnational History 1945–1965* (Houndmills: Palgrave, 2012); Jaume Sastre-Juan and Jaume Valentines-Álvarez, "Fun and Fear: The Banalization of Nuclear Technologies through Display. Special Issue," *Centaurus* 61, no. 1–2 (2019): 2–90.
26. John Krige, "The Peaceful Atom as Political Weapon: Euratom and American Foreign Policy in the Late 1950s," *Historical Studies in the Natural Sciences* 38, no. 1 (Winter 2008): 5–44; Sonja D. Schmid, "Nuclear Colonization? Soviet Technopolitics in the Second World," in *Entangled Geographies: Empire and Technopolitics in the Global Cold War*, ed. Gabrielle Hecht (Cambridge, MA: MIT Press, 2011), 125–54.
27. Radkau and Hahn, *Aufstieg und Fall der deutschen Atomwirtschaft*; Joachim Radkau, *Aufstieg und Krise der deutschen Atomwirtschaft 1945–1975: Verdrängte Alternativen in der Kerntechnik und der Ursprung der nuklearen Kontroverse* (Reinbek: Rowohlt, 1983).
28. Wolfgang Rüdig, *Anti-Nuclear Movements: A World Survey of Opposition to Nuclear Energy* (London: Longman, 1990), 63.
29. Joseba de la Torre and Mar Rubio-Varas, *La Financiación Exterior Del Desarrollo Industrial Español a Través Del IEME (1950–1982)* (Estudios d. Madrid: Banco de España, 2015), http://www.bde.es/f/webbde/SES/Secciones/Publicaciones/PublicacionesSeriadas/EstudiosHistoriaEconomica/Fic/roja69.pdf; Joseba de la Torre and Mar Rubio-Varas, "Nuclear Power for a Dictatorship: State and Business Involvement in the Spanish Atomic Program, 1950–85," *Journal of Contemporary History* 51, no. 2 (2016): 385–411.
30. Florian Bayer and Ulrike Felt, "Embracing the 'Atomic Future' in Post–World War II Austria," *Technology and Culture* 60, no. 1 (2019): 165–91; Christian Forstner, *Kernphysik, Forschungsreaktoren und Atomenergie: Transnationale Wissensströme und das Scheitern einer Innovation in Österreich* (Wiesbaden: Springer Spektrum, 2019).
31. See Hungarian, Lithuanian, Ukrainian, and Belarusian SCRs, https://hdl.handle.net/2454/38269; Zadikyan Morohov and A. Zadikyan, eds., *Nuclear Science and Technology in USSR (Moscow: Atomizdat, 1977)*, 328. See also Schmid, "Nuclear Colonization?" On the Bulgarian case, for example, see V. Sichev, Yu Krohmalev, K. Menzel, V. Tolpigo, V. Shtregober, and A. Bilbao, "Collaboration between Comecon Member States in the Nuclear Power Field," in *Nuclear Power Experience: Proceedings of an International Conference on Nuclear Power Experience Held by the International Atomic Energy Agency in Vienna, 13–17 September 1982, Volume V* (Vienna: International Atomic Energy Agency, 1983), 495–509; Ivaylo T. Hristov, *The Communist Nuclear Era: Bulgarian Atomic Community During the Cold War, 1944–1986* (Amsterdam: Amsterdam University Press, 2014), https://doi.org/10.6100/IR770869.
32. Helena Flam, ed., *States and Anti-Nuclear Movements* (Edinburgh: Edinburgh University Press, 1994).

33. Helmuth Trischler and Robert Bud, "Public Technology: Nuclear Energy in Europe," *History and Technology* 34, no. 3–4 (2018): 187–212. See also chapter 8.
34. Hein-Anton van der Heijden, "The Great Fear: European Environmentalism in the Atomic Age," in *A History of Environmentalism: Local Struggles, Global Histories*, ed. Marco Armiero and Lise Sedrez (London: Bloomsbury, 2014), 185–86.
35. Arne Kaijser and Jan-Henrik Meyer, " 'The World's Worst Located Nuclear Power Plant': Danish and Swedish Perspectives on the Swedish Nuclear Power Plant Barsebäck," *Journal for the History of Environment and Society* 3 (2018): 71–105.
36. Andrew Tompkins, "Grassroots Transnationalism(s): Franco-German Opposition to Nuclear Energy in the 1970s," *Contemporary European History* 25, no. 1 (2016): 117–42; Andrew Tompkins, *Better Active than Radioactive! Anti-nuclear Protests in 1970s France and West Germany* (Oxford: Oxford University Press, 2016).
37. Claire Le Renard, "The Superphénix Fast Breeder Nuclear Reactor—Cross-Border Cooperation and Controversies," *Journal for the History of Environment and Society* 3 (2018): 107–44.
38. Luise Schramm, *Evangelische Kirche und Anti-AKW-Bewegung: Das Beispiel der Hamburger Initiative kirchlicher Mitarbeiter und Gewaltfreie Aktion im Konflikt um das AKW Brokdorf 1976–1981* (Göttingen: V&R, 2018); Augustine, *Taking on Technocracy*, 126–60.
39. Astrid M. Eckert, *West Germany and the Iron Curtain: Environment, Economy and Culture in the Borderlands* (Oxford: Oxford University Press, 2019), 201–44; Astrid Mignon Kirchhof, "East-West German Transborder Entanglements through the Nuclear Waste Sites in Gorleben and Morsleben," *Journal for the History of Environment and Society* 3 (2018): 145–78.
40. Astrid Mignon Kirchhof and Helmuth Trischler, "The History behind West Germany's Nuclear Phase-Out," in *Pathways into and out of Nuclear Power in Western Europe: Austria, Denmark, Federal Republic of Germany, Italy, and Sweden*, ed. Astrid Mignon Kirchhof (Munich: Deutsches Museum, 2019), 124–69.
41. Bernd-A. Rusinek, "Wyhl," in *Deutsche Erinnerungsorte*, vol. 2, ed. Hagen Schulze and Etienne François (Munich: C. H. Beck 2003), 652–66; Natalie Pohl, *Atomprotest am Oberrhein: Die Auseinandersetzung um den Bau von Atomkraftwerken in Baden und im Elsass (1970–1985)* (Stuttgart: Steiner, 2019).
42. Janine Gaumer, *Wackersdorf: Atomkraft und Demokratie in der Bundesrepublik 1980–1989*. Munich: Oekom, 2018).
43. Alexander Glaser, "From Brokdorf to Fukushima: The Long Journey to Nuclear Phase-out," *Bulletin of the Atomic Scientists* 68, no. 6 (2012): 10–21.
44. Michael Schüring, *"Bekennen gegen den Atomstaat": Die evangelischen Kirchen in der Bundesrepublik und die Konflikte um die Atomenergie 1970–1990* (Göttingen: Wallstein, 2015). See also Michael L. Hughes, "Civil Disobedience in Transnational Perspective: American and West German Anti-Nuclear-Power Protesters, 1975–1982," *Historical Social Research* 39, no. 4 (2014): 236–53.
45. Christian Salm, *Transnational Socialist Networks in the 1970s: European Community Development Aid and Southern Enlargement* (Basingstoke: Palgrave, 2016).
46. Viktor Klima and Thomas Klestil, "Bundesverfassungsgesetz für ein atomfreies Österreich" [Constitutional Act for a Nuclear-Free Austria], *Bundesgesetzblatt für die Republik Österreich* 1 (August 13, 1999), 149, https://www.ris.bka.gv.at/Dokumente/BgblPdf/1999_149_1/1999_149_1.pdf.

47. Birgit Müller, "Anti-Nuclear Activism at the Czech-Austrian Border," in *Border Encounters: Asymmetry and Proximity at Europe's Frontiers*, ed. Jutta Lauth Bacas and William Kavanagh (New York: Berghahn, 2013), 68–89.
48. Cf. Herbert Kitschelt, "Political Opportunity Structures and Political Protest: Anti-Nuclear Movements in Four Democracies," *British Journal of Political Science* 16, no. 1 (1986): 57–85.
49. There was a certain overlap membership between antinuclear and *antifranquist* activism, including Mario Gaviria, Pedro Costa Morata, José Manuel Naredo, Juan Serna, and especially José Allende. M. del Mar Rubio-Varas et al., "Spain Short Country Report (version 2018)," in *History of Nuclear Energy and Society (HoNESt) Consortium Deliverable no. 3.6* (2019), 16 (fn. 54), 76, https://hdl.handle.net/2454/38269; Pedro Costa Morata, *Nuclearizar España* (Barcelona: Los libros de la frontera, 1976).
50. See also Maria del Mar Rubio-Varas, António Carvalho, and Joseba de la Torre, "Siting (and Mining) at the Border: Spain-Portugal Nuclear Transboundary Issues," *Journal for the History of Environment and Society* 3 (2018): 33–69; Stefania Barca and Ana Delicado, "Anti-Nuclear Mobilisation and Environmentalism in Europe: A View from Portugal," *Environment and History* 22, no. 4 (2016): 497–520.
51. Natasha Zaretsky, *Radiation Nation: Three Mile Island and the Political Transformation of the 1970s* (New York: Columbia University Press, 2018).
52. John G. Kemeny et al., *The Need for Change: The Legacy of TMI* (Washington, DC: The President's Commission on the Accident at Three Mile Island, 1979).
53. Bösch, "Taming Nuclear Power."
54. Karena Kalmbach, *The Meanings of a Disaster: Chernobyl and Its Afterlives in Britain and France* (New York: Berghahn, 2020); Katrin Jordan, *Ausgestrahlt: Die mediale Debatte um "Tschernobyl" in der Bundesrepublik und in Frankreich 1986/87* (Göttingen: Wallstein, 2018).
55. Kaijser and Meyer, " 'The World's Worst Located Nuclear Power Plant,' " 88.
56. Karena Kalmbach, *Tschernobyl und Frankreich: Die Debatte um die Auswirkungen des Reaktorunfalls im Kontext der französischen Atompolitik und Elitenkultur* (Frankfurt: Peter Lang, 2011).
57. Tetiana Perga, "The Fallout of Chernobyl: The Emergence of an Environmental Movement in the Ukrainian Soviet Socialist Republic," in Kirchhof and McNeill, *Nature and the Iron Curtain*, 55–72.
58. Jane Dawson, *Econationalism* (Durham, NC: Duke University Press, 1996).
59. Kate Brown, *Manual for Survival: A Chernobyl Guide to the Future* (New York: W.W. Norton, 2019); Serhii Plokhy, *Chernobyl: History of a Tragedy* (London: Penguin, 2019).
60. On Ignalina, see also Anna Storm, *Post-Industrial Landscape Scars* (Basingstoke: Palgrave, 2014).
61. Hendrik Munsberg, "Abschied vom Atomstrom," *Der Spiegel* 52 (December 20, 1998): 22–26.
62. "Germany: Nuclear Power Plant to Close by 2022," *BBC*, May 30 2011, http://www.bbc.com/news/world-europe-13592208.
63. European Commission, Directorate-General for Research, *Benefits and Limitations of Nuclear Fission for a Low-Carbon Economy. 2012 Interdisciplinary Study Synthesis Report and Compilation of the Experts' Reports: 2013 Symposium Agenda and Speakers'*

Symposium Speeches Delivered on 26–27 February 2013 (Brussels: Publications Office of the European Union, 2014).

64. European Commission, Directorate-General for Research, *Benefits and Limitations of Nuclear Fission for a Low-Carbon Economy. Defining Priorities for Euratom Fission Research and Training (Horizon 2020): Compilation of the Experts' Reports, Background to the Synthesis Report—Study* (Brussels: Publications Office of the European Union, 2014).
65. European Commission, "NFRP 12-2014: Nuclear Developments and Interaction with Society," *Euratom Work Programme* 2014–2015: C(2014)5009 of 22 July 2014, revised (2014), https://ec.europa.eu/info/funding-tenders/opportunities/portal/screen/opportunities/topic-details/nfrp-12-2014.
66. Achim Brunnengräber et al., eds., *Nuclear Waste Governance: An International Comparison* (Wiesbaden: Springer VS, 2015); Achim Brunnengräber et al., eds., *Challenges of Nuclear Waste Governance: An International Comparison Volume II* (Wiesbaden: Springer VS, 2018).
67. Mark Elam and Göran Sundqvist, "Meddling in Swedish Success in Nuclear Waste Management," *Environmental Politics* 20, no. 2 (2011): 246–63.
68. Nagako Sato, "Historical Background of Nuclear Waste Policy Formation in Finland and Comparison with (West) Germany," *Asian Journal of Peacebuilding* 7, no. 1 (2019): 73–87.
69. Tatiana Kasperski, "Children, Nation and Reactors: Imagining and Promoting Nuclear Power in Contemporary Ukraine," *Centaurus* 61, no. 1–2 (2019): 51–69.
70. Gabrielle Hecht, *The Radiance of France: Nuclear Power and National Identity after World War II* (Cambridge, MA: MIT Press, 2009); Paul Josephson, *Red Atom* (Pittsburgh: University of Pittsburgh Press, 2005).

CHAPTER 2

The Changing Economic Context Influencing Nuclear Decisions

Mar Rubio-Varas

Introduction

Nuclear power stations are large and complex undertakings that are typically individually unique and built under enormous political, commercial, and social pressures. The introduction of nuclear energy was accompanied by multiple predictions about the costs of the technology.[1] Comparison with the cost of other energy sources has been a particularly complex issue because capital costs are usually high in nuclear power while operating costs are usually low.[2] Furthermore, the costs of decommissioning and managing spent fuel are hard to assess.[3] Yet during the early decades of nuclear power, backend costs were belittled "on the assumption that spent fuel would be shipped to a reprocessing plant. . . . However, the advantages of reprocessing came under international question in the late 1970s."[4] Thus, only from the 1980s did the backend costs begin to be estimated and internalized in the costs of nuclear power.[5]

While most of the economic literature dedicated to nuclear power focuses on the cost per kilowatt-hour produced,[6] this approach overlooks other crucial economic issues that eventually become the cornerstone of nuclear decision-making: the markets for electricity and nuclear reactors, the financial constraints for the utilities, the institutions required to plan and execute the nuclear programs, and the government capacity to sustain the economic effort of a nuclear program (including in areas beyond electricity generation, such as medicine and agriculture). Although the early antinuclear movements paid little attention to the economic aspects of nuclear power,[7] the decision makers did, as this chapter will demonstrate. In that sense, this chapter broadens the view beyond local protest, social movements, and the public by understanding society as a large group of people who live together in an organized way making decisions about how to do things.[8] Drawing on the evidence collected by HoNESt partners and on the economic history of the period, the chapter examines the broader economic context within which decisions were made regarding the main energy sources for electricity production. In other words, it aims at understanding the changing economic framework influencing nuclear decisions over the past fifty years.

Each of the short country reports (SCR) produced by the HoNESt consortium explicitly addresses the issue of the cost of nuclear power.[9] But they also mention elements concerning the competition with alternative technologies to produce electricity, concerns about energy independence, and other economic issues that shaped nuclear decision-making processes. By pooling together this evidence—distinct from the actual costs per kilowatt—this chapter sheds light on how wider economic aspects influenced nuclear decisions and contributed to the decision to adopt, reject, or abandon the nuclear option. In fact, from the analysis carried out by the social scientists of HoNESt, national and consumer economics emerged as two of the categories for evaluating the basis for choices about the use of nuclear electricity generation.[10]

Nuclear technology has been adopted by over thirty of the almost 200 nations of the world since the 1960s. For most countries, the decision to become a nuclear-powered nation (or not) was made well before any major accident took place.[11] In fact, in the case of almost 65 percent of the world's reactors (308 of the 478 reactors), construction began before the Three Mile Island accident in 1979, and about half of the nuclear countries had already halted nuclear orders before that date.[12] Given the average time of seven to ten years required to complete a nuclear power plant, the standstill in the worldwide growth of nuclear generation did not become apparent until some ten years later, when the plants ordered during the 1970s were eventually connected to the grid. By the 1990s, worldwide nuclear generation growth had begun to wane, and the share of nuclear power in overall electricity generation had started to decline, after reaching a maximum of 17.5 percent in 1996 (see fig. 2.1)..

The HoNESt consortium compiled historical evidence for twenty countries. Of these, fifteen chose to connect nuclear technology to their electricity grids, one of them (Italy) opting out after having produced nuclear electricity commercially for over twenty years.[13] Another five countries considered the option but eventually disregarded it.[14] The overall weight of the civil nuclear sector in the consortium's evidence is larger than the number of countries suggests: these fifteen countries represent only a half of the nuclear countries worldwide, yet they account for over two-thirds of the nuclear reactors ever built and of the world's nuclear generation capacity; together, they generated almost 90 percent of the nuclear electricity consumed in the world throughout history.

Do the nuclear-powered nations share any common economic traits? In the narrative of the great milestones of atomic history, economics seem marginal. In fact, it is fair to say that a history that includes the economic aspects, the

Fig. 2.1. Worldwide nuclear electricity generation (GWh and share on total electricity output) 1960–2014.

GWh (right) vs. % (left)

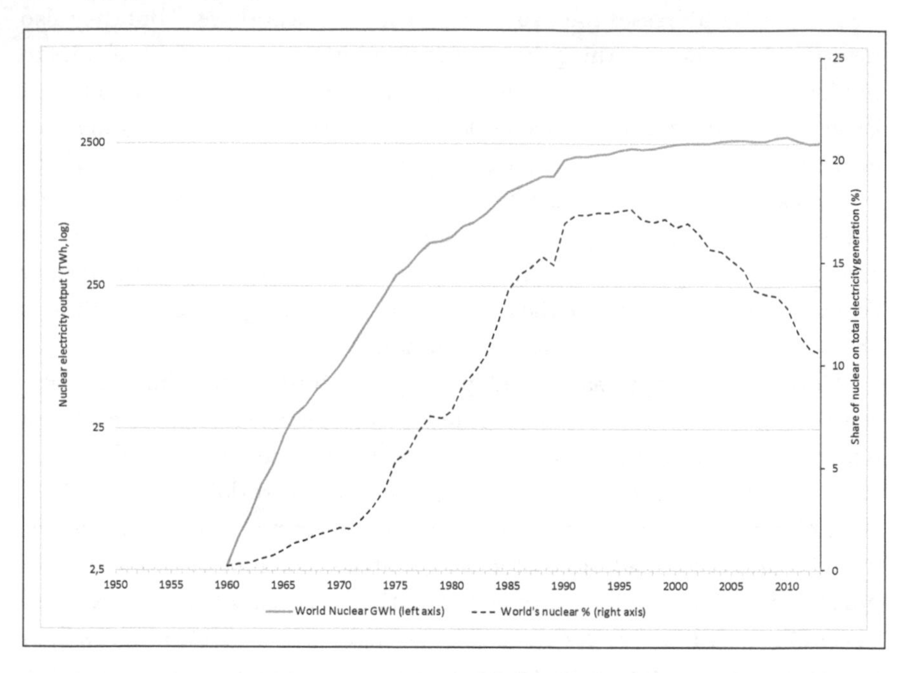

Sources: Author's elaboration from the data by International Energy Agency, IEA Statistics, OECD/IEA 2014. Data prior to 1971 compiled from Joel Darmstadter, Perry D. Teitelbaum, and Jaroslav G. Polach, *Energy in the World Economy; a Statistical Review of Trends in Output, Trade, and Consumption Since 1925* (Baltimore: Johns Hopkins University Press, 1971). Note that for ex-Soviet countries, IEA offers data only after 1990. The left axis, which reflects the total electric generation in GWh, uses a logarithm scale; thus the slope of the line is a direct indicator of the growth rate.

companies, and the finances of this technology with the tools of the economic historians has scarcely begun.[15] This absence of economic history is surprising, because from the beginning of the nuclear age there was a major concern about the financing of nuclear power plants. National and international institutions considered the economic aspects as strategic.[16] Neither is there a business history that explains the nuclear ecosystem,[17] the industrial infrastructures, the markets, the attitudes of the entrepreneurs, and the institutional influence. Nevertheless, the literature has pointed at some shared features among some of the countries running civil nuclear programs.[18] Even when the nuclear-powered nations differed in many aspects, countries opting to include nuclear technology in their electricity mix shared some basic economic traits:

> They had strong governments, and/or the sustenance of a superpower to fence the risk involved; they possessed an integrated electricity network and/or sufficient electricity demand to plug a reactor; they all had some industrial base to develop the nuclear technology and/or accommodate the technological transfer. And finally, these countries had the ability to tap into the financial resources required (whether nationally or internationally). On the business front they had dissimilar types of business organizations involved in the demand side of nuclear decision making, with standardization seeming most likely to occur in countries when lesser agents took part on the decision over the reactor choice.[19]

This chapter is structured around these issues, providing quantitative and qualitative evidence to illustrate the similarities and differences of the economic aspects that framed the nuclear decisions of the countries studied by the HoNESt consortium. It seeks to compare and contrast by grouping the countries according to the type of economic decisions they faced regarding nuclear power and to how the changing economic context affected their decisions.

First, I provide evidence on the electricity markets and their different sizes and structures. Next I look into the question that many countries faced as to whether they had a sufficient industrial base to develop their own nuclear technology or whether they could accommodate the technology transfer involved in an atomic program. Then I investigate the aspects that affected the financing of nuclear power plants, as nuclear power plants indeed rank among the largest export transactions in world commerce.[20] Finally, I analyze a key variable of capital-intensive projects: the interest rate.

The Electricity Market: How Much Electricity? Which Sources? Public vs. Private Utilities?

On May 4, 1957, a committee of three men (Louis Armand from France, Franz Etzel from Germany, and Francesco Giordani from Italy) submitted a report to the governments of the six member states of the European Coal and Steel Community (ECSC). *A Target for Euratom* reviewed Europe's needs and its resources in nuclear energy and bluntly detailed the European energy supply problems:

> Electricity consumption is growing rapidly, doubling every ten or twelve years. . . . Europe's economic growth is in danger of being seriously hampered by the lack of energy to nourish it. Being short of domestic energy supplies, our countries must turn increasingly to imports to meet their needs . . .

> Every year that is lost in constructing nuclear power stations means that conventional stations, requiring increased oil or coal imports—and which continue to consume oil or coal throughout their lifetime of twenty or thirty years—will be built instead. In view of this situation, Europe must within the limits set by the pattern of electricity production, construct nuclear stations as rapidly as possible.[21]

On these very same grounds of rapid growth of electric demand, which could not be met without importing additional oil/coal, thus endangering economic growth, nuclear programs were developed across the rest of Europe and in the United States. This idea of nuclear being crucial for industrial development and "keeping the lights on" permeated the public imaginary.[22]

This section reviews the electricity market for the countries represented in the HoNESt consortium, in terms of the evolution of the growth of demand for electricity, the actual electricity requirements, and the composition of the electric supply. There is wide consensus on the general traits of global macroeconomic history since World War II. Economic history textbooks describe the general traits shared across the world in three stages: the golden age from the end of World War II to 1971; the deceleration and crisis that followed the end of the Bretton Woods system and the oil crises (1971–1981); and the restructuring and change of the 1980s and 1990s, leading to the new challenges that closed the century.[23] Each period entailed distinct conditions for the nuclear-system decision-making. Most nuclear programs were launched in, and grew from, the 1950s to the 1970s. The developments in the nuclear system mirrored the key macroeconomic characteristics of the period: the need to match the ever-growing electricity demand, the creation of multilateral international collaboration organizations, the strong intervention by the state, and the increasing role of a few multinationals.[24] By the early 1960s, demonstration power reactors were in operation in all the leading industrial countries, although the economic competitiveness of nuclear energy was still in question, and the competition among the technological alternatives for reactor design remained unsettled.[25]

The growth of electricity demand was taken to be something akin to a law of nature during the planning phase of nuclear programs from the 1950s to the early 1970s.[26] Before 1970, the average annual growth rate of the electricity consumed in the twenty countries covered by HoNESt exceeded 9 percent (implying a doubling in less than a decade), while during the 1970s the growth of electric demand halved to 5 percent, and further slowed down during the 1980s to less than 3 percent on average. Several countries even experienced negative annual growth rates—that is, consumed less electricity than the year before—during the oil crises (1974 and 1979) and in Eastern Europe after the collapse of the Soviet Union (see figure 2.2). The growth rates of electricity demand never recovered the speed of the 1960s.

For a short while, the oil crisis of 1973–1974 looked like an opportunity to foster nuclear power as a solution for the now onerous oil imports.[27] But this was belied by the economic crisis that unfolded in its wake. As the economy slowed down, so did the demand for electricity. By 1975, the curve of nuclear orders had already passed its peak. Furthermore, over two-thirds of all nuclear plants ordered after January 1970 were eventually canceled.[28] In the US, nuclear investments stalled in the late 1970s, with no new plants ordered after 1978. The second oil crisis, in 1979, hardened the world's economic outlook and ushered in a definitive change in energy policy as efforts concentrated on reducing energy consumption. The uncertainties over the world economy

Fig. 2.2. Annual growth rates of total electricity consumption in HoNESt's twenty countries.

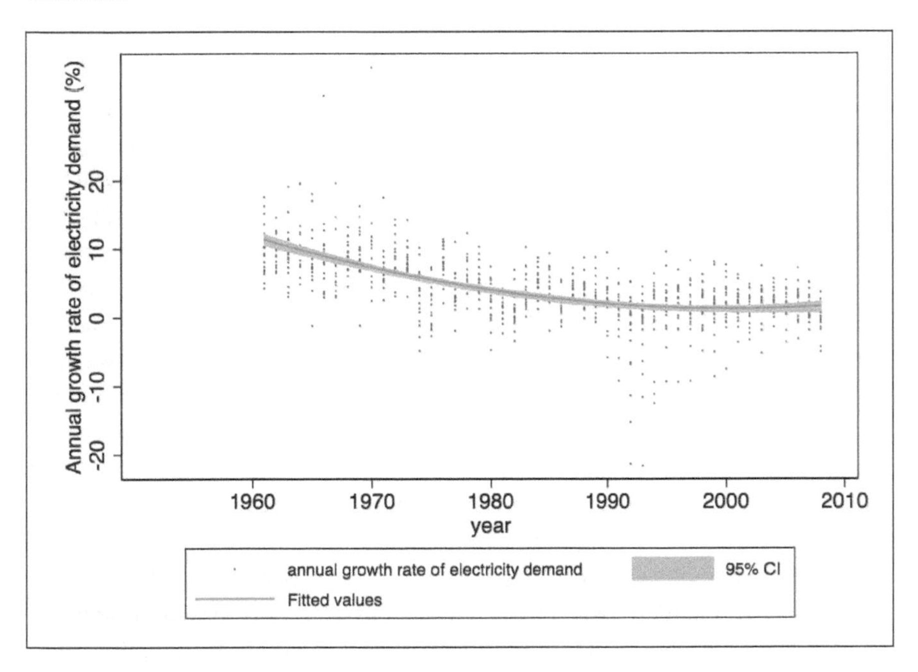

Sources: Author's elaboration from the data by International Energy Agency, IEA Statistics, OECD/IEA 2014. Note that data for former Soviet bloc countries is only available after 1990.

translated into falling energy consumption in many countries. In parallel, the 1970s saw a surge in environmentalism, resulting in new environmental legislation, environment ministries, and, in several countries, the founding of formal Green political parties, all antinuclear.[29]

Besides the obvious changes to the economy brought about by the oil crises, other fundamental long-term shifts altered the demand for electricity from the 1970s onward, such as the increase in energy efficiency and the onset of deindustrialization (to which we will return later in this chapter). Planners had overestimated countries' electricity needs, assuming that demand would increase steadily at a rate exceeding that of economic growth. Forecasters extrapolated the trends of the fifteen or twenty years before 1970, and consequently overestimated future electricity requirements. Figure 2.3 makes this evident by comparing the actual consumption of electricity in 1969 and 1989, with the forecasted consumption for 1989 had the projections of doubling every decade or less held. In 1989, each country consumed an average of about 50 percent less electricity than had been forecasted. Yet there were great variations: while the Portuguese consumption was 20 percent smaller than expected, the British consumption fell short of the forecasts made two decades earlier by some 80 percent.

In many countries, nuclear power came into question because it had to face the slowdown in the growth of electricity demand in the long run, which had justified its emergence in the first place. This is reflected in the SCRs of Western countries in the 1980s as well as in the Eastern ones, particularly after 1989.[30]

Figure 2.3 introduces a second aspect that influenced the decision whether to opt for nuclear technology: the amount of electricity required. While there was a common trend toward slowing down in the growth of the electricity demand, the actual amount of electricity required varied massively across countries. The US consumed as much electricity in 1969 as the next ten largest consumers of the Western world together;[31] that is to say, it consumed more electricity than the remainder of the Western countries studied by the HoNESt consortium taken as a whole, and more than three times the electricity consumed by Russia. Some regional US utilities generated more electricity than some of the small electricity-consuming nations on our list.

Economies of scale is an economic aspect that was very much at the fore of nuclear development. Power scaling is well known in the energy sector. As units of a similar design increase in size, their cost is expected to rise more slowly—and thus the cost per kWh produced is expected to fall. In theory, this effect could make the unit capital costs of a 200 MWe more than double those of a 1,000 MWe reactor.[32] Such reasoning dovetailed with the fact that early nuclear technology was developed by the countries with the largest electricity demands

Fig. 2.3. Actual electricity consumption in 1969 and actual vs. forecasted electricity consumption by 1989 of HoNESt consortium countries.

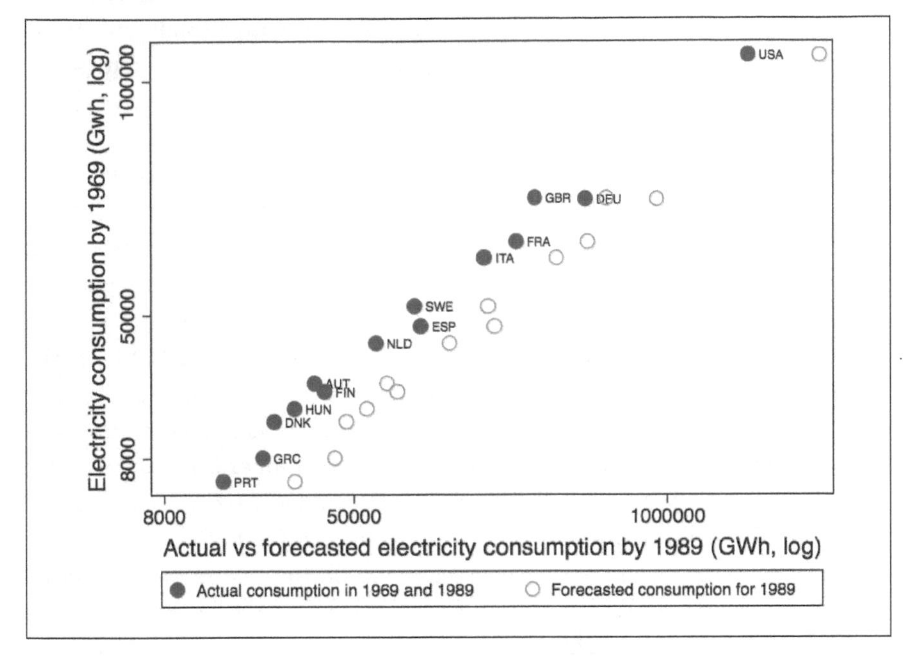

Sources: Author's elaboration from the data by International Energy Agency, IEA Statistics, OECD/IEA 2014. Extrapolated backward for Bulgaria, Hungary, and Russia. Note that data for Belarus, Lithuania, and Ukraine are only available after 1990. Expected consumption calculated as the double of the consumption a decade earlier (the usual assumption of 7 percent growth). Note the logarithm scale in both axes.

(the US, USSR, Canada, and France). Thus, reactor size steadily increased from a few hundred MWe in the early 1960s to 1,500 MWe and more today. Nuclear electric-generating plants typically operate throughout the year as baseload generation sources. Building larger reactors implied having to accommodate larger continuous flows of electricity.[33]

The trend toward larger reactors made the nuclear option poorly adapted to the needs of the many nations and utilities that had irregular or insufficient electricity demand and inadequate grid connections.[34] They simply could not take advantage of an uninterrupted supply of large amounts of electricity, even allowing for potential exports to neighboring countries to complement the small domestic market. In fact, siting nuclear installations at borders, hoping to share the electricity with neighboring countries, was fairly common in the early days of the industry.[35] However, international exchanges of electricity have remained limited, and today, even for the three largest electricity exporters

(Germany, Canada, and France), exports represent only 10–12 percent of their domestic generation of electricity.[36] The exports alone would hardly justify the installation of a nuclear power plant in a country with a small grid. Those managing electricity grids advise that "a grid should not be subjected to power variations in excess of 10 percent of the total grid capacity. So, 1 GWe plants cannot be deployed in grids of 10 GWe or less."[37]

To show that the trend toward larger reactors left out many nations that had insufficient electricity demand, figure 2.4 plots the estimated nuclear generation from a hypothetical program of up to three average-sized reactors against the actual electric demand of the countries studied by the HoNESt consortium.

Nuclear manufacturers, flooded with orders for larger plants, showed little enthusiasm for pursuing the production of smaller reactors.[38] In fact, the world's largest nuclear manufacturing country, the US, recognized in the early 1980s that "if the rational economic development of the customer nations was to be considered, only a few Asian, African and Latin American countries had power grids large enough to distribute electricity produced by even the smallest commercially available US reactor."[39] Those countries (and utilities) that were in the market for nuclear power had to have sufficiently large and integrated electrical networks to accommodate nuclear plants of standard sizes—a steady flow of 3,500 to 7,000 GWh from a single reactor in a year, corresponding to the lower bound of the gray area in figure 2.4.[40] Portugal and Greece had insufficient electric demand to host even one average-sized reactor in the 1960s, and even when they expected a rapid increase in demand, neither the expected nor the actual electricity demand of the following twenty years would have justified the installation of an average-sized reactor. For Bulgaria and Denmark, three average-sized reactors would have more than satisfied their entire domestic electricity demand almost until 1980. These figures, however, may be misleading since Denmark is neither unified electrically nor isolated—East Denmark (Zeeland) and West Denmark (Jutland and Funen) are not part of the same major grid system, and are connected only by a 500 MWe link. East Denmark is part of the Nordic grid, while West Denmark belongs to the main continental zone.[41] Thus, although typically analyzed as single units, national electric markets are not necessarily unified.

Although the general message of figure 2.4 remains strong (large electricity consumers would be more likely to opt for nuclear technology than small electricity consumers), there are noticeable exceptions. Lithuania stands out as a small consumer with nuclear power. However, the decision to build nuclear power was made while Lithuania belonged to the Soviet Union. In contrast, Italy stands out as a large consumer without nuclear power. After much debate and

Fig. 2.4. Actual electricity consumption and hypothetical electricity generation by up to three average-size reactors for HoNESt consortium countries.

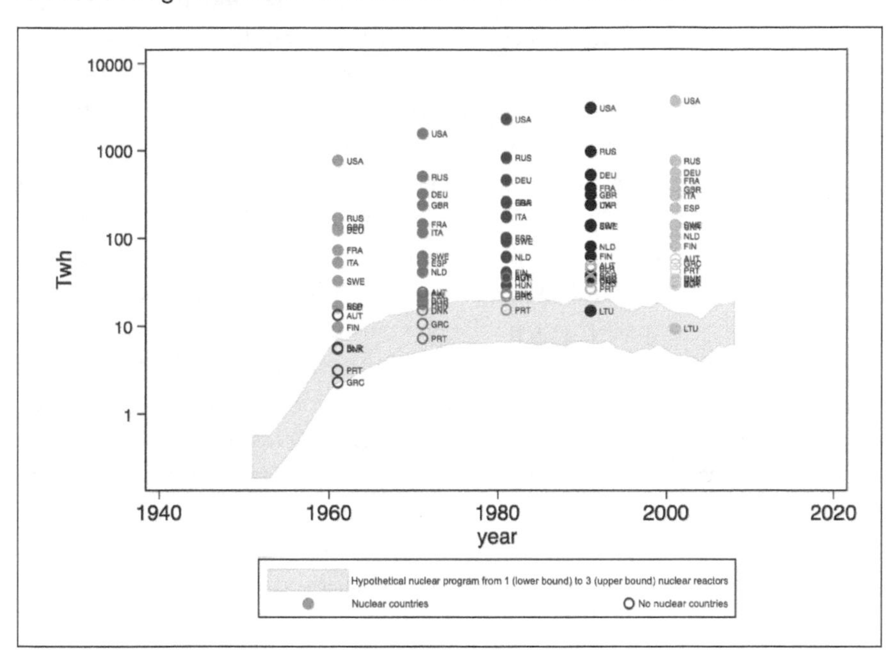

Source: Author's elaboration from the data on total electricity consumption from IEA Statistics, OECD/IEA 2014. Data about size of reactors being built from PRIS-IAEA database.

a 1987 referendum, Italy opted out of nuclear power. Hence, one must observe other variables besides the mere need to provide electricity to explain the decision to opt for nuclear technology.

The scale issue was crucial for the utilities too. The size of the reactor determined the needed upfront investment, which in turn meant that not all utilities could undertake nuclear projects. In the US and Spain, where the utility industry was fragmented among a number of private utilities, few had large enough markets and deep enough pockets to undertake projects with more than two reactors at a single site. To overcome these issues, privately owned utilities created joint ventures and consortia for specific nuclear projects. In Spain, most reactors were built and operated by joint ventures of privately owned utilities. In other countries, ownership expanded beyond the electricity sector, inviting co-ownership of the nuclear plant by other industries (different from electricity companies) and/or public bodies (for instance, municipalities partially own nuclear power plants in Sweden, Finland, the Netherlands, and Germany). In

Sweden, six of the twelve nuclear power plants were owned by consortia. In contrast, European nations where the electricity sector remained under the control of a single state-owned utility (see table 2.1) could accommodate larger reactors in groups of four or more.[42] Their electricity markets were integrated into one large single market, rather than unconnected regional/local networks too small to accommodate the standard nuclear technology.

Beyond the size/scale effect, the type of ownership has a crucial impact on nuclear projects. While for private companies profitability constitutes the primary decision criterion, a public investor may value other benefits for society (job creation, energy independence, industrial development, etc.) over business profitability. One aspect that a public planner took into consideration while pondering the nuclear option was the electricity mix. It is worth investigating the electricity-generation alternatives available for the different countries. For the three dozen nuclear-powered countries, nuclear electricity has very different weights in their respective electricity mixes, ranging from less than 2 percent to almost 80 percent of the total electric consumption from 1960 to 2014.[43] Nuclear electricity's mean share of the world's total electricity consumption was 12.5 percent in the same period. Variation is wide also between the countries analyzed by HoNESt (see figure 2.5).

In the twenty countries studied by HoNESt, only in France did nuclear manage to become the principal source of electricity. In Bulgaria, Finland, Sweden, and Ukraine, nuclear matched the contribution of the second major type of generation (either fossil fuels or renewables). Hydroelectricity (the largest renewable provider) made significant contributions in Austria and Sweden, while its relative contribution declined in Italy, Portugal, and Spain. In the Netherlands and the United Kingdom, new discoveries of natural gas helped to keep fossil fuels as the primary source for electricity generation. It seems that in Italy, the pressure from the fossil fuel lobby contributed to the decision to abandon nuclear.[44] In fact, and despite the oil crises, fossil fuels (coal and gas, but hardly any oil in the last thirty-five years) have remained the preferred technology for electricity generation in fifteen of the twenty countries, regardless of whether nuclear was added to the mix.

This analysis shows that (1) the key economic questions underpinning the nuclear choice differed for large and small electricity consumers (even though all of them faced decreasing growth in the demand for electricity); (2) types of ownership (public, private, mixed) of the nuclear projects affected decision-making; and (3) the choices regarding the electricity mix depended on endogenous endowment with other energy sources.

Table 2.1. Types of ownership reported in HoNESt consortium's cases for commercial nuclear reactors

Nuclear producers in HoNESt	State owned	Private utilities	Mixed
Bulgaria	*		
France	*		
Finland			*
Germany (Democratic Rep.)	*		
Germany (Federal Rep.)		*	*
Hungary	*		
Italy	*	*	
Lithuania	*		
Netherlands			*
Russia	*		
Spain		*	*
Sweden	*		*
Ukraine	*		
United Kingdom[a]	*	*	
United States		*	
No nuclear countries (projected ownership)			
Austria	*		
Belarus	*		
Denmark		*	*
Greece	*		
Portugal	*		

Note: When more than one type of ownership is reported for a country, it implies that there are nuclear power plants with distinct types of ownership. In Italy, for instance, one reactor was state owned, another privately owned by utilities.

"State owned" indicates at least one reactor reported as fully owned by the central government at any point in time.

"Private utilities" indicates at least one reactor reported as fully owned by one or more private utilities at any point in time.

"Mixed" includes reactors reported as having equity owned by public and private bodies at the same time.

[a] In the UK, nuclear power plants were state owned until the 1990s and privatized thereafter, except the Magnox reactors that remained under public ownership. Author's elaboration from HoNESt SCRs.

Fig. 2.5. Electricity market shares of different technologies for HoNESt's consortium countries.

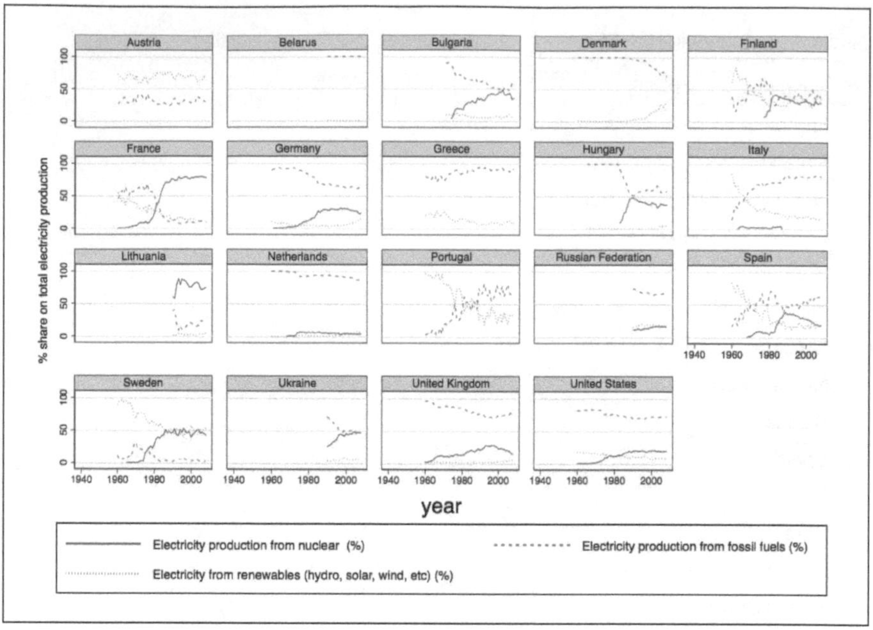

Source: Author's elaboration from the data by International Energy Agency, IEA Statistics, OECD/IEA (2014). Note that data for former Soviet bloc countries is only available after 1990.

Industrial Base and Nuclear Markets: Domestic or Imported Nuclear Technology?

Choosing nuclear required not only a sufficient demand for electricity but also an industrial base allowing the country to develop its own nuclear technology and/or make use of imported technology. As Euratom stated in 1957, "Scientific and technological knowledge can be borrowed, but industrial capacity one must create oneself."[45] Historically, there has been a strong positive correlation between electricity consumption and industrial development (figure 2.6). Industry has been the primary driver for increasing electricity demand: while in the 1960s industry consumed over 75 percent of the electricity generated, in the 1990s it still represented 46 percent of total final consumption of electricity in the European Union (27 members).[46]

It is useful to distinguish between the "ability to use" and the "ability to produce" technology.[47] The construction and commissioning of a nuclear power

Fig. 2.6. Industrial added value and electricity consumption in the world (1995–2008), nuclear vs. nonnuclear countries.

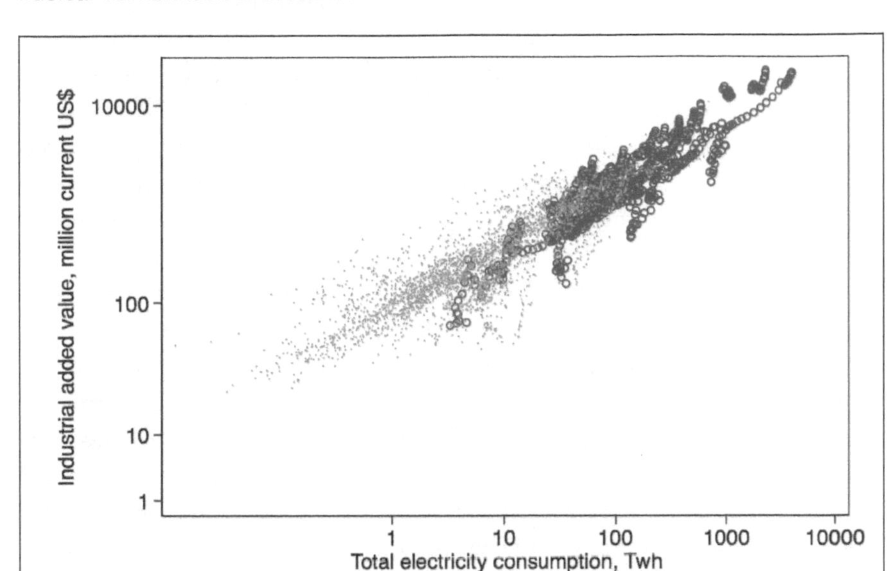

Source: Author's elaboration from World Bank Database.

plant is a clear example of this distinction. To receive the technology transfer, a host country had to have a trained scientific community and a group of companies capable of taking on the nuclear challenge.[48] Accordingly, the countries in figure 2.6 could be classified as follows:

1. Too small (industrially and electricity-wise) to opt for nuclear power (figure 2.6, bottom left)
2. Large industrial countries, all of which adopted nuclear and also become exporters of the technology (figure 2.6, upper right)
3. The intermediate group, which had a sufficient electricity demand and industrial base to accommodate a nuclear program. If they opted for nuclear, most ended up importing the technology from the countries in the top right, despite their initial aspirations to build their own reactor.[49] This group includes some countries that eventually developed strong nuclear capabilities, starting from imported technology and licenses.

On the one hand, technology exporters (i.e., US, UK, and USSR and later France, Germany, and Sweden) had to decide whether to invest public money to foster the export of nuclear technology. While some (the US, the USSR, France, and Germany) did so without hesitation, others either had fewer resources to do so (Sweden) or were more hesitant to do so (the UK). On the other hand, for technology importers (Southern and Eastern Europe, but also the Netherlands and Finland, for example), acquiring foreign nuclear technology required specialized knowledge of the alternatives available on the market. Engineers and scientists were at the core of all commercial negotiations and transactions involving nuclear technology.[50] To understand the issues at stake, a brief look at the global market for nuclear reactors in the 1950s–1980s is required.[51]

The global sales of nuclear reactors constituted a tight market. Only about 100 reactors were ordered internationally between 1955 and 1980, while the rest were built domestically (see table 2.2). By the early 1950s, it was clear that nuclear science could not remain an American monopoly and that its spread was inevitable. By taking an active role in assisting foreign nuclear programs, the US government influenced other nations' nuclear policies, shared their technological developments, and obtained guarantees on safeguarding nuclear materials.[52] Across the Iron Curtain, the Soviet Union followed a similar active strategy to launch civil nuclear programs in the countries under its influence.[53]

The light-water reactor, fueled by low-enriched uranium and cooled and moderated by ordinary water, was the US alternative to the more expensive gas-cooled reactors built by the Europeans in the 1950s.[54] Two US manufacturers, Westinghouse and General Electric, became major developers of light-water reactors, specializing in pressurized and boiling-water reactors, respectively. By the early 1960s, demonstration power reactors were in operation in all leading industrial countries, and expectations were high. In December 1963, the US introduced the idea of "turnkey" plants, with a bid for the construction of a plant in Oyster Creek, New Jersey.[55] The main advantage was that turnkey plants were offered at a guaranteed fixed price, set in advance, competitive with coal and oil-fired alternatives. In fact, before turnkey projects came about, international nuclear sales only came drop by drop. By 1964, the US had received seven international orders for nuclear reactors, while the UK sold two reactors (to Japan and Italy[56]) and the USSR one to East Germany (see table 2.2).[57]

Even though they proved to be bad business for manufacturers in the long term,[58] turnkey projects propelled domestic and international sales of nuclear reactors. With the help of US economic diplomacy and the financial assistance of the Export–Import Bank (see below), by the mid-1970s General Electric and Westinghouse had captured about 80 percent of the international sales of

Table 2.2. Global nuclear export orders (number of reactors) 1955–1980

		1955–1964	1965–1970	1971–1973	1974–1980	Total export orders
Global nuclear export orders*		**9**	**23**	**27**	**36**	**95**
Suppliers	Canada		3	2	1	6
	France		1		9	10
	Germany (Democratic Rep.)		2	2	8	12
	Sweden			1	1	2
	UK	2				2
	US	7	17	22	17	63
	Of which EXIM financed	*3*	*11*	*20*	*16*	*50*
	USSR	1	8	28	37	
Ratios (percent)						
	US /world*	78%	74%	81%	47%	66% (48% incl. USSR)
	EXIM/US	43%	65%	91%	94%	79%

Sources: M.d.M. Rubio-Varas and J. De la Torre, "Spain—Eximbank's Billion Dollar Client: the Role of the US financing the Spanish Nuclear program" in *Electric Worlds: Creations, Circulations, Tensions, and Transitions, 19th–21st Centuries*, ed. A. Beltran, L. Laborie, P. Lanthier, and S. Le Gallic (London: Peter Lang, 2016), 245–70.

nuclear reactors to the Western countries. During the second half of the 1970s, other Western manufacturers, who had been gaining experience by building nuclear plants in their countries, came to compete in the international market. Among these were the German Kraftwerk Union, the French Framatome, and the Canadian AECL. The British mostly failed in their attempts to export their Magnox reactor. Meanwhile, the Soviet Union built another fifty reactors in its territory and in the countries within its political orbit. On top of this, the virtual US monopoly on enrichment services ended abruptly in 1974 with the Soviet Union's decision to sell enriched uranium to Western countries.[59] By 1975, the curve of orders for nuclear reactors had already passed its peak. Furthermore, over two-thirds of all nuclear plant orders placed after January 1970 were eventually canceled.[60] By the early 1980s, a dozen companies competed with the US multinationals supplying the core elements of the reactor. Few, however, could export entire nuclear projects beyond their own borders, and few are able to do so still today. To understand the reasons behind this short supply of international tenders for nuclear power, we have to look at the financial constraints involved.

Financial Constraints: The Borrowing Challenge and the Onset of Neoliberalism

As with any capital-intensive investment, a major challenge in developing a new nuclear project is mobilizing the resources to finance it. The financing payback periods for new-build nuclear projects are typically relatively long, which creates a need for an equally long-term financing (which may not be readily available).[61] Given the economics of nuclear new-build's sensitivity to the cost of capital, it is crucial that project developers secure as low a cost of capital as they possibly can.

Depending on the ownership and commercial structure that is chosen to develop a new nuclear project, the financing may come directly from the government (in other words, from taxpayers and public debt) or, if undertaken by the private sector, from a combination of equity—that is, the amount of the funds contributed by the owners—and (long-term) loans from national and international sources. Crucially, capital costs differ according to the kind of promoter in question (public, private, or mixed), as explained in a study for the European Commission:

> Public investors (say governments or state-owned institutions) have access to cheap capital when borrowing money. The interest rates on government bonds are usually relatively low compared to interest charged for financial loans taken up by private actors. For that reason, the financing of public projects is basically done though the issue of bonds, and this for 100 percent. In the electricity generation business in Europe, this public investor situation may apply to the largely state-owned companies such as Electricité de France (EDF) and Vattenfall (Sweden). Private investors in liberalized markets operate in an uncertain environment, and their interest rates are the highest, depending on the rating of the company and the type of project. . . . The Finnish example with large customers

acting as coinvestors is an intermediate case that lowers the cost of capital for the investors.[62]

The pioneering countries of nuclear technology were able to tap into the financial resources required domestically; in addition, most of them helped to finance their exports of domestically manufactured reactors to third parties (including the Soviet exports to Eastern Europe): "With only a few exceptions, public financing institutions of the principal supplier nations undertook external financing for the export of nuclear power projects."[63] The US Export–Import Bank (EXIM bank) played this role in the US, contesting other government-sponsored export-financing institutions.[64] Almost all industrialized countries had official export credit agencies through the 1970s and 1980s.[65] In fact, by the mid-1970s, according to a US report that analyzed the global nuclear markets, "financing by the supplier's government became more important to customers than the overall cost evaluation of the project."[66]

Between 1955 and 1985, the EXIM bank financed more than half of all the international sales of nuclear reactors to the Western world, as shown in table 3.2. By the late 1970s, all US reactor exports but one—sold to Switzerland—came with an EXIM bank financial package.[67] One can therefore hypothesize that the US government's decision to pump public money into exporting nuclear facilities explains a great deal its share in the global nuclear market before the 1980s. To the dismay of British, French, and German manufacturers that were also bidding for the international nuclear market, their American counterparts systematically won most of the bids for nuclear reactors and technical assistance launched by European utilities until the mid-1970s. The US financial assistance, both directly through EXIM bank loans and indirectly by guarantees to private loans for the nuclear projects, made it impossible for European manufacturers to compete with the Americans.[68]

The tide began to turn in the mid-1970s, and even more so after the Three Mile Island accident in 1979. Critics of nuclear power questioned the environmental, developmental, and diplomatic consequences of EXIM bank's assistance to exporters of nuclear power plants. Investigations were launched within the bank but also in US governmental agencies, the Senate, and the Congress.[69] In fact, EXIM bank's mere existence came under attack as a new economic paradigm began to gain foothold: neoliberalism. As a particular regime of liberalism, capitalism, and democracy, neoliberalism has been globalized since the 1970s in the form of an active state promotion of market and competition principles.[70] The switch in US policy away from subsidies and a smaller role for the state in the economy implied a smaller role for EXIM bank. Nuclear loans by the EXIM bank became more exceptional until their total cessation by 1985.[71] In other words, the financial facilities provided by the historically largest exporter of nuclear reactors came to an end by the mid-1980s, before Chernobyl. Neoliberalism affected, to varying degrees, the economic role of the state across the Western world.[72] Thus, the number of nations willing to support the exports of nuclear technology and services with public money declined. The existence of the credit (subsidized or not) needed to be matched on the buyer side, with companies of sufficient size and solvency willing to accept the required financial commitments for ten to fifteen years. EXIM bank did not concern itself with the wisdom of the nuclear purchases from the buyer's point of view.[73]

Interest Rates and Nuclear Investments

"The economics of new nuclear plants are heavily influenced by their capital cost, which accounts for at least 60 percent of their levelized cost of electricity. Interest charges and the construction period are important variables for determining the overall cost of capital."[74] When interest rates are high, projects with high initial capital costs, such as nuclear, are disadvantaged in comparative financial appraisals with alternative technologies. A 2015 NEA report made the important point regarding interest rates and the competitiveness of nuclear projects:

> At a 3 percent discount rate, nuclear is the lowest cost option for all countries. However, consistent with the fact that nuclear technologies are capital intensive relative to natural gas or coal, the cost of nuclear rises relatively quickly as the discount rate is raised. As a result, at a 7 percent discount rate the median value of nuclear is close to the median value for coal [but lower than the gas in CCGTs], and at a 10 percent discount rate the median value for nuclear is higher than that of either CCGTs or coal. These results include a carbon cost of $30/tonne, as well as regional variations in assumed fuel costs.[75]

In other words, nuclear projects can only compete when the cost of capital is below 10 percent.[76] Given the crucial role of interest rates in the viability of nuclear projects, the last section of the chapter analyzes their evolution. The issue again is that countries in different positions will face different issues regarding the capital costs. In those countries where nuclear was developed domestically by a state-owned company and financed from within their own market, the relevant interest rate would be the one applied to government bonds. Private promoters had to accept the higher lending interest rates offered in the market, including a risk premium (i.e., a higher interest rate) since private lenders were reluctant to finance nuclear projects due to their size and long maturity. For nuclear projects, a premium of up to three additional percentage points is not unusual. The real interest rate discounts from the nominal interest rate the effect of inflation.

Figure 2.7 helps explore the relationship between real interest rates and nuclear projects. It describes the evolution of the real interest rates in the countries that contributed most to financing nuclear projects both domestically and abroad: the United States, Germany, France, and the United Kingdom. These interest rates reflect the average market conditions for credit in these countries: public-related projects faced lower interest rates to these ones, but private promoters of nuclear projects certainly confronted higher rates than the ones reflected in figure 2.7. Against the interest rates we have also plotted the number of new reactor projects started in any given year in the countries studied within HoNESt, classified by the type of promoter/ownership at the inception..

Until 1977, real interest rates remained very low, thus making nuclear projects attractive for private investors as well as for many governments. Real interest rates began to increase in the mid-1970s in the US and Germany, and in the mid-1980s in France and the UK. From the mid-1980s until the end of the 1990s, interest rates remained on the edge of what would allow nuclear to remain attractive for the private sector. No private nuclear project could be competitive at the very high interest rates of the early 1980s. In countries where the state backed up the nuclear projects with taxpayer money and public debt rather than private loans, the financial cost was less onerous but still much higher than a decade earlier. The neoliberal wave meant a decreasing role for the state, and it became increasingly difficult for state-owned companies to defend nuclear projects. Not surprisingly, a nuclear renaissance seemed to begin when interest rates fell from 5 percent at the beginning of the twenty-first century (in the US, the UK, and France, but not Germany). The Fukushima accident in 2011 again made nuclear new-build difficult, if not impossible, due to its impact on risk perceptions.

If financing a nuclear project became increasingly difficult for those countries that could raise the capital within their own borders, the evolution of the exchange rates made it nearly impossible for those that had to import the technology and search for funding in international markets. It is worth remembering that up to the suspension of the convertibility of the dollar in August 1971, exchange rates were semifixed under the Bretton Woods system. The end of the convertibility implied a new uncertainty as the values of currencies varied according to market conditions. In the case of the US, the devaluation of

Fig. 2.7. Reactors in HoNEST partner countries, by year of construction start and ownership at inception (piled bars), and evolution of real interest rates 1960–2002 (lines)

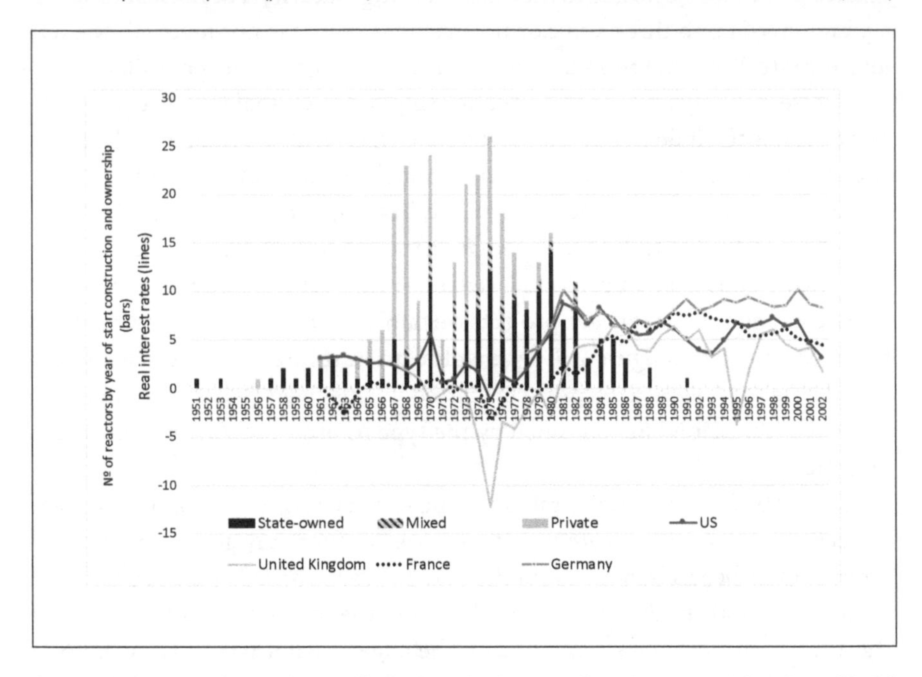

Source: Author's elaboration from PRIS database for the year of starting construction. See table 3.1 for ownership and World Development Indicators, World Bank for real interest rates. Note that the real interest rate is the lending interest rate adjusted for inflation. Deposit and lending rates are collected by the International Monetary Fund as representative interest rates offered by banks to resident customers. A negative real interest rate indicates a loss in the purchasing power of the principal.

the dollar made US imports cheaper for the rest of the world until 1979, when the dollar began a rally that made it much more expensive over the following decade. Other major currencies responded in varying ways to the wide swings in the value of the dollar. In a little more than four years starting in 1977, the British pound appreciated by one-third in nominal terms and by an astonishing (and unprecedented for a reserve currency) 87 percent in real terms. The German mark and the French franc were stable in real terms until the second half of the 1980s. Most (between 65 and 80 percent) international bank loans were denominated in dollars, as a consequence of which all those countries, companies, or institutions that had to pay back their international loans suddenly found that the costs in their national currencies had more than doubled. This was particularly problematic for electric utilities, which were also facing a slowdown in their business given the collapse in electricity demand, as we saw in the first section of the chapter.

For Eastern Europe, we can only apply this same logic for the period since the early 1990s. The transition to market economies led to astronomical real interest rates in most of the Eastern countries studied within HoNESt (close to 100 percent). Even if these countries tried to secure financing by resorting to their national financial markets only (if they had a sufficiently wide capital market), the interest rates were subject to heavy oscillations. Such uncertainty obviously made it very difficult to plan long-term projects.

Throughout the 1970s, nuclear plant safety regulations were tightened, something that began before the Three Mile Island accident in 1979, but which was certainly amplified further after the accident. Some speak in this context of "regulatory ratcheting." [77] As a result, the increasing complexity of the plants made construction more expensive, and redesigns and adaptations had to take place during construction. In addition, construction periods tended to become much longer, and this in a time of high interest rates as shown by figure 2.7. Thus, "the interest during construction became one of the killing factors in the cost of nuclear power plants." [78]

Conclusions

The economic context, which varied across time and space, influenced nuclear decisions. The development of a civil nuclear power plant was subject to a number of very basic economic conditions requiring sufficient electricity demand to justify the construction of an average-sized reactor. It also required a minimum industrial base to develop the nuclear technology and/or accommodate the technological transfer. Some countries did not fulfill these two initial preconditions, which helps us to better understand why they opted out of nuclear. The context was also different for countries with large domestic endowments of energy sources, which had to face a different set of issues from those highly dependent on imports of energy from abroad. Technology exporters also had to solve a different set of economic questions than the countries that imported nuclear technology. Moreover, the countries where the nuclear endeavor depended on the private sector faced distinct economic issues (e.g., the possibility of bankruptcy) compared to those countries in which a public utility constructed and operated nuclear power plants (financed by electricity consumers but backed up by taxpayers).

In addition to a sufficiently large domestic market for electricity and the necessary industrial base for developing a nuclear program, additional economic/

industrial conditions needed to be met, such as low interest rates and a strong financial position of the country and/or its electricity utilities, all accompanied by a sufficiently developed financial market. From the end of the 1970s and into the early 1980s, nuclear power faced a perfect storm, which undermined its economics. Two reasons have frequently been highlighted: the costs of increasingly stringent safety measures, and, from the late 1970s onward, the inclusion of the costs of waste management. These two factors are largely attributable to public pressure and safety concerns. This chapter adds two concurrent factors that further undermined the economics of nuclear projects: the fact that financing became more expensive and more difficult to obtain, and the slowdown in the growth of electricity demand. The complications that affected the financing of nuclear projects came from two key fronts: increased interest rates, and the advent of neoliberalism—and thus a reduced role for the state, which among other things meant abandoning government financing of nuclear power (exemplified by the end of funding from the US Export–Import Bank, the world's single largest nuclear financier).

This chapter has emphasized that economic aspects were prominent in framing nuclear decisions, but sociopolitical aspects (e.g., public trust, quality of democracy) were of course also important in the decision-making processes, as many of the chapters in this volume make evident.

Notes

Institute for Advanced Research in Business and Economics, Universidad Pública de Navarra (INARBE-UPNA). Contact: mar.rubio@unavarra.es. The Spanish Ministry of Economy and Competitiveness and the European Regional Development Fund (ERDF) (project ref. HAR2017-86086-R (AEI/FEDER, UE)) financed parts of the meetings and research required to complete this chapter. The bulk of the research received funding from the Euratom research and training program 2014–2018 under grant agreement N°662268. Despite this, the views expressed here, as well as any errors and/or omissions, are entirely the responsibility of the author.

1. Warren Young, *Atomic Energy Costing* (New York: Springer US, 1998); Steve Cohn, "The Political Economy of Nuclear Power (1945–1990): The Rise and Fall of an Official Technology," *Journal of Economic Issues* XXIV, no. 3 (1990): 781–811; Brian Balogh, *Chain Reaction: Expert Debate and Public Participation in American Commercial Nuclear Power, 1945–1975* (Cambridge: Cambridge University Press, 1991).
2. Steve Thomas, "Is the Real Cost of Nuclear Power Plants Increasing?" *Energy Policy 102* (2017): 639; Noah. J. D. Lucas, "Economies of Scale in Nuclear Power Plant," *International Journal of Energy Research* 3, no. 3 (1979): 297–300; Ari Rabl, "External Costs of Nuclear: Greater or Less Than the Alternatives?" *Energy Policy* 57, no. C (2013): 575–84.
3. Pabrita L. De and Edward G. Delaney, "Overview of Cost Estimates and Financing

Practice," *IAEA Bulletin* (Winter 1985): 13–15, https://www.iaea.org/sites/default/files/27405091315.pdf.
4. IAEA, *Costing of Spent Fuel Storage* (IAEA Nuclear energy series No. NF-T-3.5, 2009), 3, https://www.iaea.org/sites/default/files/costingfuel0909.pdf.
5. A review of the bibliography listed in the IAEA *Cost Analysis Methodology of Spent Fuel Storage* (Technical report series no. 361, 1994) supports this assertion. For an analysis of the historical evolution of the reactions of society to spent-fuel management, see chapter 6 in this volume.
6. Usually calculated using the levelized cost of energy (LCOE), which estimates the average lifetime cost of power production per MWh. The LCOE typically include investment costs, fuel costs, operation and maintenance costs, environmental externalities, system costs, and heat revenue for combined heat and power plants. Some recent comparative estimations can be found at OECD-NEA, *The Full Cost of Electricity Provision* (NEA no. 7298, 2018).
7. We are aware of the wide differences across what are known as antinuclear movements, which are discussed in chapter 3 of this volume.
8. According to the first entry under "society" in the *Cambridge Academic Content Dictionary* (2008).
9. All the twenty SCR are available at https://hdl.handle.net/2454/38269.
10. Wilfried Konrad et al., "Comparative Cross-Country Analysis on Preliminary Identification of Key Factors Underlying Public Perception and Societal Engagement with Nuclear Developments in Different National Contexts," *HoNESt Consortium Deliverable D.4.2* (2018). See also chapter 5.
11. Mar Rubio-Varas and Joseba De la Torre, "Seeking the Perennial Fountain of the World's Prosperity," in *The Economic History of Nuclear Energy in Spain* (Cham: Springer International Publishing, 2017), 1–32, doi:10.1007/978-3-319-59867-3_1.
12. Nathan Hultman and Jonathan Koomey, "Three Mile Island: The Driver of US Nuclear Power's Decline?" *Bulletin of the Atomic Scientists* 69, no. 3 (May 1, 2013): 63–70, doi:10.1177/0096340213485949.
13. The nuclear countries included in HoNESt are Bulgaria, France, Finland, East and West Germany, Hungary, Italy, Lithuania, the Netherlands, Russia, Spain, Sweden, Ukraine, the United Kingdom, and the United States.
14. The countries studied in HoNESt that considered nuclear but renounced it are Austria, Belarus, Denmark, Greece, and Portugal.
15. A first attempt has been the panel "The Atomic Business: Industrial, Financial and Economic Issues of the Development of Nuclear Power over the 20th Century" at the World Economic History Congress in Boston, MA, in August 2018. Program available at: http://wehc2018.org/the-atomic-business-industrial-financial-and-economic-issues-of-the-development-of-nuclear-power-over-the-20th-century/.
16. OEEC, *European Nuclear Energy Agency, The Industrial Challenge of Nuclear Energy. Stresa Conference, Vol. III Survey of European Programmes. Economics of Nuclear Power and Financing Programmes* (New York: OEEC, 1959); I.R. Maxwell, P.W. Mummery, and Philip Sporn, eds., *Progress in Nuclear Energy: The Economics of Nuclear Power* (New York: Pergamon Press, 1959); Federal Power Commission, *The 1970 National Power Survey* (Washington, DC: US Government Printing Office, 1971), chapters 2, 6, and 20.
17. Jonathan Scurlock, "A Concise History of the Nuclear Industry Worldwide," in

Nuclear or Not? Does Nuclear Power Have a Place in a Sustainable Energy Future?, ed. David Elliott (New York: Palgrave Macmillan, 2007), 24–33.

18. Scott Victor Valentine and Benjamin K. Sovacool, "The Socio-Political Economy of Nuclear Power Development in Japan and South Korea," *Energy Policy* 38, no. 12 (December 2010): 7971–79, doi:10.1016/j.enpol.2010.09.036. Comparing Japan and South Korea, the authors find the following shared features: (1) strong state involvement guiding economic development; (2) centralization of national energy policymaking; (3) campaigns to link technological progress with national revitalization; (4) influence of technocratic ideology on policy decisions; (5) subordination of challenges to political authority; and (6) low levels of civic activism.
19. Rubio-Varas and De la Torre, "Seeking the Perennial Fountain of the World's Prosperity."
20. United States Comptroller General's Report to the Congress, *U.S. Nuclear Non-Proliferation Policy: Impact on Exports and Nuclear Industry Could Not Be Determined* (Washington, DC: US General Accounting Office, 1980), 38, https://www.gao.gov/assets/140/130475.pdf.
21. Louis Armand, Franz Etzel, and Francesco Giordani, *A Target for Euratom: Report submitted by Mr. Louis Armand, Mr. Franz Etzel and Mr. Francesco Giordani at the request of the governments of Belgium, France, German Federal Republic, Italy, Luxembourg and the Netherlands* (May 1957), 13–40, https://www.cvce.eu/content/publication/1997/10/13/e72917a4-3c9d-48b1-b8cb-41307736731e/publishable_en.pdf.
22. See chapter 8.
23. Angus Maddison, *Monitoring the World Economy, 1820–1992* (Paris: OECD Development Centre Studies, 1995).
24. This chronology matches the one presented in other chapters in this volume. See chapter 1. The evolution and role of the international organizations in the nuclear realm are also discussed in chapter 4.
25. There was an array of possible designs for nuclear reactors. The basic differences depended on the fuel, the cooling, and the moderating elements, and until the late 1960s there were no clear advantages to any of them. Eventually, three types of reactors were commercialized internationally: (1) the light-water nuclear power reactor, using low-enriched uranium as its fuel and ordinary water as its coolant and moderator, built originally to a US design in Western countries and to a similar Soviet design in the USSR and Eastern European countries; (2) the gas graphite reactor using natural uranium as its fuel, moderated by graphite and cooled by carbon dioxide—a technological design favored by Britain and France; (3) a quite different nuclear power reactor, marketed by Canada, using natural uranium as its fuel and heavy water as its coolant and moderator. Within each main type there were further design categories. For instance, among the light-water reactors, pressurized (PWR) and boiling water (BWR) reactors were developed in the West, while the Soviets developed the VVER series and the RBMK, the latter made conspicuous by Chernobyl. David Fischer *History of the International Atomic Energy Agency: The First Forty Years* (Vienna: The Agency, 1997), 149.
26. Most of HoNESt's SCRs refer to the electricity needs forecasts made during the 1960s in identical terms: doubling every decade or so.
27. US embassies throughout the world received orders to explain the advantages of

nuclear power to their host economies. The United States embassy in Madrid responded to the request for opportunities in the following terms on 8 March 1974: "Embassy suggests following areas presenting either new opportunities for US exports or good prospects for increased US export share of Spanish market, as a result of higher petroleum and raw material prices and supply shortages: 1. Technical assistance and equipment sales for Spain's expanding nuclear energy and thermal energy industry. . . . While Westinghouse and GE as well as US design firms, e.g. Bechtel, Foster Wheeler, have played major role in Spanish nuclear energy program, US industry must continue to be energetic in meeting other competition sources in order exploit increasingly attractive possibilities" Source: Telegram from the US Embassy in Madrid to the Secretary of State, Washington, Subject: Trade and investment opportunities arising from energy situation. Document Number 1974MADRID01553, National Archives at College Park, College Park, MD.

28. Steve Cohn, *Too Cheap to Meter: An Economic and Philosophical Analysis of the Nuclear Dream* (Albany: State University of New York Press, 1997), 127.
29. The 1970s were a decade of significant opposition to nuclear power in Europe. In 1978 Austria rejected nuclear power in a general referendum; in the late 1970s opposition stopped Irish and Portuguese attempts at nuclear development; and in 1980 Swedish voters approved a referendum to phase out the country's operating nuclear power plants. See Christian Forstner, "Austria Short Country Report (version 2018)," in *History of Nuclear Energy and Society (HoNESt) Consortium Deliverable no. 3.6* (2019), https://hdl.handle.net/2454/38269; Arne Kaijser, "Sweden Short Country Report (version 2018)," in *History of Nuclear Energy and Society (HoNESt) Consortium Deliverable no. 3.6* (2019), https://hdl.handle.net/2454/38269.
30. "In the situation of an economic crisis in which the GDR was, the development of the nuclear energy economy was terminated." Bernd Helmbold, "German Democratic Republic (GDR) Short Country Report (version 2018)," in *History of Nuclear Energy and Society (HoNESt) Consortium Deliverable no. 3.6* (2019), 49, https://hdl.handle.net/2454/38269. "Starting from the period of Independence the national economy and energy systems faced the problem of managing the oversupply of energy in the context of the decreasing energy demand, economic crises, and changing economic and political relations in the post-Soviet region." Andrei Stsiapanau, "Lithuania SCR Short Country Report (version 2018)," in *History of Nuclear Energy and Society (HoNESt) Consortium Deliverable no. 3.6* (2019), 19, https://hdl.handle.net/2454/38269.
31. The second to eleventh positions on the list of the Western world's largest electricity consumers in 1965 were (in decreasing order of consumption) United Kingdom, Japan, Germany, Canada, France, Italy, Sweden, Norway, Poland, Australia, and Spain.
32. Tony Roulstone, "Economies of Scale vs. Economies of Volume," *Nuclear Engineering International* (August 2015), https://www.neimagazine.com/features/featureeconomies-of-scale-vs-economies-of-volume-4639914/. Yet the issue has been widely discussed, as larger reactors did not in fact prove to be capable of reducing the cost per kWh. Building larger reactors to achieve economies of scale led to longer construction times, offsetting the cost savings of larger reactors. See Mark Cooper, *Policy Challenges of Nuclear Reactor Construction, Cost Escalation and Crowding out Alternatives* (2010), http://citeseerx.ist.psu.edu/viewdoc/download?doi=10.1.1.1

75.2423&rep=rep1&type=pdf. It is a trait shared by other technologies. See chapter 9 in Nuno Luis Madureira, *Key Concepts in Energy* (New York: Springer), 181–203.

33. For example, a generator with 1 megawatt (MW) capacity that operates at that capacity consistently for one hour will produce 1 megawatt per hour (MWh) of electricity. If the generator operates at only half of that capacity for one hour, it will produce 0.5 MWh of electricity. Many generators do not operate at their full capacity all the time.
34. This is often the point of departure of studies investigating the introduction of small and medium reactors and small modular reactors (SMRs). IAEA, *Deployment Indicators for Small Modular Reactors*, IAEA-TECDOC-1854 (2018), https://www-pub.iaea.org/MTCD/Publications/PDF/TE-1854web.pdf.
35. See chapter 9.
36. Comparative data on electricity exports obtained from the *CIA Factbook* (https://www.cia.gov/the-world-factbook/field/electricity-exports/country-comparison), which can then be divided by the data on electric generation from BP statistics or an alternative source.
37. Mario D. Carelli et al., "Economic Comparison of Different Size Nuclear Reactors," *Proceedings of the International Joint Meeting (IJM) of the Latin American Section of the American Nuclear Society* (2007): 654, https://www.researchgate.net/publication/228463939_Economic_comparison_of_different_size_nuclear_reactors.
38. Fischer, *History of the International Atomic Energy Agency*, 166.
39. George Holliday, *Eximbank's Involvement in Nuclear Exports* (Washington, DC: Congressional Research Service, GPO, March 2, 1981), 20. Box L1, Folder 277. Ex-Im Bank Archives.
40. Standard size of a reactor of 500–1000 MW(e), assuming they would work at 80 percent of the hours of a year, multiplied by 7000 hours.
41. World Nuclear Association, *Nuclear Energy in Demark (updated February 2017)*, http://www.world-nuclear.org/information-library/country-profiles/countries-a-f/denmark.aspx.
42. Anthony C. Krautmann and John L. Solow, "Economies of Scale in Nuclear Power Generation Economies of Scale in Nuclear Power Generation," *Southern Economic Journal* 55, no. 1 (1988): 70–85.
43. Averages calculated over the data by International Energy Agency, IEA Statistics, OECD/IEA (2014), http://www.iea.org/data-and-statistics.
44. Matteo Gerlini, "Italy Short Country Report (version 2017)," in *History of Nuclear Energy and Society (HoNESt) Consortium Deliverable no. 3.6* (2019), https://hdl.handle.net/2454/38269.
45. Armand, Etzel, and Giordani, *A Target for Euratom*.
46. European Environment Agency, "Final Electricity Consumption by Sector EU-27," last modified March 19, 2013, https://www.eea.europa.eu/data-and-maps/figures/final-electricity-consumption-by-sector-5.
47. Frederik Tell et al., eds., *Managing Knowledge Integration Across Boundaries* (Oxford: Oxford University Press, 2017).
48. Richard G. Hewlett and Jack M. Holl, *Atoms for Peace and War, 1953–1961: Eisenhower and the Atomic Energy Commission* (Berkeley: University of California Press, 1989).
49. Countries such as Denmark, Portugal, Spain, and Sweden cherished the idea of a

nuclear program relying on domestically mined natural uranium. As elsewhere, the idea was abandoned in the mid-1960s, when turnkey American light-water reactors relying on enriched uranium entered the market. Mar Rubio-Varas et al., "Siting (and Mining) at the Border: Spain-Portugal Nuclear Transboundary Issues," *Journal of History Environment and Society* 3: 33–70, https://doi.org/10.1484/J.JHES.5.116794.

50. Joseba De la Torre et al., "Engineers and Scientist as Commercial Agents of the Spanish Nuclear Programme," in *Technology and Globalisation. Networks of Experts in World History*, ed. David Pretel and Lino Camprubí (London: Palgrave Macmillan, 2018), 313–40.
51. Much of the following paragraphs derives directly from Mar Rubio-Varas and Joseba De la Torre, "Spain—Eximbank's Billion Dollar Client: The Role of the US financing the Spanish Nuclear program," in *Electric Worlds: Creations, Circulations, Tensions, and Transitions, 19th–21st Centuries*, ed. A. Beltran et al. (London: Peter Lang, 2016), 245–70.
52. US Comptroller General, *U.S. Nuclear Non-Proliferation Policy*, 4.
53. Matthew Adamsom and Gabor Pallo, "Hungary Short Country Report (version 2017)," in *History of Nuclear Energy and Society (HoNESt) Consortium Deliverable no. 3.6* (2019), https://hdl.handle.net/2454/38269.; Bernd Helmbold, "German Democratic Republic (GDR) Short Country Report (version 2018)," in *History of Nuclear Energy and Society (HoNESt) Consortium Deliverable no. 3.6* (2019), https://hdl.handle.net/2454/38269.; Ivan Tchalakov and Ivaylo Hristov, "Bulgaria Short Country Report (version 2018)," in *History of Nuclear Energy and Society (HoNESt) Consortium Deliverable no. 3.6* (2019), https://hdl.handle.net/2454/38269.
54. US Comptroller General, *U.S. Nuclear Non-Proliferation Policy*, 8–9.
55. International Atomic Energy Agency, *50 Years of Nuclear Energy* (last modified September 24, 2004), https://www.iaea.org/sites/default/files/gc/gc48inf-4-att3_en.
56. Mauro Elli, *Atomi per l'Italia: La vicenda politica, industriale e tecnologica della centrale nucleare di Latina 1956–1972* (Milano: Unicopli, 2011); Mauro Elli, "British First Nuclear Export: ENI's Atomic Power Station at Latina and Anglo-Italian Nuclear Cooperation," *Annales historiques de l'électricité* 9 (2011): 27–42.
57. The US sales went to Belgium, Italy, Japan, West Germany, India, France, and Spain. The latter was the first turnkey project exported.
58. H. Stuart Burness et al., "The Turnkey Era in Nuclear Power," *Land Economics* 56, no. 2 (1980): 188–202.
59. Later, two consortia formed for the same purpose: France, Italy, Belgium, Iran, and Spain formed EURODIF, and the United Kingdom, the Netherlands, and West Germany formed URENCO. US Comptroller General, *U.S. Nuclear Non-Proliferation Policy*, 10.
60. Cohn, *Too Cheap to Meter*, 127.
61. International Atomic Energy Agency, *IAEA Nuclear Energy Series Managing the Financial Risk Associated with the Financing of New Nuclear Power Plant Projects* (Vienna: International Atomic Energy Agency, 2017), https://www-pub.iaea.org/MTCD/Publications/PDF/P1765_web.pdf.
62. William D. D'haeseleer, *Synthesis on the Economics of Nuclear Energy* (Study for the European Commission Directorate-General for Energy, Contract no. ENER/2012

/NUCL/Si2.643067, 2013), 25–26, https://www.mech.kuleuven.be/en/tme/research/energy_environment/Pdf/wpen2013-14.pdf.

63. *Eximbank Programs in Support of Nuclear Power Projects* (Washington, DC, 1970), 3. Box J11, Folder 2347. Ex-Im Bank Archives.
64. Eximbank, Press Release, February 17, 1970, Bound Press Releases, January 6, 1970–June 30, 1970, J6g, 2275, Ex-Im Bank Archives.
65. Robert S. Rendell, "Export Financing and the Role of the Export-Import Bank of the United States," *Journal of International Law and Economics* 91 (1977): 91–146.
66. US Comptroller General, *US Nuclear Non-proliferation Policy*, 10.
67. *Eximbank Programs in Support of Nuclear Power Projects* (Washington, DC, 1970), 3. Box J11, Folder 2347. Ex-Im Bank Archives
68. See Rubio-Varas and De la Torre, "Spain—Eximbank's Billion Dollar Client."
69. Holliday, *Eximbank's Involvement in Nuclear Exports.* In 1975 EXIM Bank was already under scrutiny. See Arthur B. Laffer, "Testimony of the Export-Import Bank, prepared for the Subcommittee on International Security and Scientific Affairs on Nuclear Proliferation of the Committee on International Relations, October 28, 1975." Box G26, Folder 991. Ex-Im Bank Archives.
70. One has to be aware that "any discussion of the relationship between liberalism and neoliberalism must start by recognizing the contested and nebulous nature of both categories." Sean Phelan and Simon Dawes, "Liberalism and Neoliberalism," *Oxford Research Encyclopedia of Communication*, February 6, 2018, https://oxfordre.com/communication/view/10.1093/acrefore/9780190228613.001.0001/acrefore-9780190228613-e-176.
71. William H. Becker and William M. McClenahan Jr., *The Market, the State and the Export-Import Bank* (Cambridge: Cambridge University Press, 2003), Appendix B. No nuclear credit was authorized in 1986, and a tiny credit amount of $8,900 was authorized in 1987, with none thereafter.
72. The effect of neoliberalism on nuclear energy has been studied by Shasi van de Graaff, "Much Ado About Nothing? Rhetoric and Reality of the Nuclear Renaissance," (PhD thesis, Queensland University, 2016). It implied, among other things, the reduction of the public expenditure in R&D in nuclear technologies in the UK and the US but also in most of the Western world, and the privatization of previously state-owned business related to the nuclear realm.
73. Holliday, *Eximbank's Involvement in Nuclear Exports*, 20.
74. World Nuclear Association, *Nuclear Power Economics and Project Structuring* (2017), 3, http://www.world-nuclear.org/getmedia/84082691-786c-414f-8178-a26be866d8da/REPORT_Economics_Report_2017.pdf.aspx.
75. OECD/IEA NEA, *Projected Costs of Generating Electricity* (Organization for Economic Co-operation and Development/International Energy Agency: Paris, 2015), https://www.oecd-nea.org/jcms/pl_14756.
76. Almost all studies on the matter come to similar conclusion, with the break-even line for nuclear to be competitive ranging from 7 to 10 percent.
77. Bernard Cohen, *Before It's Too Late: A Scientist's Case for Nuclear Energy* (New York: Springer, 1983), 217–47; E. Berndt and D.P. Aldrich, "Power to the People or Regulatory Ratcheting? Explaining the Success (or Failure) of Attempts to Site Commercial US Nuclear Power Plants: 1954–199,." *International Journal of Energy Research* 40, no. 7 (2016): 903–23.
78. D'haeseleer, *Synthesis on the Economics of Nuclear Energy*, 25–26.

PART 2

Actors

CHAPTER 3

One Movement or Many? The Diversity of Antinuclear Movements in Europe

Albert Presas i Puig and Jan-Henrik Meyer

Introduction

The history of movements against nuclear power is frequently told as a progressive story of one single movement. The story is often narrated in the following manner: Starting at the end of the 1960s, many European countries began to build numerous nuclear power plants. Ambitious nuclear programs were implemented to ensure the security of energy supply, based on expectations of high levels of economic growth and continuously rising energy consumption.[1] In the context of the emerging environmentalism of the 1970s, the antinuclear movement in Western Europe challenged individual nuclear plants and their construction sites locally.[2] The movement also criticized nuclear power more generally, highlighting the risks of radiation or accidents, and more fundamentally as a symbol of a society primarily devoted to economic growth. Young citizens, influenced by the events of 1968, articulated new demands for more democratic forms of participation and objected to technocratic top-down decisions imposed by an older generation. This led to the founding of Green parties, many of which continue to represent antinuclear and environmentalist positions in parliaments throughout Europe.[3] Such an account reflects the conventional and often linear story of the rise and political establishment of a notionally single antinuclear movement in Europe as part and parcel of a wider environmental movement.[4] Yet this version of events is not adequate to capture the true origins and characteristics of European movements against nuclear power in their full complexity and diversity.

Since the early 1970s, social scientists have studied the societal critique of nuclear programs in a range of detailed or comparative studies.[5] Historians first began to examine the history of social, environmental, and antinuclear movements in the early 2000s.[6] Yet most historical analyses have involved individual, local, and regional, at most national, case studies,[7] and have rarely compared countries.[8] Only the most recent historical research has also emphasized transnational connections and transfers between movements, notably between France and Germany.[9]

Many social science studies have been comparative in scope, engaging in the comparison of national cases.[10] But the purpose of most of these comparisons has been to draw out common features rather than to account for differences.[11] Social scientists' disciplinary ambition has chiefly been to explain which factors determined the success of social movements. Pursuing the ideal of parsimony in theorizing, political sociologists and social movement scholars have tended to draw on a limited range of factors in the relevant theories, thus drastically reducing complexity. Antinuclear movements have featured as prototypical examples for devising (and testing) new theories, notably theories of "new social movements."[12] The emphasis on common features tended to induce researchers to underplay the great diversity of actors, ranging from local groups and regional networks to larger movements. Traditional methodological nationalism in comparative research led many researchers to conceptualize movements along national lines, downplaying regional and local diversity. Such studies tend to understate relevant political, economic, or ideological contexts—which account for such diversity—as well as the varying relation to the wider environmental movements.

In this chapter we seek to critically revisit such conventional views of the antinuclear movement as a single phenomenon, arguing that antinuclear movements existed in the plural. Thus, we will do greater justice to the enormous variety of protest and opposition to the "peaceful uses of the atom" in the antinuclear heyday of the 1970s and 1980s, and account more clearly for its contexts. For this purpose, we will consider various experiences of societal engagement with and opposition to nuclear power in Western Europe. Empirically, this chapter is primarily based on examples of nuclear-societal relations presented in the short country reports drawn up within the HoNESt, the History of Nuclear Energy and Society project.[13] We examine the experience of antinuclear movements in five West European countries: Spain, West Germany, Austria, Denmark, and Sweden. All have seen substantial opposition to nuclear power and political decisions to stop the expansion of nuclear power—even if at different points in time and involving different modes of decision-making. They differ, however, with regard to their political systems in the postwar period, an important contextual factor regarding social movements. While Denmark and Sweden had long-standing experience of a well-functioning democratic system, democratic traditions were more fragile in West Germany and Austria, which only reinstated democracy after National Socialism. As late as the mid-1970s, Spain emerged from more than thirty years of the Franco dictatorship. Nascent antinuclear protest on the Iberian Peninsula started earlier, but became much stronger during the transition to democracy. All of these countries had been economically transformed into relatively affluent consumer societies in the postwar boom, even if Spain lagged behind, and all were affected by the economic crises of the 1970s.[14] At that time, only West Germany and Denmark were part of the European Communities (EC), with its promotion of nuclear via Euratom (see chapter 4). Denmark and Sweden considered themselves in many ways to be Nordic, rather than European, countries and shared close ties in the Nordic Council,[15] which also facilitated cooperation on nuclear issues.[16] Sweden and Austria were neutral in the Cold War, while West Germany and Denmark were NATO powers, and thus more exposed in the Cold War conflict. Spain only joined NATO after the end of the dictatorship, in 1982, four years before joining the EC in 1986.

Our goal is to demonstrate that what we tend to lump together under the heading "antinuclear movements" are indeed very different entities. They differ notably with regard to their origins and members, their perception of the conflict and ideological orientation, and the strategies and protest behavior they used. They also differed—and this is somewhat a cross-cutting category—in their geographical scope, ranging from local and regional to national and transnational. Researchers have increasingly come to appreciate such geographical distinctions in the wake of the "spatial turn" in history and the humanities and social sciences more generally.[17] The related phenomenon of "transboundary relations" between social movements is treated in chapter 9.[18]

First, we provide a short introduction to the conceptualization of antinuclear movements as social movements and some of the leading approaches on this issue in social sciences. Then we discuss the impressive diversity among social movements regarding the four criteria outlined in the previous paragraph. Finally, we explore what this means for how we can conceptualize and understand antinuclear movements, what explains the great variety of antinuclear movements, and to what extent this apparent variety potentially also reflects some more fundamental similarities.

Conceptualizing Antinuclear Movements as Social Movements

The critique of nuclear energy became politically relevant when it ceased to be voiced solely by single individuals.[19] Once smaller or larger groups have stated their opposition collectively and publicly, we can speak of antinuclear movements, of social mobilization, and of antinuclear movements as social movements.[20] What are social movements? For purposes of definition, we will draw on recent social movement research in the social sciences.

Social movement scholars have defined their object of research differently over time, depending on their goals and purpose. Generally speaking, they perceive such movements as forces of social and political change: they try to understand why and how actors mobilize within society and what kind of change they demand—in policy or law, or more fundamentally in structural features of state and society. Their interest lies in why such groups emerge, what accounts for their success, and how they manage to influence their society (or fail to do so).

Antinuclear movements' attitude toward social, political, and economic change was at times ambiguous: antinuclear activists did not agree with the promises of the advocates of nuclear power, who promoted the technology as a force of progress, of economic and technological modernization. Many antinuclear activists' visions of the future were different, in many ways more conservative. They opposed this technology exactly because they objected to changes in their local environments and economies, their vineyards, farms, and livelihoods.[21]

Social movements are usually defined by structural features: what they look like and who is part of them, what holds them together and what they oppose, and what means and methods they use in conflict. In a recent handbook entry, Donatella della Porta, a leading social movement researcher, defined them as networks: "Social movements are constituted by networks of informal relations between a plurality of individuals and groups, which are more or less structured from an organisational point of view."[22]

As opposed to associations and political parties or other formally organized institutions, social movements are hence characterized by their informal and noninstitutionalized nature. Social movements usually do not issue membership cards, keep a register, or charge a membership fee, even if some of them do develop into more formalized bodies—for instance, the environmental (and antinuclear) non-governmental organization Greenpeace, which started out from social movement structures.[23] At the same time, within their network structure social movements may well include organizations and associations.

By contrast, social movements are held together as networks by common goals and a "system of beliefs that nourish solidarity and collective awareness," often including "alternative world visions and value systems"—or what has traditionally been called ideologies.[24] These visions, values, and belief systems emerge and are constructed in the process of mobilization around conflicts or "contentious themes."[25] Classical, often Marxist-inspired, social movement theorizing used to focus on given, objective socioeconomic conflicts, such as class conflict. Early social movement theories argued that the relative deprivation of workers led them to voice their grievances, which in turn brought about the rise of the labor movement.[26] Studies of the new social movements springing up since the 1960s—the civil rights movement, the women's movement, the environmental movement, the antinuclear movement—alerted social movement researchers to conflicts that emerged around nonmaterial issues, which in turn constituted social movements that seemed novel and distinctly different from previous ones.[27] In many respects, however, antinuclear movements dealt with a problem with clear material implications: a source of energy and pollution involving significant risks. Accidents and radiation could have health impacts and become life-threatening. And, in the form of nuclear waste, nuclear was to have a lasting impact on the environment for future generations.

Mobilizing around societal conflicts, the main instrument social movements use to voice their grievances is protest. Researchers have distinguished between different kinds of protest—conventional ones, such as letter writing, and unconventional ones, such as taking to the streets. The style and mode of protest very much depend on a social movement's access to the given political systems, which has differed greatly in Europe.[28]

Building on such a definition, we address first not only the structure but also the "members" of the network. From a historical perspective, we focus in particular on the origins of the respective antinuclear movements and the context of their emergence, which accounts for much of the diversity between such movements in Europe. Second, this definition of social movements encourages us to enquire into the shared values and beliefs of members of the antinuclear movement, their identity and their sense (or invocation) of solidarity and their cultural manifestation and strategic use. It also alerts us to think critically about the social and political construction of the nuclear conflict and its link with other conflicts—such as regionalism—or ideologies. In the eyes of those protesting and establishing a movement, what exactly was the conflict about? Was it about a threat to material livelihoods? In the 1970s, for instance, vintners near the construction site of a new nuclear power plant in the village of Wyhl in Baden-Württemberg, Germany, worried that the steam from cooling towers would block the sunlight and prevent their grapes from ripening. This would endanger their harvest, and thus their economic livelihood. For many locals, this concern was initially more important than the less tangible risk of radiation.[29] Nuclear power plants meant different things to different people. Hence, the conflict about nuclear power in Europe was in fact an assembly of different conflicts in different countries and localities, and existing research on the patterns of and arguments within antinuclear debates provides ample evidence for this.

Third, we can distinguish between different methods and repertoires of protest, ranging from conventional letter writing to the unconventional occupation of sites. Some protests involved methods of bearing witness to wrongdoing, in the tradition of Gandhi and the peace movement.[30] Some protestors even resorted to violence against themselves to bear witness in this sense. The German radical antinuclear activist Hartmut Gründler went on hunger strike at Wyhl in 1975 and burned himself to death in Hamburg in 1977. This act of self-immolation was considered so extreme and radical that it went largely unreported at the time, undermining its intended impact.[31] Protest strategies involving violence against others were used more frequently in conflict with the police: demonstrations turned violent, such as in Brokdorf and Malville.[32] The most extreme example of violence was the outright terrorism and murder committed by the Basque terrorist organization ETA at the Lemoníz site in the Basque Country and by the terrorists of the radical left Red Army Faction (RAF) in West Germany.[33]

With regard to our historical approach, we may also look at how these repertoires changed over time and differed between places. Have social movement activists learned from each other and/or cooperated across borders?[34]

Finally, we will consider the movements' geographical scope: Did they primarily involve local groups fighting against local plants? Or did the networks extend throughout a region or even nationally? Were there transnational contacts, and to what extent did they matter?

In trying to assess what accounts for the movements' success, social movement researchers from the social sciences have developed different approaches that highlight the role of different causal factors. With a view to antinuclear movements, three approaches have featured prominently. First, the political opportunity structures approach, which emphasizes the importance of the political and institutional conditions within which social movements operate, notably whether they could access the political system.[35] A case in point is the unwillingness of so-called "closed" political systems to take a critique of nuclear power seriously. This fueled the protests especially in France and West Germany for a long time and may have triggered frustration and contributed to violence, but also led to the founding of Green, antinuclear political parties to have a voice within the political system.[36]

Second, the resource mobilization approach highlights the fact that movements rely on resources—notably, moral, cultural, socio-organizational, human, and material ones—in order to make a difference.[37] With its strategy of bearing witness and mobilizing counter-expertise, the US antinuclear movement in the 1970s, for instance, was particularly adept at mobilizing moral and cultural resources.[38]

Third, the framing approach emphasizes the importance of cultural features and narratives in the perception of societal problems for social movement success.[39] A recent study comparing the 2011 Fukushima disaster's varying resonance in Japan, Germany and the United Kingdom illustrates this convincingly. While in Germany Fukushima strengthened existing antinuclear discursive frames and narratives and weakened already rather defensive pronuclear narratives, Fukushima had much less of an impact in the UK and even in Japan, where pronuclear narratives continued to predominate.[40] Elements of each of these three approaches will be used throughout to strengthen our analysis.

Origins and Members

Historically, a more comprehensive social mobilization against nuclear energy in Western Europe indeed only started after the massive building of nuclear power plants began in the late 1960s. Throughout Western Europe, antinuclear movement groups and organizations became more numerous in the mid-1970s, as knowledge of the consequences and risks of nuclear power became more widespread.

Nevertheless, it is important to be aware of the movements' antecedents in the much smaller-scale protests against nuclear weapons in the 1950s and 1960s, in countries such as Sweden, the UK, and West Germany, such as the appeal of the Göttingen Eighteen, nuclear researchers protesting in 1957 against government plans to develop nuclear arms capability in West Germany.[41] Within a number of movements against nuclear power, there were important continuities of people and ideas with the movement against nuclear weapons, notably regarding risks of proliferation. In Denmark, for instance, peace groups from within the Protestant church were central among those founding the Danish social movement Organization for Nuclear Information (OOA).[42] In some countries, such as West Germany, links and overlaps existed between the movements against nuclear power and the movement against nuclear weapons in the context of NATO's double-track decision in the first half of the 1980s and returned to the revived antinuclear energy protest after Chernobyl.[43]

In many cases social mobilization against nuclear power started and generally manifested itself at specific construction sites, most of which were in the countryside. This was true in particular of France, West Germany, Austria, and

Spain. The local antinuclear movement at Wyhl was itself inspired by protests on the French side of the Rhine in Fessenheim in Alsace,[44] and in turn kick-started the broader movement across West-Germany.[45]

Rather than the stereotypical long-haired and bearded young leftist, it was a local landowner and mill owner, the clean-shaven Siegfried Göpper, from the neighboring Protestant and largely Social Democratic village of Weisweiler, who became the key figure in starting local antinuclear opposition against Wyhl. Göpper had learned about the risks of nuclear fission at a physics lecture at the Technical University of Karlsruhe and was informed early on about the plans for a nuclear power plant at Wyhl. He organized an information meeting in 1971, involving his local networks, a local Protestant pastor, the church elders, a leftist pub owner, the Social Democratic mayor, and the local landowners' association, conventionally in charge of distributing the local hunting fees. These citizens from the village of Weisweiler who were initially at the heart of the movement were slightly off the region's conservative, Christian Democratic and Catholic mainstream.[46] They soon mobilized the region's vintners, who worried about the plant's impact on their grape harvest and were not given sufficiently credible reassurances at a 1972 public hearing organized by the authorities.[47] The nascent movement gained much credibility and symbolic capital from its roots in the rural population. It also tapped into the existing protest networks across the Rhine, which had emerged amid opposition to a lead factory in Marckolsheim and the Fessenheim nuclear power plant. Once the local protests commenced in 1974, students from Freiburg and leftist circles started to support the originally rural resistance. Unsurprisingly, the striking diversity of the activists, with their very different attitudes, habits, ways of life, and styles of political engagement, gave rise to numerous conflicts within the movement.[48]

Generally the political left militancy dominated the antinuclear movements in West Germany and elsewhere in Western Europe, even if mainstream Social Democratic governments fostered nuclear expansion and Communists were only against nuclear power in capitalist countries. However, opposition to nuclear power also hailed from the other end of the political spectrum. On the Upper Rhine as well as in Austria, adherents of eugenics, with deep roots dating back to National Socialism, were among the first to alert citizens of the potential impact of radiation on human genes. Günther Schwab, an Austrian forester and "old fighter" of the Nazi Party, founded the World Federation for the Protection of Life (Weltbund zum Schutze des Lebens), which warned against the detrimental effects of radiation for the survival of the planet's "biological elite of the human species" at events on the Swiss, French, and German sides of the Upper Rhine in the early 1970s.[49] In 1971, they supported local protest by the Austrian

Nature Protection League (Naturschutzbund) in rural Vorarlberg, Austria's westernmost state, opposing the Rüthi nuclear power plant in neighboring Switzerland. Protest marches across the border to the Swiss site continued between 1972 and 1975. Already in 1970, the Bund für Volksgesundheit (Union for Public Health)—a eugenic association dating back to 1926—organized early protests against the planned Zwentendorf nuclear power plant in Lower Austria. The group cooperated closely with Schwab's World Federation, with which they shared many views. Schwab's writings resonated widely, and also impressed scientific critics of nuclear energy, such as Peter Weish, a young scientist and leading activist in the Austrian antinuclear struggle, who was also well connected transnationally.[50]

Science-based criticism in fact first began in Austria in 1969 in Lower Austria, the federal state around Vienna, where the controversial Zwentendorf nuclear power plant was to be built. The Lower Austrian Chamber of Physicians issued a protest memorandum against nuclear power, warning of the impact on human health and demanding that the medical profession be consulted in the decision-making. By 1978, Zwentendorf had become the symbolic focus of the antinuclear struggle when the Austrian antinuclear movement mobilized at the national level for a referendum.

Protest against the second planned Austrian nuclear power plant in Stein/St. Pantaleon in Upper Austria also started early and involved a diversity of groups, also beyond the local level, ranging from nature protection groups (Naturschutzbund) to the eugenicist World Federation for the Protection of Life (Weltbund) and from Maoist-oriented communists to the network of scientists led by Peter Weish in Vienna. All of these groups joined a broader regional upper Austrian network, which soon became a national one. The local groups around Stein/St. Pantaleon founded a citizen action group (Bürgerinitiative gegen Atomgefahren).[51]

Citizen action groups are a very open and only lightly structured form of social movement organization that was characteristic of the new social movements of the 1970s and 1980s, when environmental and antinuclear citizen action groups flourished. In some countries, the state intervened and helped to structure the nascent organizations. In 1972, West Germany's liberal minister of the interior, Hans-Dietrich Genscher, responsible for the new policy of the environment, sponsored the creation of a national umbrella organization for environmental citizen action groups, the Bundesverband Bürgerinitiativen Umweltschutz (BBU). The minister's intention had been to establish an environmental interest group as an ally within the political system and to link the new environmental issue more firmly to the liberal Free Democratic Party (FDP), which at the time included numerous left-leaning and pro-environment voices. The first chairman of the BBU, Hans-Helmuth Wüstenhagen, was a member of the Free Democratic Party, until he was forced to resign in 1977 by radical left-wingers.[52] By that time, the antinuclear issue dominated the BBU agenda. Within the BBU, antinuclear—and transnationally connected—activists such as Petra Kelly and Jo Leinen played a leading role, including in the debates on moving from a movement organization to founding a Green party.[53]

The Spanish antinuclear movement had already emerged before the end of the Franco dictatorship in 1975, in the form of local and rural protests against nuclear power plants, such as the Ascó plant in Catalonia. By that time, the Spanish commercial nuclear program was well under way. The Ascó plant was intended to reinforce the economic modernization of a predominantly agricultural area. Though construction was announced early in 1970, it was not until 1974 that the village inhabitants became aware of the potential risks of the installation. The first voices of opposition arose during the time of the dictatorship among Christian groups gathering within the Catholic parish. Local clerics considered it their duty as Christians to show solidarity with those who, they believed, were exploited by the big electric companies, a position not supported by the church hierarchy. By interacting with city dwellers, the local discourse against nuclear power plants broadened. Ascó activists looked for support among scientists and lawyers, established links with the universities in Barcelona, and also forged connections with the anti-Franco opposition. At the beginning of 1975, various professional associations joined the protest. In 1977, the movement went regional, and the ensuing Antinuclear Committee of Catalonia (CANC) brought together the various small antinuclear groups that had formed since the end of the 1960s.[54]

After Franco's death, the democratic political system seemed to offer new opportunities for antinuclear protest at the local level and at the regional level where a new Catalan autonomous government was established. Mixed commissions between government and local representatives were formed to address territorial questions. However, different levels of government had different preferences regarding nuclear power. The first democratically elected mayor of Ascó was elected on an antinuclear platform, having risen to prominence in the struggle against the power plant. Antinuclear groups demanded compliance with municipal laws and regulations and reliable information about the plant, but they were unable to stop the ongoing construction.

At both the national and regional levels, the newly democratic political and party system remained deaf to the demands of the rural movement against the Ascó nuclear power plant, contradicting its vision of modernization.

Nuclear power stood for modernity, whereas renewable energy sources symbolized the past. Jordi Pujol, president of the new autonomous government, declared, "*We cannot return to the times of the windmills.*"[55] In response, the antinuclear movement organized large-scale demonstrations in Barcelona in 1978 with more than 100,000 participants. Frustrated by the lack of reaction to their demands for a nuclear moratorium, some groups radicalized, engaging in violence and boycotts of the building work. This considerably delayed the project.[56]

Unlike in West Germany, where Wyhl was stopped by the courts, even in an alliance with the local government the movement was unable to benefit from the opportunities afforded by the judicial system. The plants are still in operation, with their licenses extended.[57]

In Sweden, an individual scientist played a central role in putting the nuclear issue on the agenda and paving the way for an antinuclear movement. Hannes Alfvén was not only charismatic but also a highly respected physicist and winner of the 1970 Nobel Prize in Physics. He had, moreover, been closely involved in research for atomic energy. In the late 1960s, while conducting research in California, Alfvén met American researchers and engineers who had become increasingly critical of the safety of nuclear power plants, the persistent problems of radioactive waste, and the risks of proliferation. In the early 1970s, Alfvén started to voice his critique of nuclear energy in public, in newspapers and vis-à-vis politicians. His public criticism left a lasting impression on Thorbjörn Fälldin, the leader of the Swedish Center Party, which traditionally represented agricultural constituencies and adopted an antinuclear stance in the subsequent elections.[58] Alfvén's impact on Fälldin proved very important for Swedish politics and the antinuclear movement because it opened up the Swedish party system for the political representation of antinuclear views, which was not the case in countries like Austria, Germany, France, or Spain until the rise of Green parties.[59] Alfvén was also an important point of reference for Sweden's nascent environmental movement and became an internationally known activist; the Danish antinuclear movement, for instance, invited him for talks.[60]

As the host country of the 1972 United Nations Conference on the Human Environment, Sweden became the center of international environmental politics. The event had an impact on the country itself and its new environmental movement.[61] Even if Alfvén was not allowed to give a talk at the UN conference, the counter-conferences and events on its margins facilitated the transmission of antinuclear ideas, mostly from the United States, toward the often-youthful environmental activists.[62] During the conference, the network of Friends of the Earth—which had split away from the Sierra Club in the US on the issue of nuclear power—published a daily newspaper that strongly criticized Swedish nuclear power and its safety.[63] Just one month before the Stockholm conference, the king had opened Sweden's first commercial nuclear power plant, celebrating it as the mark of "a new epoch in our country's energy supply" and "a milestone in our country's technical development."[64]

Similar to the situation on the Upper Rhine, early Swedish antinuclear protest benefited from transnational and transboundary links among environmentalists. In the context of a pan-Nordic summer camp in southern Sweden in August 1976, environmentalists from Sweden, Denmark, and Norway marched against the Barsebäck nuclear power plant for the first time. Located directly opposite Copenhagen, the plant attracted transboundary protest well into the 1980s, as discussed in chapter 9.[65]

In 1978, the Folkkampanjen mot Atomkraft (FMA—People's Campaign against Atomic Power) was founded to create a national umbrella organization for a very diverse and politically divided hodgepodge of environmental and political groups and activists ranging from environmentalists and the communist left to the conservative-agricultural spectrum. Expertise played an important role within the movement. Among FMA's leaders were some of those experts who had been involved in a government-sponsored public inquiry in 1976–1977. FMA mobilized massively in the campaign for the 1980 referendum. When the FMA lost the referendum, the movement quickly weakened and disintegrated. Only local initiatives, such as those against a nuclear waste repository, persisted and continued to mobilize.[66]

By contrast, the Danish antinuclear movement, which emerged in 1973–1974, was national in scope right from the start. Opposition to nuclear power arose even before any decision on a specific building site had been finalized, and this contributed strongly to its success in preventing nuclear power plants in the country. The OOA captured the limelight with its first press conference in Copenhagen in January 1974, in direct response to the submission of plans for the licensing of commercial nuclear power plants in Denmark. The movement arose from different sources. Young Christians got together in June 1973 at the Danish section of the International Fellowship of Reconciliation (IFOR)[67] in Copenhagen to discuss urgent contemporary issues relating to peace and the fight against global inequality. They singled out growing energy consumption and the plans for nuclear power as particularly problematic developments, and they decided to gather information and eventually to campaign against nuclear power. The group began organizing and involving other groups critical of nuclear power from the new Danish environmental organization NOAH, members of the Danish section of the Women's International League for Peace and Freedom (WILPF), and the Danish War Resisters International (WRI). To enhance its credibility and to ensure broad appeal, the group chose a name that was deliberately neutral: Organization for Nuclear Information, or more literally, for "enlightenment about atomic power" (Organisationen til Oplysning om Atomkraft). The OOA reached out to antinuclear groups across the country, including local protest groups that had sprung up at Gyllingnæs near Aarhus, where a nuclear power plant was to be erected.[68] All of these groups were subsequently represented within the national movement organization, through very informal, light-touch consensus-based decision-making at national meetings (*landsmøder*). At the same time, the OOA maintained a very effective secretariat in Copenhagen, where some activists, including theology student Siegfried Christiansen, one of the founding members, were devoted to the movement—including its transnational networking—full-time.[69]

The movement started out as a national one from the outset, because it responded to national politics and decision-making at the national level. The Danish utility Elsam had submitted an application for the licensing of new nuclear power plants. According to the law in force at the time, dating back to the 1960s, the minister of research and education was competent to issue such a license. This was what the OOA criticized at their first press conference in Copenhagen on January 31, 1974. The organization not only challenged the nuclear option but also called for an assessment of alternative energy sources. The OOA's press release warned against what it considered an undemocratic and hastily taken decision, accusing the minister of making a "*panikbeslutning*" (panic-induced decision).[70] The OOA called for a period of more general reflection on energy policy. After the 1973 landslide elections, the Danish parliament had numerous new members and a fragmented party system, and thus proved very open to such a critique. The minister of commerce actually started the Energy Information Committee (Energioplysningsudvalget), a funding instrument for organizing debates about energy policy. In this context, the OOA established itself as a critical expert and movement throughout the country.[71]

Perception of Conflict and Ideological Affiliation

Antinuclear movements in different countries and at different levels provided very different definitions of what the nuclear issue involved, and what its key problems were. *Radiation* and its potentially detrimental impact on the gene pool and public health of the "German race" were central to the eugenicists of the Union for Public Health (Bund für Volksgesundheit) and World Federation for the Protection of Life (Weltbund zum Schutze des Lebens), who were prominent in the initial phase of the German and Austrian antinuclear movements. From a scientific perspective, the problems of low-level radiation were central for Alfvén in Sweden and for the medical doctors in the Austrian movement, who based their critique on information from the United States. Radiation concerns were also important for the German (and transnational) activist Petra Kelly, one of the leaders of the West German umbrella organization BBU. Having grown up in the United States, she was also in direct contact with critical American experts.[72]

By contrast, for the local actors in Wyhl and Ascó, and other places of protest in Spain, opposition to nuclear power plants was primarily motivated by more immediate economic and material concerns about their crops and thus their livelihoods. In Wyhl, as we have seen above, the vintners feared the impact of steam from cooling towers on their vines and grapes. In Spain, nuclear power's vast requirements for cooling water meant that downstream users were to lose out on crucial water resources needed for irrigation.[73] Besides such water problems, land use was a core issue for German and Spanish local actors. Essentially, farmers were unwilling to give up their precious farmland, economic existence, and way of life for nuclear-industrial development. In both countries farmers were concerned that they might not be able to sell their produce if customers feared it was radioactively contaminated. At a more general level, rural actors in both West Germany and Spain rejected the industrialization of their homeland while not reaping its benefits, and a change of their lifestyle and environment imposed on them by government planners and utilities.

Some movements were also connected to wider political trends and ideas, notably the new *regionalism* of the 1970s, which built on the societal cleavage between center and periphery.[74] Antinuclear activists in Vorarlberg, Austria, tapped into a long history of regionalist opposition to Viennese centralism.[75] In Catalonia the antinuclear movement established itself as a regional one, even if the critique of nuclear power failed to convince the newly democratic regional authorities in charge of the region's economic development. Central to the protest against the Lemoníz plant in the Basque Country were concerns that a potential accident would wipe out the entire Basque people and with it the project of regional (or national) independence.[76] On the Upper Rhine, protesters imagined and celebrated the cross-border region Dreyeckland ("three corner country") in opposition to central government planners in Paris and in Stuttgart. In the capitals of France and Baden-Württemberg, planners aimed at the wholesale industrialization of the Upper Rhine valley, with scant regard for citizens' preferences. Similarly, in the 1980s, in the struggle over the Wackersdorf reprocessing plant, local critics also defined themselves in regionalist terms. They self-confidently asserted their claims against the state government in Munich as people from the Upper Palatinate region.[77]

Across all of the cases, the nuclear issue was linked to democracy, albeit in different ways, and to different conceptions of democracy. Emerging from a dictatorship, the Spanish antinuclear movement placed great hopes in representative democracy. Activists were thus disappointed when the newly established national and regional levels of democratic government proved unreceptive to their critique. Similarly, in Denmark, the OOA demanded that the responsibility for nuclear power be taken from the government and handed over to Parliament, as the representative of the people. They argued that the people—and their elected representatives—should arrive at an enlightened decision, based on sound and thorough information and a period of reflection and discussion in the public sphere, in line with a Habermasian understanding of deliberative democracy.[78]

In Wyhl, and West Germany more generally, activists were much more skeptical of the institutions of representative democracy that had imposed the plant on them, and demanded more participatory democracy. They asserted their right to resist, protest, and voice public dissent against what they perceived as merely formally democratic decisions. Thus, they considered direct action, including the occupation of the building site, to be justified by a right to democratic opposition or "resistance."[79]

In Austria and Sweden, antinuclear movements were equally skeptical of representative democracy by political parties, elected representatives, and governments. On the fateful issue of nuclear power, they believed it was necessary to directly listen to the will of the people and hold a referendum.

In a number of European countries, leftist groups sought to attach their ideas on class struggle and societal change to the antinuclear cause while typically remaining uncritical of nuclear power in the Soviet Union. In Austria and West Germany, even Maoists joined the protest. Maoist groups played a role in the protest against the Brokdorf nuclear power plant in northern Germany. Such communist splinter groups interpreted rural protest in ideological terms as a peasant revolution.[80]

Many members of antinuclear movements understood the issue of nuclear power as a moral one. From the perspective of their faith, Christians emphasized their responsibility for the environment (as created by God), people's health and livelihoods, the interests of future generations (concerning nuclear waste), and the prevention of war. These issues mattered for Christians in Denmark, for Protestant pastors and (in the 1980s) Catholic priests in West Germany, and local Catholic clerics in Spain. Despite their denominational differences, remarkably, activist theologians all entered into conflict with the more senior and more conservative higher echelons of their churches' hierarchy.[81] In West Germany, the strong involvement of Protestant pastors was based on moral arguments relating to the country's National Socialist past. Many antinuclear Protestant pastors felt morally obliged to speak out against nuclear power, to confess and resist, because their parents' generation had failed to oppose National Socialism.[82] This diversity of motivations for antinuclear protest was also reflected in the strategies used in this conflict.

Protest Strategies

Protest strategies ranged from traditional letter-writing campaigns, contributions to newspapers, and signature collection to more unconventional strategies that involved breaking the law by site occupation or even violence. Different movements pursued different strategies at different times, depending on the opportunities afforded by the political system in which they operated. For instance, some governments, such as those in Sweden and Austria, granted a public vote or referendum. Many governments also financed public information campaigns in which movement actors could participate, as in the cases of Austria, Denmark, and the Netherlands.[83]

Any effective critique of nuclear power required first and foremost the collection of often very complex information about nuclear technology, its risks, implications, and consequences—a major challenge in the days before the internet. Advocates of nuclear power tended to dismiss criticism by suggesting that antinuclear activists were arguing on a purely emotional basis and lacked sufficient technical and scientific knowledge.[84] In fact, antinuclear activists did amass substantial knowledge and information. Groups built up and shared counter-expertise, also by including "converts" from the nuclear sector, such as the German nuclear expert Klaus Traube or the Austrian radiation protection scientist Peter Weish.[85] The OOA prioritized information collection in its founding phase in 1973. The first books in Danish on nuclear energy targeting a nonspecialist audience weren't published until 1974,[86] and thus, the OOA gathered information from Swedish, American, German, and other sources to build up an archive. Hannes Alfvén's efforts at popularizing nuclear critique—which placed scientific information on the risks of nuclear power in the Swedish public sphere—also proved very useful for the Danes, because linguistic similarity facilitated information transfer.

Inviting counter-experts as guest speakers was an important strategy for antinuclear movements to popularize critical knowledge and arguments. The OOA organized an entire series of such events, with important counter-experts from the US and Sweden, using the opportunities of the Energy Information Committee, a government-funded public information campaign. The OOA also advised public libraries in Denmark, providing lists of books they should purchase to help inform the public.[87]

The so-called people's high school of the Wyhl Forest, a self-organized educational institution set up on the occupied territory, served to popularize knowledge by inviting counter-experts. Siegfried Göpper's first attempts to mobilize against Wyhl in 1971 involved a talk by a retired atomic physicist on the risks of nuclear power.[88] Movements in Spain organized information lectures too.[89]

Many movements, such as that against Wyhl, started with simple letter writing to the authorities. A letter sent by the landowners' association of the village of Weisweiler to West German Chancellor Willy Brandt in 1972 protested the originally planned nuclear power plant site at Breisach.[90] By 1974, the movement on the Upper Rhine had scaled up its protest with a petition and the collection of 90,000 signatures, which were submitted to the local county authorities.[91] Signature collection demanding a referendum in Sweden started in March 1979.[92] In Denmark, local OOA groups such as at Gyllingnæs and Vendsyssel not only collected signatures from 87 to 90 percent of the local population but also used a signature collection drive to mobilize in the wake of the Three Mile Island accident. Backed up by 312,000 signatures, the OOA demanded that the Danish government address the issue of Barsebäck with the Swedish government. After the Chernobyl disaster in 1986, the OOA collected some 160,000 signatures directed at nuclear power plants in three neighboring countries. In a protest march in Copenhagen, the OOA took the signatures to the Swedish, East German, and West German embassies and actually traveled to these countries to demand nuclear phaseout.[93] In Spain, the Basque movement against the Lemoníz nuclear power plant collected more than 150,000 signatures and organized large-scale demonstrations with more than 50,000 participants.[94]

Information collection, signature collection, and street protest involving demonstrations in front of power plant construction sites or in cities were core instruments across all of the antinuclear movements of the five countries under consideration and beyond. Nevertheless, movements behaved very differently with regard to site occupations and violence. Site occupation was pioneered in France, first in the rural Larzac in southern France in opposition to the extension of a military site, then in Alsace at the Marckolsheim lead factory.[95] Successful site occupation in West Germany not only established a model for neighboring Kaiseraugst in Switzerland but instantly achieved an almost mythical status.[96] At the construction site of the Brokdorf plant, the protesters followed the Wyhl model very closely. Ironically, this allowed the police, who had also studied the Wyhl protests, to prepare accordingly, and enabled them to intercept the occupation at Brokdorf.[97]

Whereas protests in Denmark, Sweden, and Austria remained peaceful, the situation was much more confrontational in West Germany, France, and Spain. Violent clashes between police and protesters attracted media attention, which did not necessarily benefit either side. In West Germany, at Wyhl, the police's forceful removal of protesters—many of them women and elderly people—backfired. Broadcast on television nationwide, the images of the police action came across as an excessive use of executive powers in favor of the state-owned utility. This created a wave of sympathy for the Wyhl protest.[98] Later, in the late 1970s and early 1980s, the increasingly violent protests at reactor sites at Brokdorf or Grohnde undermined the credibility of the antinuclear cause.[99] Nevertheless, in the latter half of the 1980s, protests against the Wackersdorf reprocessing plant were equally overshadowed by violence on both sides.[100] In France, violent clashes between the armed police and French, Swiss, Italian, and German activists at the Malville fast breeder site in 1977 left one protester dead. This event significantly delegitimated and disillusioned the antinuclear movement in France, and is considered to have contributed to its decline.[101]

In Spain and West Germany, the conflict even went beyond violence at demonstrations, and involved terrorism and bloodshed. In the Basque Country, the Basque separatist group ETA exploited the antinuclear cause for its own ends. In 1981 ETA kidnapped and assassinated the chief engineer of Lemóniz, and killed his successor the following year. These killings contributed to the Spanish authorities' decision not to press forward with Lemóniz.[102] Similarly, in July 1986, in the wake of Chernobyl and in the context of conflict over Wackersdorf and Brokdorf, the radical left-wing terrorists of the West German Red Army Faction (RAF) murdered Karl Heinz Beckurts, who was a leading nuclear physicist and advocate of nuclear power. He was director of research at the Siemens company, an important supplier of nuclear technology. The West German government remained unimpressed by the attacks, and allowed Brokdorf to go critical in autumn 1986. When Wackersdorf was stopped two years later, this was primarily for economic reasons.[103]

Breaking the law and using violence during protests were controversially discussed within antinuclear movements, notably in France and West Germany. The terrorist groups ETA and RAF, which carried out these murders, sought to attach themselves to the antinuclear cause. However, within the antinuclear movement there was no support for a violent strategy such as that carried out by the terrorists.[104]

The varying presence of violence may be explained by political opportunity structures. Where the political system proved open to antinuclear arguments and/or held referendums, as in Denmark, Austria, and Sweden, there were political options for resolving the conflict. Violence in Spain and West Germany to some extent resulted from frustration about the unwillingness of the political system to take antinuclear claims seriously. The resort to physical force and bloodshed in these countries may ultimately also reflect the unresolved legacy of dictatorships that was common to both.

By contrast, referendums demonstrate a certain openness of the political system to activist demands. They also provided important political opportunities for the movements to mobilize beyond the local level. Given the high stakes of the referendum decision, movements in Austria and Sweden were able to muster substantial support. However, depending on the outcome of the referendum, such mobilization was often rather short-lived. The referendum's failure in Sweden effectively depleted the movement of much of its support base. For the Austrian activists, however, success in the referendum campaign emboldened them to proselytize and support antinuclear causes in other countries. In 1980, the Austrian movement distributed postcards to Austrian citizens, inviting them to send the postcards to Swedish friends and call upon them to follow the Austrian example and vote against nuclear power.[105] Similarly, in the 1990s, the Austrian antinuclear movement supported the Czech antinuclear movement. This help backfired, as it undermined the Czech movement's domestic credibility (see chapter 9).[106]

Conclusions: Antinuclear Movements in Space and Time

Antinuclear activism in Western Europe in the 1970s and 1980s came in different forms and guises—with respect to origins, membership, perception of the nuclear conflict, ideological concerns, as well as to protest instruments and strategies. This chapter suggests that we should speak of antinuclear movements in the plural rather than essentializing the "antinuclear movement" as one single homogeneous phenomenon. Antinuclear groups formed their movement structures and networks at different levels, from the local to the national, and at different points in time. Which kinds of structures emerged depended strongly on contextual conditions of their respective countries and nuclear sites. For a number of reasons, in most countries national movement organizations, networks, or campaign organizations emerged, connecting the various local groups. Local specificities, however, continued.

Transnational connections mattered, notably for the exchange of scientific and technological knowledge and protest strategies within Europe and across the Atlantic. Occasionally, movement representatives from different countries cooperated transnationally on common projects. In 1978, for instance, Austrian, West German, Danish, and other antinuclear activists formed the World Information Service on Energy (WISE) to collect information on nuclear and to advocate alternative energies.[107] Such cooperation, however, was often very fragile, required specific skills and resources, and involved only a very small and select number of members of the various movements.[108] Clearly, such cooperation does not allow us to emphatically declare the existence of a single transnational antinuclear movement.

When looking at this diversity of antinuclear movements across space and time, we can observe certain shared patterns and differences. In terms of space, we can distinguish between movements that originated from the local level, often organized in direct opposition to a construction site, and movements that started almost immediately at the national level. Clearly, in Spain and West Germany—building on earlier experiences in neighboring France—movements emerged locally. They usually encompassed a wide variety of local citizens—ranging from farmers and vintners concerned about their crops to clergy, local elites, housewives, and, subsequently, city folk. In both countries, as well as in the Austrian Vorarlberg, protest attached itself to regionalism. Hence, regionalism reinforced antinuclear activism ideologically and as an established pattern of center-periphery conflict. Drawing on the concept of political opportunity structures, the Spanish regions and the German states (Länder) should have benefited the antinuclear movements, by exploiting conflicts between the different levels of government.[109] However, this proved impossible because the relevant regional governments in both countries continued to be committed to nuclear power as a means of economic modernization well into the 1980s.

Political opportunity structures provided by national referendums spurred the establishment of national movements in Austria and Sweden, which, however, faced the challenge of integrating a diversity of groups from across the political spectrum. Similarly, in early 1974, the nascent OOA began to operate at the national level right from the start, because it feared that the national government would make a hasty decision on the licensing of nuclear power plants without involving the Danish parliament. While maintaining a very effective national central office in Copenhagen, the OOA had a strong presence throughout the country and near potential building sites. Traditional centralism and the existence of the Swedish Barsebäck power plant across the sea may also have strengthened the focus of protest in the Danish capital.

Antinuclear movements differed to an astonishing degree with a view to their perception of the conflict and problems posed by nuclear power. This also changed over time. Initially, Austrian and West German eugenicists with National Socialist roots worried about radiation's effect on the human genome, and physicians and doctors worried about health effects. Local farmers and vintners in Spain and West Germany were concerned about crops, water, and land use. In Sweden and Denmark, and among some of the Austrian activists, scientific concerns about accidents, nuclear waste, and radiation, transferred from the US and popularized by experts like Alfvén, mattered early on. Nuclear was frequently perceived as a challenge to democracy and led to the demand for participation in decision-making, albeit by different means, ranging from the assertion of parliamentary rights to calls for direct democracy. National political culture, trust in the state and authorities, norms, and ethical and religious traditions account for the great variety of interpretations of the nuclear conflict outlined above.

Such contexts also explain the varying use of means of protest—ranging from the traditional and conventional collection and dissemination of information, letter writing, and signature collection to more unconventional methods such as site occupation and the use of violence. As argued above, the use of violence at demonstrations in Spain and West Germany (and France, to some extent) may be explained in terms of the legacy of dictatorship, lack of trust in the authorities, and very unfavorable opportunity structures. Terrorists from existing terrorist groups, notably the Basque ETA and the left-extremist West German RAF—tried to piggyback on the popular antinuclear cause.

Over time, antinuclear movements changed. Some officially or effectively dissolved, once the conflict was settled, such as in Denmark or Sweden. In West Germany, France, Sweden, and Austria, the movements indeed helped form Green parties or shifted toward protesting rearmament and NATO's double-track decision in the 1980s. Some persisted, notably local movements in places where the conflict remained unresolved, such as in Gorleben,[110] which until 2020 was still on the map for the deep geological disposal of Germany's nuclear waste, or in Spain (and Portugal), where uranium mining and waste disposal are still on the agenda.[111]

The comparison across five West European countries has elicited common patterns across some of the antinuclear movements that help us to better understand why certain movements emerged or behaved in certain ways. Nevertheless, varying contexts have led antinuclear movements toward very distinct trajectories. Antinuclear movements have been and continue to be highly diverse phenomena across time and space—within and across nations.

Notes

1. Hendrik Ehrhardt, "Energiebedarfsprognosen. Kontinuität und Wandel energiewirtschaftlicher Problemlagen in den 1970er und 1980er Jahren," in *Energie in der modernen Gesellschaft. Zeithistorische Perspektiven*, ed. Hendrik Ehrhardt and Thomas Kroll (Göttingen: Vandenhoeck & Ruprecht, 2012), 193–222.
2. This chapter only addresses cases from Europe west of the Iron Curtain, where democratic political systems allowed for public antinuclear protest. Spain is a special case in this respect, because its dictatorship ended only in the mid-1970s. Hence, early movements had to operate under conditions of an authoritarian regime. In this respect, the Spanish antinuclear movement resembles that of Socialist countries as discussed in chapter 1.
3. Stephen Milder, "Between Grassroots Protest and Green Politics: The Democratic Potential of the 1970s Antinuclear Activism," *German Politics and Society* 33, no. 4 (2015): 25–39, https://doi.org/10.3167/gps.2015.330403.
4. E.g., Andrew Jamison, Ron Eyerman, and Jacqueline Cramer, *The Making of the New Environmental Consciousness: A Comparative Study of the Environmental Movements in Sweden, Denmark and the Netherlands* (Edinburgh: Edinburgh University Press, 1990).
5. E.g., Alain Touraine, *La prophétie anti-nucléaire* (Paris: Seuil, 1980); Dieter Rucht, *Von Wyhl nach Gorleben: Bürger gegen Atomprogramm und nukleare Entsorgung* (Munich: C.H. Beck, 1980); Herbert Kitschelt, *Kernenergiepolitik: Arena eines gesellschaftlichen Konflikts* (Frankfurt: Campus, 1980).
6. E.g., for Germany: Jens Ivo Engels, *Naturpolitik in der Bundesrepublik. Ideenwelt und politische Verhaltensstile in Naturschutz und Umweltbewegung 1950–1980* (Paderborn: Schöningh, 2006), 344–76; Ute Hasenöhrl, *Zivilgesellschaft und Protest. Eine Geschichte der Naturschutz- und Umweltbewegung in Bayern 1945–1980* (Göttingen: Vandenhoeck & Ruprecht, 2011), 405–71.
7. E.g., Luis Sánchez-Vázquez and Alfredo Menéndez-Navarro, "Nuclear Energy in the Public Sphere: Anti-Nuclear Movements vs. Industrial Lobbies in Spain (1962–1979)," *Minerva* 53, no. 1 (2015): 69–88, https://doi.org/10.1007/s11024-014-9263-0; Stefania Barca and Ana Delicado, "Anti-Nuclear Mobilisation and Environmentalism in Europe: A View from Portugal," *Environment and History* 22, no. 4 (2016): 497–520; Dolores L. Augustine, *Taking on Technocracy: Nuclear Power in Germany, 1945 to the Present* (New York: Berghahn, 2018); Janine Gaumer, *Wackersdorf: Atomkraft und Demokratie in der Bundesrepublik 1980–1989* (Munich: Oekom, 2018).
8. E.g., Henrik Kaare Nielsen, "Youth and the Antinuclear Power Movement in Denmark and West Germany," in *Between Marx and Coca-Cola: Youth Cultures in Changing European Societies, 1960–1980*, ed. Axel Schildt and Detlef Siegfried (New York: Berghahn Books, 2006), 203–23.
9. Andrew Tompkins, *Better Active than Radioactive! Antinuclear Protests in 1970s France and West Germany* (Oxford: Oxford University Press, 2016); Stephen Milder, *Greening Democracy: The Anti-Nuclear Movement and Political Environmentalism in West Germany and Beyond, 1968–1983* (Cambridge: Cambridge

University Press, 2017); Natalie Pohl, *Atomprotest am Oberrhein: Die Auseinandersetzung um den Bau von Atomkraftwerken in Baden und im Elsass (1970–1985)* (Stuttgart: Steiner, 2019).

10. E.g., Lutz Mez, *Der Atomkonflikt: Atomindustrie, Atompolitik und Anti-Atom-Bewegung im internationalen Vergleich* (Berlin: Olle & Wolter, 1979); Herbert Kitschelt, *Politik und Energie: Eine vergleichende Untersuchung zur Energie-Technologiepolitik in den U.S.A. der Bundesrepublik, Frankreich und Schweden* (Frankfurt: Campus, 1983); Dorothy Nelkin and Michael Pollak, *The Atom Besieged: Extraparliamentary Dissent in France and Germany* (Cambridge, MA: MIT Press, 1981); Helena Flam, ed., *States and Anti-Nuclear Movements* (Edinburgh: Edinburgh University Press, 1994); Wolfgang C. Müller and Paul W. Thurner, eds., *The Politics of Nuclear Energy in Western Europe* (Oxford: Oxford University Press, 2017).
11. Wolfgang Rüdig, *Anti-Nuclear Movements. A World Survey of Opposition to Nuclear Energy* (London: Longman, 1990), 8.
12. Dieter Rucht, *Modernisierung und neue soziale Bewegungen: Deutschland, Frankreich und USA im Vergleich* (Frankfurt: Campus, 1994); Hanspeter Kriesi et al., "New social Movements and Political Opportunities in Western Europe," *European Journal of Political Research* 22, no. 2 (1992): 219–44, https://doi.org/10.1111/j.1475-6765.1992.tb00312.x.
13. This chapter draws on evidence from the short country reports (SCRs) of the HoNESt project (https://hdl.handle.net/2454/38269). These reports were designed to assemble in an accessible manner information and research results on the history of the relations between nuclear energy and society for nineteen European countries and the United States. Five country studies based on the SCRs were published in Astrid Mignon Kirchhof, ed., *Pathways into and out of Nuclear Power in Western Europe: Austria, Denmark, Federal Republic of Germany, Italy, and Sweden* (Munich: Deutsches Museum Verlag, 2019).
14. Niall Ferguson, "Crisis, What Crisis? The 1970s and the Shock of the Global," in *The Shock of the Global: The 1970s in Perspective*, ed. Niall Ferguson et al. (Cambridge, MA: Belknap Press, 2010), 1–21.
15. Lene Hansen and Ole Waever, *European Integration and National Identity: The Challenge of the Nordic States* (London: Routledge, 2001); Melina Antonia Buns, "Green Internationalists: Nordic Environmental Cooperation 1967–1988" (PhD dissertation, University of Oslo, 2020).
16. Arne Kaijser and Jan-Henrik Meyer, " 'The World's Worst Located Nuclear Power Plant': Danish and Swedish Perspectives on the Swedish Nuclear Power Plant Barsebäck," *Journal for the History of Environment and Society* 3 (2018): 71–105, https://doi.org/10.1484/J.JHES.5.116795.
17. Matthias Middell and Katja Naumann, "Global History and the Spatial Turn: From the Impact of Area Studies to the Study of Critical Junctures of Globalization," *Journal of Global History* 5 (2010): 149–70.
18. See also Arne Kaijser and Jan-Henrik Meyer, "Siting Nuclear Installations at the Border. Special issue," *Journal for the History of Environment and Society* 3 (2018): 1–178, https://www.brepolsonline.net/doi/abs/10.1484/J.JHES.5.116793.
19. Markku Lehtonen, "France Short Country Report," in *History of Nuclear Energy and Society (HoNESt) Consortium Deliverable no. 3.6* (2019), https://hdl.handle.net/2454/38269.
20. Pedro Ibarra, "¿Qué son los movimientos sociales? Anuario de Movimientos

sociales," in *Anuario de Movimientos sociales: Una mirada sobre la red*, ed. Elena Grau and Pedro Ibarra (Barcelona: Icaria Editorial y Getiko Fundazioa, 2000), 9–26.

21. Ute Hasenöhrl, "Just a Matter of Habituation? The Contentious Perception of (Post)energy Landscapes in Germany, 1945–2016," *Environment, Place, Space* 10, no. 1 (2018): 63–88.
22. Donatella della Porta, "Social Movements," in *International Encyclopedia of Political Science*, ed. Bertrand Badie (Los Angeles: Sage Publications, 2011), 2431. Emphasis in the original.
23. Frank Zelko, "The Umweltmulti Arrives: Greenpeace and Grass Roots Environmentalism in West Germany," *Australian Journal of Politics and History* 61, no. 3 (2015): 397–413, https://doi.org/10.1111/ajph.12110; Frank Zelko, *Make It a Green Peace! The Rise of Countercultural Environmentalism* (Oxford: Oxford University Press, 2013).
24. Della Porta, "Social Movements," 2431.
25. Della Porta, "Social Movements," 2431.
26. Steven M. Buechler, "The Strange Career of Strain and Breakdown Theories of Collective Action," in *Blackwell Companion to Social Movements*, ed. David A. Snow, Sarah A. Soule, and Hanspeter Kriesi (Oxford: Oxford University Press, 2007), 47–66.
27. Doris Wastl-Walter, "Social Movements: Environmental Movements," in *International Encyclopedia of the Social and Behavioral Sciences*, ed. Neil J. Smelser and Paul B. Baltes (Oxford: Pergamon, 2001), 14352–57; Claus Offe, "New Social Movements: Challenging the Boundaries of Institutional Politics," *Social Research* 52, no. 4 (1985): 817–68.
28. Herbert P. Kitschelt, "Political Opportunity Structures and Political Protest: Anti-Nuclear Movements in Four Democracies," *British Journal of Political Science* 16, no. 1 (1986): 57–85.
29. Rucht, *Von Wyhl nach Gorleben*, 78; Milder, *Greening Democracy*, 40–42.
30. Michael Hughes, "Civil Disobedience in Transnational Perspective: American and West German Anti-Nuclear-Power Protesters, 1975–1982," *Historical Social Research* 39, no. 1 (2014): 236–53.
31. Alex Raack, "Selbstverbrennung eines Umweltaktivisten" Feuerzeichen gegen 'atomare Lügen,' " *Spiegel Online*, November 21, 2017, https://www.spiegel.de/geschichte/hartmut-gruendler-selbstverbrennung-des-umweltaktivisten-1977-a-1178242.html; Peter Brügge, "Waldeslust und Widerstand," *Der Spiegel*, July 21, 1975, 41–43, https://www.spiegel.de/spiegel/print/d-41458244.html; Tompkins, *Better Active than Radioactive!*, 181–82.
32. Augustine, *Taking on Technocracy*, 126–60; Claire Le Renard, "The Superphénix Fast Breeder Nuclear Reactor: Cross-Border Cooperation and Controversies," *Journal for the History of Environment and Society* 3 (2018): 107–44, https://doi.org/10.1484/J.JHES.5.116796.
33. M.d. Mar Rubio-Varas et al., "Spain Short Country Report," *History of Nuclear Energy and Society (HoNESt) Consortium Deliverable no. 3.6* (2019), https://hdl.handle.net/2454/38269.
34. For an excellent example of transnational learning, see Hughes, "Civil Disobedience in Transnational Perspective."
35. Hanspeter Kriesi, "Political Context and Opportunity," in *Blackwell Companion to*

Social Movements, ed. David A. Snow, Sarah A. Soule, and Hanspeter Kriesi (Oxford: Oxford University Press, 2007), 69.

36. Kitschelt, "Political Opportunity Structures and Political Protest," 83.
37. Bob Edwards and John D. McCarthy, "Resources and Social Movement Mobilization," in *Blackwell Companion to Social Movements*, ed. David A. Snow, Sarah A. Soule, and Hanspeter Kriesi (Oxford: Oxford University Press, 2007), 118–25.
38. Steven E. Barkan, "Strategic, Tactical and Organizational Dilemmas of the Protest Movement against Nuclear Power," *Social Problems* 27, no. 1 (1979): 22, https://doi.org/10.2307/800014; Dieter Rucht, "Gegenöffentlichkeit und Gegenexperten: Zur Institutionalisierung des Widerspruchs in Politik und Recht," *Zeitschrift für Rechtssoziologie* 9, no. 2 (1988): 290–305.
39. Rhys H. Williams, "The Cultural Contexts of Collective Action: Constraints, Opportunities, and the Symbolic Life of Social Movements," in *Blackwell Companion to Social Movements*, ed. David A. Snow, Sarah A. Soule, and Hanspeter Kriesi (Oxford: Oxford University Press, 2007), 105f.
40. Lukas Hermwille, "The Role of Narratives in Socio-Technical Transitions: Fukushima and the Energy Regimes of Japan, Germany, and the United Kingdom," *Energy Research and Social Science* 11 (2016): 237–46, https://doi.org/10.1016/j.erss.2015.11.001.
41. Arne Kaijser, "The Referendum that Preserved Nuclear Power and Five Other Critical Events in the History of Nuclear Power in Sweden," in *Pathways into and out of Nuclear Power in Western Europe: Austria, Denmark, Federal Republic of Germany, Italy, and Sweden*, ed. Astrid Mignon Kirchhof (Munich: Deutsches Museum, 2019), 242–244; Astrid Mignon Kirchhof and Helmuth Trischler, "The History behind West Germany's Nuclear Phase-Out," in Kirchhof, *Pathways into and out of Nuclear Power in Western Europe*, 130; Robert Lorenz, *Protest der Physiker: Die "Göttinger Erklärung" von 1957* (Bielefeld: Transcript, 2011).
42. Jan-Henrik Meyer, " 'Atomkraft—Nej tak': How Denmark Did not Introduce Commercial Nuclear Power Plants," in Kirchhof, *Pathways into and out of Nuclear Power in Western Europe*, 74–123.
43. Stephen Milder, "From Antinuke to Ökopax: 1970s Anti-Reactor Activism and the Emergence of West Germany's Mass Movement for Peace," in *Nature and the Iron Curtain: Environmental Policy and Social Movements in Communist and Capitalist Countries, 1945–1990*, ed. Astrid Mignon Kirchhof and John R. Mc Neill (Pittsburgh: University of Pittsburgh Press, 2019), 87–101; Silke Mende and Birgit Metzger, "Eco-pacifism: The Environmental Movement as a Source for the Peace Movement," in *The Nuclear Crisis: The Arms Race, Cold War Anxiety, and the German Peace Movement of the 1980s*, ed. Christoph Becker-Schaum et al. (New York: Berghahn, 2016), 119–37; Luise Schramm, *Evangelische Kirche und Anti-AKW-Bewegung: Das Beispiel der Hamburger Initiative kirchlicher Mitarbeiter und Gewaltfreie Aktion im Konflikt um das AKW Brokdorf 1976–1981* (Göttingen: V&R, 2018).
44. Andrew Tompkins, "Grassroots Transnationalism(s): Franco-German Opposition to Nuclear Energy in the 1970s," *Contemporary European History* 25, no. 1 (2016): 117–42, https://doi.org/10.1017/S0960777315000508.
45. Bernd A. Rusinek, "Wyhl," in *Deutsche Erinnerungsorte II*, ed. Etienne François and Hagen Schulze (München: Beck, 2001), 652–66.

46. Milder, *Greening Democracy*, 71–75.
47. Milder, *Greening Democracy*, 41.
48. Brügge, "Waldeslust und Widerstand"; Kirchhof and Trischler, "The History behind West Germany's Nuclear Phase-Out"; Jens Ivo Engels, " 'Politischer Verhaltensstil': Vorschläge für ein Instrumentarium zur Beschreibung politischen Verhaltens am Beispiel des Natur- und Umweltschutzes," in *Natur- und Umweltschutz nach 1945: Konzepte, Konflikte, Kompetenzen*, ed. Franz-Josef Brüggemeier and Jens Ivo Engels (Frankfurt: Campus, 2005), 184–202.
49. Milder, *Greening Democracy*, 29–31.
50. Christian Forstner, "The Failure of Nuclear Energy in Austria: Austria's Nuclear Energy Programmes in Historical Perspective," in *Pathways into and out of Nuclear Power in Western Europe: Austria, Denmark, Federal Republic of Germany, Italy, and Sweden*, ed. Astrid Mignon Kirchhof (Munich: Deutsches Museum, 2019), 36–73, here: 56, 62; Jan-Henrik Meyer, " 'Where Do We Go from Wyhl?' Transnational Anti-Nuclear Protest Targeting European and International Organisations in the 1970s," *Historical Social Research* 39, no. 1 (2014): 214, 225.
51. Forstner, "The Failure of Nuclear Energy in Austria," 57–58.
52. Frank Uekötter, *The Greenest Nation? A New History of German Environmentalism* (Cambridge, MA: MIT Press, 2014), 90–91.
53. Stephen Milder, "Between Grassroots Activism and Transnational Aspirations: Anti-Nuclear Protest from the Rhine Valley to the Bundestag, 1974–1983," *Historical Social Research* 39, no. 1 (2014): 191–211; Meyer, " 'Where Do We Go from Wyhl?' "
54. Rubio-Varas et al., "Spain Short Country Report," 39–41.
55. Antonio Ribes Serrano, *Memòries d'un regidor de l'Ajuntament d'Ascó* (Barcelona: Viena Edicions, 2008), 31.
56. Rubio-Varas et al., "Spain Short Country Report," 41–43.
57. Rubio-Varas et al., "Spain Short Country Report," 43–44.
58. Kaijser, "The Referendum that Preserved Nuclear Power," 247–49, 257–58.
59. Kitschelt, "Political Opportunity Structures and Political Protest"; Lutz Mez and Birger Ollrogge, *Energiediskussion in Europa: Berichte und Dokumente über die Haltung der Regierungen und Parteien in der Europäischen Gemeinschaft zur Kernenergie* (Villingen: Neckar-Verlag, 1979).
60. Meyer, " 'Atomkraft—Nej tak,' " 101.
61. Anna Kaijser and David Larsson Heidenblad, "Young Activists in Muddy Boots. Fältbiologerna and the Ecological Turn in Sweden, 1959–1974," *Scandinavian Journal of History* 43, no. 3 (2018): 301–23, https://doi.org/10.1080/03468755.2017.1380917; David Larsson Heidenblad, "Mapping a New History of the Ecological Turn: The Circulation of Environmental Knowledge in Sweden 1967," *Environment and History* 24, no. 2 (2018): 265–84, https://doi.org/doi.org/10.3197/096734018X15137949591936.
62. Kaijser, "The Referendum that Preserved Nuclear Power," 247–49; Wake Rowland, *The Plot to Save the World: The Life and Times of the Stockholm Conference on the Human Environment* (Toronto: Clarke, Irwin & Co., 1973).
63. The Ecologist/Friends of the Earth, "Swedish Cover-Up on Nuclear Safety," *Stockholm Conference Eco*, June 7, 1972, 1, 8.
64. Kaijser, "The Referendum that Preserved Nuclear Power," 246.
65. Kaijser and Meyer, " 'The World's Worst Located Nuclear Power Plant,' " 82–84;

Jan-Henrik Meyer, "To Trust or Not to Trust? Structures, Practices and Discourses of Transboundary Trust around the Swedish Nuclear Power Plant Barsebäck near Copenhagen," *Journal of Risk Research* 24 (2021).

66. Kaijser, "The Referendum that Preserved Nuclear Power," 258–63, 272–74, 277–81.
67. IFOR is a nongovernmental organization founded in 1914 in response to the horrors of war in Europe. It was also instrumental in the early days of the antinuclear movement in the Upper Rhine valley. Milder, *Greening Democracy*, 32.
68. Meyer, " 'Atomkraft—Nej tak,' " 84; Henry Nielsen et al., *Til samfundets tarv—Forskningscenter Risøs historie* (Risø: Forskningscenter Risø, 1998), 301.
69. Meyer, " 'Atomkraft—Nej tak,' " 81–83; *Interview with Siegfried Christiansen, founding member of OOA, conducted by Jan-Henrik Meyer, 28 September 2017* (Copenhagen); Siegfried Christiansen, "Den danske oplysningskampagne om atomkraft 1974–1977," text prepared for International Seminar on Training for Nonviolent Action, July 6–27, 1977, Guernavaca, Mexico, https://danmarkshistorien.dk/leksikon-og-kilder/vis/materiale/sigfred-christiansen-den-danske-oplysningskampagne-om-atomkraft-1974-1977-juni-1977/.
70. OOA, "Pressemeddelse, 31.1.1974" (Folder: Energiprojekt → OOA, Planlægningsmøder, June 5, 1973–January 31, 1974, in "Mødereferater med bilag fra tirsdagsmøder 1974–1995," *Rigsarkivet, Copenhagen* OOA 10451, no. 40 (1974).
71. Meyer, " 'Atomkraft—Nej tak,' " 81–83, 92–94.
72. Meyer, " 'Where Do We Go from Wyhl?' " 213.
73. Rubio-Varas et al., "Spain Short Country Report," 18.
74. Seymour Martin Lipset, and Stein Rokkan, "Cleavage Structures, Party Systems and Voter Alignments. An Introduction," in *Party Systems and Voter Alignments: Cross-National Perspectives*, ed. Seymour Martin Lipset and Stein Rokkan (New York: The Free Press, 1967), 9–13.
75. Forstner, "The Failure of Nuclear Energy in Austria," 62–64.
76. Rubio-Varas et al., "Spain Short Country Report," 38–44, 47.
77. Jan-Henrik Meyer, "Nature: From Protecting Regional Landscapes to Regionalist Self-Assertion in the Age of the Global Environment," in *Regionalism and Modern Europe: Regional Identity Construction and Regional Movements from 1890 until the Present*, ed. Xosé M. Núñez Seixas and Eric Storm (London: Bloomsbury, 2019), 73–76; Gaumer, *Wackersdorf*, 203.
78. Bernhard Peters, "Deliberative Öffentlichkeit," in *Die Öffentlichkeit der Vernunft und die Vernunft der Öffentlichkeit: Festschrift für Jürgen Habermas*, ed. Lutz Wingert and Klaus Günther (Frankfurt: Suhrkamp, 2001), 655–77.
79. Milder, "Between Grassroots Protest and Green Politics." Similarly, Michael Schüring, "West German Protestants and the Campaign against Nuclear Technology," *Central European History* 45, no. 4 (2012): 744–62, https://doi.org/10.1017/S0008938912000672.
80. Augustine, *Taking on Technocracy*, 140–41.
81. Jan-Henrik Meyer, "Kernkraft, Gesellschaft und Demokratie in den 1970er- und 1980er-Jahren," *H-Soz-Kult*, June 14, 2019, http://www.hsozkult.de/publicationreview/id/rezbuecher-29811.
82. Schüring, "West German Protestants"; Michael Schüring, *"Bekennen gegen den Atomstaat": Die evangelischen Kirchen in der Bundesrepublik und die Konflikte um die Atomenergie 1970–1990* (Göttingen: Wallstein, 2015); Thomas Kroll, "Protestantismus und Kernenergie: Die Debatte in der Evangelischen Kirche der

Bundesrepublik Deutschland in den 1970er und frühen 1980er Jahren," in *Energie in der modernen Gesellschaft. Zeithistorische Perspektiven*, ed. Hendrik Ehrhardt and Thomas Kroll (Göttingen: Vandenhoeck & Ruprecht, 2012), 93–118.

83. Erik Berkers, "Netherlands Short Country Report," in *History of Nuclear Energy and Society (HoNESt) Consortium Deliverable no. 3.6* (2019), 52–58, https://hdl.handle.net/2454/38269.
84. Rucht, *Von Wyhl nach Gorleben*, 79; European Parliament, "Debates on 'Community Policy on the Siting of Nuclear Power Stations' (13 January 1976)," *Official Journal of the European Communities, Annex: Debates of the European Parliament. Report of Proceedings from 12 to 15 January 1976*. No. 198 (1975/76 Session): 36, 53, 64.
85. Sezin Topçu, "Confronting Nuclear Risks: Counter Expertise as Politics within the French Nuclear Debate," *Nature and Culture* 3, no. 3 (2008): 225–45; Rucht, "Gegenöffentlichkeit und Gegenexperten"; Uekötter, *The Greenest Nation?*, 97; Forstner, "The Failure of Nuclear Energy in Austria," 56–57.
86. The first book published by promoters of nuclear power highlights this fact in its introduction: Uffe Korsbech and P. L. Ølgaard, *Atomkraft i Danmark, fordele og ulemper* (Copenhagen: Lindhardt og Ringhhof, 1974).
87. Meyer, " 'Atomkraft—Nej tak,' " 101–7.
88. Milder, *Greening Democracy*, 74.
89. Rubio-Varas et al., "Spain Short Country Report," 47.
90. Milder, *Greening Democracy*, 74.
91. Milder, *Greening Democracy*, 76.
92. Kaijser, "The Referendum that Preserved Nuclear Power," 250.
93. Meyer, " 'Atomkraft—Nej tak,' " 84; Kaijser and Meyer, " 'The World's Worst Located Nuclear Power Plant.' "
94. Rubio-Varas et al., "Spain Short Country Report," 47.
95. Robert Gildea and Andrew Tompkins, "The Transnational in the Local: The Larzac Plateau as a Site of Transnational Activism since 1970," *Journal of Contemporary History* 50, no. 3 (2015): 581–605, https://doi.org/10.1177/0022009414557909.
96. Patrick Kupper, *Atomenergie und gespaltene Gesellschaft: Die Geschichte des gescheiterten Projektes Kernkraftwerk Kaiseraugst* (Zürich: Chronos, 2003).
97. Milder, *Greening Democracy*, 137.
98. Milder, *Greening Democracy*, 110–17.
99. Augustine, *Taking on Technocracy*, 126–60.
100. Gaumer, *Wackersdorf*, 170–92, 212–33.
101. Renard, "The Superphénix Fast Breeder Nuclear Reactor"; Rucht, *Modernisierung und neue soziale Bewegungen*; Andrew Tompkins, "Transnationality as a Liability? The Anti-Nuclear Movement at Malville," *Revue Belge de Philologie et d'Histoire* 89, no. 3–4 (2011): 1365–79.
102. Rubio-Varas et al., "Spain Short Country Report," 44–48.
103. Gaumer, *Wackersdorf*, 286–87.
104. Tompkins, *Better Active than Radioactive!*, 183.
105. NN, "Nein zur Atomenergie, Österreich 5. November 1978. Nej till Kärnkraft, Sverige, 23 Mars 1980," (postcards from Austria to Sweden, 1980), in "Materiale vedr. Barsebäck-aktiviteter: OOA's Barsebak gruppe 1986–1988," *Rigsarkivet, Copenhagen* OOA 10451, no. 25 (1980).
106. Birgit Müller, "Anti-Nuclear Activism at the Czech-Austrian Border," in *Border*

Encounters: Asymmetry and Proximity at Europe's Frontiers, ed. Jutta Lauth Bacas and William Kavanagh (New York: Berghahn, 2013), 68–89.

107. Meyer, " 'Where Do We Go from Wyhl?' " 229. On current WISE activities, see https://www.wiseinternational.org.
108. Astrid Mignon Kirchhof and Jan-Henrik Meyer, "Global Protest Against Nuclear Power. Transfer and Transnational Exchange in the 1970s and 1980s," *Historical Social Research* 39, no. 1 (2014): 181, https://doi.org/10.12759/hsr.39.2014.1.165-190.
109. Kitschelt, "Political Opportunity Structures and Political Protest," 58–60.
110. Astrid M. Eckert, *West Germany and the Iron Curtain. Environment, Economy and Culture in the Borderlands* (Oxford: Oxford University Press, 2019), 201–43.
111. Maria del Mar Rubio-Varas, António Carvalho, and Joseba de la Torre, "Siting (and Mining) at the Border: Spain-Portugal Nuclear Transboundary Issues," *Journal for the History of Environment and Society* 3 (2018): 33–69, https://doi.org/10.1484/J.JHES.5.116794.

CHAPTER 4

International Organizations and the Atom: How Comecon, Euratom, and the OECD Nuclear Energy Agency Developed Societal Engagement

Paul Josephson and Markku Lehtonen

Introduction

The postwar period was an era of institutionalized multilateralism: various international organizations were created after World War II, in pursuit of economic growth and prosperity through modern technology and international cooperation among governments.[1] While designed to promote cooperation between their member states, international organizations typically also assumed a life of their own.

Nuclear power was a significant element in the postwar vision of progress. Hence, it is hardly surprising that key postwar European organizations designed to advance economic collaboration between governments set up their own organizations to promote and regulate nuclear energy. The worldwide interest in the so-called Atoms for Peace program gave further impetus to the creation of intergovernmental organizations (IGOs) in the area. The program followed US President Dwight D. Eisenhower's 1953 eponymous speech to the United Nations, and took shape notably through the Geneva conferences on the peaceful uses of atomic energy. The first of these conferences, held in 1955, paved the way for the creation of the International Atomic Energy Agency (IAEA, 1957), European Atomic Energy Community (Euratom, 1957), and the European Nuclear Energy Agency (ENEA, 1958). Euratom was established to accompany the European Communities (European Coal and Steel Community of 1952 and European Economic Community of 1957), while the ENEA emerged under the auspices of the Organisation for European Economic Co-operation (OEEC). Originally established in 1948 to coordinate the US Marshall aid, in 1961 OEEC became the OECD (Organisation for Economic Co-operation and Development) as Canada and the US joined the organization. With Japanese membership in 1972, the ENEA also expanded beyond Europe and was renamed the OECD Nuclear Energy Agency (NEA). On the other side of the Iron Curtain, the socialist civilian nuclear programs were placed under the umbrella of the Eastern Bloc economic trade organization the Council for Mutual Economic Assistance (Comecon, 1949). Nuclear research cooperation at Comecon was controlled by the USSR through the Joint Institute for Nuclear Energy (JINR, 1957).

Nuclear IGOs came into existence against the backdrop of the burgeoning Cold War and fears over proliferation of nuclear know-how, weapons, and materials. They managed these fears and materials and successfully promoted a variety of nuclear applications both in Europe and across the world. But international cooperation was constrained by member countries' desires to retain their sovereignty in a strategically sensitive policy area, notably in countries with nuclear weapons programs, such as France and the UK.[2] Generally, as chapter 1 highlights, apart from major accidents and their transboundary impacts, nuclear programs and events are administered and negotiated at the national level.

In this chapter, we examine how three intergovernmental organizations—Comecon, Euratom, and NEA—have engaged with the broader public beyond government and industry stakeholders. We seek to illustrate the evolution of public engagement work within these IGOs, in the context of tensions between contrasting objectives among member governments. In particular, we examine the role of these three IGOs in the so-called participatory turn, which emerged in the 1990s in nuclear waste management,[3] in governance in various policy areas more broadly,[4] and to some extent also in nuclear energy policies.

In focusing on IGOs created in the immediate postwar period, our analysis excludes organizations whose mandate covers only a specific aspect of nuclear governance (for example, the European Nuclear Safety Regulators Group [ENSREG] and the Western European Nuclear Regulators Association [WENRA]) as well as nongovernmental organizations such as Greenpeace, the World Nuclear Association, and the World Association of Nuclear Operators (WANO). Given our focus on civilian nuclear energy, Europe, and public engagement, we also exclude the IAEA, for three reasons. First, as a truly global organization and part of the United Nations family, the IAEA has retained the prevention of proliferation of nuclear weapons—along with the promotion of nuclear energy—as its *raison d'être*. Second, its remit far exceeds the concerns of European countries. Third, the IAEA has been late to initiate work on public engagement.[5]

While Comecon disappeared in 1991 with the fall of the Iron Curtain, and therefore never had a chance to develop any significant public engagement, it crucially shaped nuclear development in the socialist countries, many of which were integrated into the Western nuclear governance structures in the 1990s and 2000s, including Euratom and the NEA. Russian-European interaction in nuclear energy continues today through bilateral agreements, especially in Central and Eastern Europe. Similarly, although the NEA's membership today is global, as an organization with West European roots, it played a particular role in the evolution of the European nuclear sector.

Types of Public Engagement

To characterize the activities and evolution of the ways in which our three IGOs have interacted with the public, this chapter follows a typology by Gene Rowe and Lynn Frewer, who distinguish between three types of public engagement, from simple toward more advanced forms:[6]

1. *Communication* describes a one-way information flow, from a sponsor or a promoter to the public. No feedback from the public is sought.
2. *Consultation* is a step toward greater engagement. Information still flows in one direction only but can take the form of questions from the public or answers/information from a sponsor or a promoter.
3. *Participation* is the highest level of engagement in this typology, consisting of two-way exchange of information and dialogue between the two parties, which can ideally transform the opinions of either party and lead to greater understanding of each other's views, if not necessarily greater agreement.

Whether the more advanced forms of engagement also enhance the democratic quality of policymaking depends strongly on a host of contextual factors—not least on power relations among the actors. The more advanced forms can therefore be seen as a necessary yet insufficient condition for enhancing the democratic quality of policymaking.

A further category—public-initiated engagement—is only addressed in passing in this chapter.[7] As our examples show, this type of activity by NGOs and other civil society actors had a significant yet only indirect effect on the operation of the IGOs, spurring these to more actively engage with the public (see also chapter 9).[8]

We first briefly describe the birth of the three IGOs, and then examine their public engagement activities throughout three periods. Over time, the NEA's and Euratom's relations with the public indeed largely followed a succession from mere communication toward more advanced forms of engagement.

Communication dominated until the late 1960s, consultative approaches emerged as a response to growing public concerns and changes in the international landscape from the early 1970s, and participatory practices were increasingly adopted from the 1990s onwards. Comecon did not live long enough to integrate the more advanced forms of engagement. In the concluding section, we explore reasons for the differences in the ways the three organizations have engaged with the public and highlight topics for further research.

The Birth of Nuclear-Sector IGOs

During the first wave of development of nuclear governance, our three IGOs were designed to promote the peaceful use of the atom in the context of the emerging Cold War.

Comecon and the Soviet and East European Atom

Created in 1949, Comecon was a foreign policy tool of the USSR to integrate the socialist nations of Central and Eastern Europe and a response to the US Marshall Plan, which helped postwar reconstruction and modernization in Western Europe.[9] Comecon and the Warsaw Treaty Organization were created in 1955 as a counterbalance to NATO. They were intended to integrate the socialist world by harnessing the industrial-military system of the Soviet Union, its vast natural resources, and its enormous market potential as a basis for the economic systems of other socialist countries.[10] Comecon sought to foster economic development by applying the principles of Soviet planning at an international level, via a division of labor between its member countries that was designed to bring about rationality and efficiency.

After Stalin's death in 1953, socialist scientists sought new and expanded contacts, especially within the socialist world, to end economic and scientific autarky. The authority of Soviet physicists enabled them to open up heretofore secret research institutes connected with Minsredmash (the Ministry of Medium Machine Building, responsible for nuclear weapons and peaceful program developments). When they involved European and American scientists, the visits also helped build mutual confidence in the area of arms control.[11]

In January 1955, the USSR Council of Ministers announced its willingness to share nuclear knowledge within the socialist world, out of which grew the Joint Institute for Nuclear Energy (JINR), in 1957. JINR was founded at the Institute of Nuclear Problems and the Electrophysical Laboratory of the Academy of Sciences (EFLAN), in Dubna, USSR, with access to world-leading particle accelerators.[12] Initially, the USSR invited China, Poland, Czechoslovakia, the German Democratic Republic (DDR), and Romania to work with Soviet-designed accelerators and experimental reactors to acquire the nuclear materials needed for research and develop scientific exchanges.[13] Later, in 1955 and 1956, Hungary, Bulgaria, Yugoslavia, and Egypt signed similar agreements with the USSR. Claiming that its peaceful programs preceded those of the US, the USSR now engaged in a kind of nuclear imperialism over the socialist countries.[14] Eventually, the JINR involved virtually the entire socialist world: Albania, Bulgaria, China, Czechoslovakia, the DDR, Hungary, Mongolia, North Korea, Poland, Romania, and the USSR.

The UN Geneva Conferences on the Peaceful Uses of Nuclear Energy were crucial in opening Soviet and East European research institutes and their scientists to the international arena.[15] As its participants noted, the first conference, held over two weeks in August and September 1955, was an outstanding scientific success. The scientists learned nothing deeply new about their research topics, yet in hundreds of sessions they made acquaintances, lifted the veil of secrecy, created a genuinely international, exciting occasion to talk and exchange ideas, gauged the vitality of each other's research programs—and also engaged in scientific intelligence gathering.[16] The conferences emboldened socialist scientists to push for the declassification of basic research and promote exchanges. The 1956 visit by Igor Kurchatov, head of the Soviet atomic bomb project, and the Soviet Premier Nikita Khrushchev to the UK's Atomic Energy Research Establishment at Harwell signaled the determination of Soviet physicists to become members of the international community.[17] Kurchatov came to play a major role in fostering East-West research collaboration and opening up the hitherto closed Soviet nuclear research to "capitalist" scientists.[18]

Euratom: Nuclear Energy in the Service of European Integration?

In part with the support of the US government, the growing American civilian nuclear industry offered the world its cooperation to build nuclear power stations, including in Europe.[19] This was the backdrop to discussions in 1955 to create a European Community for Atomic Energy—an idea that gained support among a number of prominent diplomats and engineers.[20] In the "Benelux Memorandum" (1955), the governments of Benelux, Italy, France, and West Germany agreed to work together toward what would become the Treaties of Rome (March 25, 1957) establishing the EEC and Euratom.[21] Euratom would foster joint West European nuclear research to develop the industry and have a supply system of fissile fuel through a dedicated agency.[22]

Pressure primarily from the French government led to the exclusion of military applications from Euratom, which would concentrate on the promotion of nuclear energy, nuclear safeguards, and health and safety regulations.[23] National industries were to remain independent but under a common supranational framework. For its founders, Euratom was also a tool of postwar European integration.[24] By obliging its member states to collaborate in developing this new, emerging technology, Euratom would foster political integration and address the pressing economic problems of the still young and fragile European Community.[25]

The US government endorsed Euratom as the culmination of West Germany's integration into the Western coalition and as a means of spreading the peaceful atom, tying the emerging European nuclear industry to the American one, and safeguarding it from diversion for military uses.[26]

Unlike the two other founding treaties of the European Communities, the Euratom Treaty has remained essentially unchanged since its adoption, retaining its original arrangement that largely excluded European Parliament involvement and scrutiny.[27] At a time when nuclear power was expected to provide a major solution to the continent's energy problems, the treaty gave considerable centralized powers to the European Commission. These concerned the ownership and control of the supply of all fissile materials, as well as the distribution of patent rights and production licenses for technologies to be developed by the Joint Nuclear Research Center (JNRC). European research was seen as vital for building nuclear capabilities, while international agreements would allow European Community members to gain access to fissile materials and technologies.[28] The four specialized research centers—Ispra (Italy), Petten (the Netherlands), Geel (Belgium), and Karlsruhe (Germany)—together absorbed up to 50 percent of the Euratom budget, with Ispra the largest of the four. Euratom was committed to ensuring that nuclear information would be shared among all the actors active in the sector. A 1958 agreement with the United States gave Euratom the additional tool of a joint research program with a budget of US$100 million.[29] Crucially, Euratom was granted no competencies in radiological protection and nuclear safety.[30]

The OECD Nuclear Energy Agency (NEA): Soft Persuasion via Cooperation

The NEA has its origins in the Organisation for European Economic Cooperation (OEEC), created on April 16, 1948, to administer the Marshall Plan. Energy questions, notably those relating to the oil, electricity, and coal sectors, figured prominently in OEEC work.[31] In 1953, Robert Marjolin, the French secretary-general of the OEEC, highlighted the rising cost of energy in Europe as a major impediment to postwar recovery. The OEEC commissioned Luis Armand, a French engineer who assisted in the creation of Euratom and became its first director, to report on possible solutions. The 1955 Armand report stressed the significant potential of nuclear energy, and urged Europeans to pool their resources to develop this new energy source.[32] Bringing together seventeen European countries,[33] the European Nuclear Energy Agency was born a few months after Euratom and the International Atomic Energy Agency (IAEA)—with which it had "distinct but overlapping roles, missions, and memberships."[34] From the beginning, establishing its own identity, not least in relation to the IAEA, was a major challenge for the NEA and its numerous standing committees, each with its own tasks, remit, and work culture.[35] Relations with Euratom were not simple either: for example, the British first strongly supported the creation of the NEA as an alternative to Euratom but quickly lost interest after failing to prevent the establishment of Euratom.[36]

The NEA soon adopted the working methods still in practice in the agency. Devoid of regulatory power over its member countries, the NEA relied on "soft persuasion" and joint research projects. This would take place via work under specialized standing technical committees that in turn supervised the work of numerous permanent and ad hoc expert groups consisting of member country representatives. The groups provide a venue for exchange of information, development of joint opinions on technical and policy issues, identification of areas for further work, and organization of joint research projects among interested countries.[37]

Public Engagement in the Early Years: 1950–1970

Founded in the heyday of technological and technocratic nuclear enthusiasm, none of the three IGOs examined in this chapter had stakeholder involvement as a founding principle. In the spirit of the Atoms for Peace initiative, they nevertheless greatly advanced exchange among experts, also across the Iron Curtain. In the early postwar years of nuclear development, these IGOs were largely promotional, engaging in public communication that extolled the virtues of the "peaceful atom," especially its contribution to economic growth, well-being, and European economic recovery. Dissemination of popular media images designed to serve this end was strikingly transnational in character, as similar images, reporting on the same events, circulated across the globe.[38] Legal regulation likewise served the purpose of promoting the new industry.[39]

Laypeople were seen as a target of one-way communication, rather than as potentially active partners and interlocutors.

Comecon: Extolling the Virtues of Socialist Atomic Science

While the Soviet atomic bomb project was shrouded in secrecy, Comecon served economic and strategic goals: helping to overcome economic despair in the countries of the Central and Eastern Europe (CEE) after the Second World War while ensuring Soviet control over their substantial mineral resources, including uranium. The Soviets forced the CEE countries to adopt the USSR's development model based on five-year plans and industrialization. Throughout the history of Comecon, the other member nations recurrently complained about being subject to unilateral Soviet exploitation rather than true partners in cooperative ventures. The USSR represented 90 percent of Comecon's land and energy resources, 70 percent of population, and 65 percent of national income, and possessed industrial and military capacities second only to those of the US. The location of most Comecon committee headquarters in Moscow and the large number of Soviet nationals in positions of authority further accentuated the Soviet dominance.

Energy policy activities included the establishment of a unified electrical power grid; a number of bilateral and multilateral investment projects, such as the *Druzhba (friendship)* pipeline bringing oil from Russia to Eastern Europe; and scientific exchanges.[40] If economic development and military programs were paramount at first, then scientific and technical exchanges expanded in an atmosphere of greater openness that accompanied a de-Stalinization "thaw" in the late 1950s. Nuclear power programs developed in the mid-1950s continued until the breakup of the USSR.

The bilateral agreements of Comecon's nuclear research institute, the JINR, grew broader and more permanent over the years. The JINR created two new laboratories, one for nuclear reactions and the other for neutron physics with pulsed reactors (the IBR-1 and BBR-2). A Bulgarian physicist Neno Ivanchev described cooperation through Comecon as "one of the greatest conquests and advantages of the socialist system," which "especially clearly manifests itself in such a region as atomic science and technology."[41] The scientists connected with JINR constituted an "epistemic community,"[42] gathering Soviet bloc scientists in pursuit of the overarching objective of keeping up with and exceeding the achievements of the Western nations. The vast majority of information exchange through Comecon atomic programs involved scientists and graduate students, of whom there were 3,000 to 4,000 in all.

Rarely if ever included directly, the broader publics were targets of explicit communication and propaganda efforts. A series of publications was established, although at first they rarely reached nonspecialists. The Soviet *Atomnaia Energiia* (1956–) included a rubric for "short reports" that reported on developments around the world, including in Comecon, such as the signature of cooperation protocols between the USSR and other Comecon countries, the JINR Academic Council meetings,[43] joint research by scholars from countries of Comecon, and scientific exhibitions.[44]

While Atoms for Peace publicity campaigns were carried out in Eastern Europe, they seem to have been smaller in scale and to have developed later than in the USSR. Such journals as *Tekhnika-Molodezhi*, *Ogonek*, and *Nauka i Zhizn'* gave a prominent place to modern science and technology in the "glorious socialist present and future." They also had the success of Sputnik (1957) and Yuri Gagarin (1961) to tout. Dozens of books proselytized the atom.[45] The reporting hardly mentioned the risks of nuclear applications but gave great prominence to peace demonstrations abroad, conveying the message that while people in the West had to take to the streets to get their governments to act for peace, the Soviet state was inherently peaceful.[46]

In 1947 the *Znanie* (knowledge) society was created on the initiative of scientists who urged artists, writers, and others to help disseminate knowledge in support of the rebuilding of the war-damaged Soviet nation. The society spread quickly to all union republics and had an active speakers' bureau and publishing house that, over its lifetime, produced perhaps 500 titles in a total of thirty million copies.[47] A similar phenomenon unfolded in the rest of the socialist world in newspapers, journals, and books, albeit on a more modest scale, judging from an examination of holdings in Polish, Bulgarian, and Czech libraries and other searches.

The reasons for the fewer formal and informal industry-society interactions in the socialist Eastern Europe include a delay in developing a civilian nuclear program and the dominance of Comecon by Moscow, which was unwilling to engage the public. Comecon published a monthly journal for the last fifteen years of its existence, only in Russian.[48]

Euratom: Building European Identities—Communicating to the Elites or to the Public?

Euratom's communication policy evolved in the context of the European Communities' communication with a fragmented and multilingual European public sphere.[49] The basis for this work had been laid prior to the birth of the actual EC institutions. Since the Marshall Plan, public communication campaigns by numerous transatlantic, supranational, and national (especially US) institutions[50] and private companies sought not only to inform the public about the new institutions and their rationale but also to promote a sense of European identity. Next to the purely instrumental promotion of key economic objectives, the EC's early communication policy had the "explicit civil aim" of stimulating "a European civil consciousness."[51] In this context, Euratom was seen as an instrument of political integration, and hence expected to communicate a European political identity via written material, films, and documentaries (so-called information films).[52] The first president of the High Authority of the European Coal and Steel Community, Jean Monnet, considered it essential that the Communities communicate directly to all interested parties. In practice, this meant addressing the entire 160 million European population in their respective languages. In the 1950s and early 1960s, the institutions targeted the broader public via mass media, using simplified, comprehensible language, and by inviting the public to visit the Community institutions and its offices in the member states.[53]

In the early 1960s, the EC and Euratom refocused their communication efforts toward elite actors. The creation of European-wide identities and loyalties was now to be achieved by targeting opinion-leading elites, such as politicians, representatives of political associations, trade unions, economic and business organizations, journalists, academics, and youth leaders.[54] The aim was to establish and strengthen "links between EC institutions, national administrations, social actors and the media," and not so much with the public at large.[55] The shift from a public-oriented to an elite-oriented communication policy was dictated partly by resource constraints but also by the neofunctionalist belief that "rational" policymaking in Europe could only be achieved by creating a supranational community of independent experts. Including national sectoral interest groups or the public at large could only hamper the efforts of the European-wide elites to deliver the desired policy outputs in the interest of the wider society.[56] A "permissive consensus" among the general public seemed sufficient for European integration to advance.[57] Ultimately, Euratom's public engagement policy was constrained by its rather "dirigiste" and promotional founding treaty.[58]

Euratom was undergoing a more general crisis in the late 1960s, not least because of "shortcomings of the Euratom Treaty, nuclear nationalism, a lack of leadership, overinflated initial expectations, and inauspicious circumstances,"[59] especially the confrontation and mutual suspicion between West Germany and France. West Germany had sought to limit Euratom's scope and powers, which the French saw as an instrument for spreading the costs of nuclear infrastructure across the member countries and controlling the expansion of West German nuclear industry.[60] The crisis culminated in 1967, as the member countries failed to agree on a third five-year program, and the short-term plans and funding schemes no longer left room for major joint undertakings. By the end of the 1960s, the founders' hopes and expectations—that Euratom would foster political unity and transcend particular political interests—had proven to be overly optimistic.[61]

European Nuclear Energy Agency: Communication among Experts and Governments

In its early years, the NEA paid little attention to public engagement, focusing instead on expert and intergovernmental collaboration. The Health and Safety Subcommittee (HSC), established in February 1958, was to play a crucial role in advancing stakeholder engagement. Insurance and other liability questions were another key area of work in NEA's first years, as potential investors were wary in the face of legal uncertainty and fears of liability claims in the case of an accident.[62] A major NEA achievement was the drafting of the Paris Convention on Third Party Liability in the Field of Nuclear Energy, adopted by the OEEC Council on July 20, 1960. The convention established rules of responsibility and levels of compensation for damage caused by nuclear activities, seeking to ensure equal treatment across countries.[63] The US remained outside of the convention, having introduced its own, rather different, legislation in 1957—the Price-Anderson Act on Liability.

Other efforts focused on joint undertakings designed to develop experimental reactors and prototypes, safety and security, economics, training of nuclear engineers, and measurement and monitoring of radioactivity in the environment.[64] Three major joint projects were inaugurated in the late 1950s: the Eurochemic reprocessing company in Belgium, the high-temperature gas-cooled reactor Dragon, and the experimental boiling heavy water reactor in Halden, Norway. Eurochemic and Dragon were abandoned in the mid-1970s, whereas Halden operated until 2018 and gave rise to a wide international technical network coordinating the work of dozens of research institutions in numerous areas of nuclear safety.

In 1960, anticipating growing public concern, the Norwegian delegation requested work on the technical, biological, and administrative aspects of dumping radioactive waste in the North Sea. This paved the way for the NEA to take into account the broader public. Between 1965 and 1976, despite the slowly emerging public concerns, the NEA conducted more than ten disposal "campaigns"—coordinated efforts designed to "promote and supervise sea disposal under excellent safety conditions" in the Atlantic.[65]

Times of Change and Crisis: 1970–1990

The 1970s began in an atmosphere of nuclear optimism. An apparently endless growth of electricity demand was underpinning ambitious nuclear new-build plans. Now that turnkey nuclear power plants were available, the visions of the nuclear age seemed close to fulfillment. In the immediate aftermath of the first oil crisis, in 1973, nuclear power seemed economically competitive. A number of countries (e.g., Denmark, West Germany, and most notably France) launched ambitious plans to expand nuclear power to ensure energy independence and security of supply.[66]

In the West, this optimism soon faded with the onset of a crisis of public confidence, as well as an economic stagnation that ended the era of continuous economic growth, brought down electricity demand forecasts, and led many European countries to downscale their nuclear programs (see chapter 2). A 1972 report commissioned by the Paris-based Atlantic Institute of International Affairs from Achille Albonetti, a director at the Italian Committee for Nuclear Energy, illustrated the spirit of the time and foresaw problems for the IGOs. Albonetti envisaged the possibility of "rationalizing the structure of international co-operation in the nuclear field" by abolishing the NEA or merging it with other OECD directorates.[67] He also gave a rather gloomy account of Euratom's future prospects, attributing its crisis to "nationalistic tendencies" in the member countries, unable to agree on joint nuclear programs.[68]

Mass demonstrations against nuclear power programs throughout the 1970s, the 1972 UN Stockholm conference on the environment, and the accidents at Three Mile Island in 1979 and especially at Chernobyl in 1986 spurred further public engagement work at the IGOs. The failed attempts by many member countries in the late 1980s to identify willing host communities for nuclear waste repositories (see chapter 6) compounded the problems. The focus of international nuclear law development now switched from promotion toward addressing safety concerns and constraining the development of nuclear power.[69] Key multilateral initiatives that had been under consideration for several years were adopted. Nuclear power and its regulation were now seen as a truly transnational activity, which required collaboration across national boundaries, notably through IGOs. The IGOs strengthened their communication efforts, while increasingly also engaging in public consultation on issues such as risk perceptions, communication, emergency preparedness, and radioactive waste and spent nuclear fuel.

In the East, economic and public acceptance problems arrived later, in the 1980s, culminating with the Chernobyl disaster. Until then, environmental movements had remained closely controlled by the state. The economic policy based on energy-intensive industries boosted the growth of electricity demand, while the nuclear military-industrial complex occupied a privileged position in the socialist system. In the context of the glasnost period of openness, Chernobyl triggered the birth of independent environmental movements, which helped to precipitate the fall of the USSR and Comecon.

The Glory and Fall of Comecon: From International Division of Labor to Forced Opening Up

At the 1971 Comecon session, the member nations agreed to establish the Special Council Committee for Scientific and Technical Cooperation to jointly plan and coordinate research programs. Hundreds of projects were initiated, typically with a Soviet ministry and a Soviet organization responsible for supervision and coordination.[70] At the June 1975 Comecon meeting, the member nations agreed to work on fast breeder reactors. The June 1978 session announced the objective of a fantastic 110,000–130,000 MW increase in nuclear power plant capacity by the 1990s, foreseeing a 50 percent increase in electricity consumption in the European Comecon countries by 1990.[71] The nuclear targets were never close to being achieved. Nevertheless, they reflected a multinational effort to standardize power in 1,000–1,500 MW units across fifty industrial enterprises in eight nations.[72]

As part of the "international division of labor," Comecon countries maintained and built upon their pre-socialist expertise.[73] Czechoslovakia, Hungary, Poland, and the GDR were also actively involved in fusion research, in partnership with the Soviets, until the fall of the Iron Curtain.[74] Modern technology spread throughout the socialist world as a sign of progress and modernity, and transferred capital from developed to developing countries. Transfers and specialization took place between the more industrialized and trade-dependent Central European Comecon countries, such as Czechoslovakia, East Germany, and Poland on the one hand, and the less industrialized countries, such as Bulgaria and Romania, on the other. The consortium Interatominstrument was established in 1972 to provide the socialist nations with modern radioisotope equipment and arrange deliveries to other countries.[75] In this light, Comecon nuclear activities did not reflect the dominance of the USSR but helped the smaller member countries advance toward energy interdependence, economic growth, and freedom.[76] Research collaboration within Comecon contributed to these objectives but also included significant ideological components.[77] Nuclear power suited the narratives of an ideology of progress, international division of labor, and the virtues of socialism. The slogan "Let the atom be a worker, not a soldier" found its way into posters and songs and onto the side of a JINR building in Dubna.

There was little or no public protest against nuclear in the socialist nations until the Chernobyl disaster. It appears that even the glasnost openness initiated by Mikhail Gorbachev barely affected the Comecon nuclear activities directly. However, it did so indirectly by facilitating the expression of public discontent following Chernobyl, which forced the sector to greater transparency. Chernobyl constituted a major watershed not only in that it greatly assisted in the emergence of an independent environmental movement in the Soviet bloc countries but also in ultimately speeding up the downfall of the USSR itself.[78] The accident laid bare a system characterized by secrecy, lack of participation, and the belief in the infallibility of Soviet technology. The Soviet government's early attempts to cover up the disaster and belittle its true extent fueled public discontent and demand for information and transparency. In a new atmosphere of perestroika and glasnost, the civil society could now express itself more freely, and local environmental movements proliferated across the Soviet states. Environmentalist, antinuclear, and independence movements were often closely related, as many activists saw independence as a prerequisite for the prevention of further environmental degradation and nuclear accidents. The closure of nuclear power plants and halting of nuclear programs under public pressure aggravated the already serious economic problems of Comecon countries. Comecon held its final council session on June 28, 1991, in Budapest, and the USSR was dismantled less than six months later. However, the JINR survived, and continues as a key international center for nuclear physics research, especially but not exclusively in the former Soviet bloc countries.[79]

Euratom Seeking Public Acceptance for Nuclear

The abovementioned 1972 report by Achilles Albonetti reflects the spirit of Euratom thinking at the time. It identified economics, politics (diversity of supply), and "ecological considerations" as the key defining factors for the future of nuclear in Europe. Remarkably, it did not evoke public opinion—despite the rise in critical voices against nuclear power.[80] The "ever-growing needs" served to justify the "inescapability" of nuclear in Western Europe and bold forecasts of nuclear new-build.[81] The report expected nuclear power to represent by 1985 15–30 percent of the rapidly rising total electricity production in Western Europe, including over 60 percent in Germany and more than 30 percent in countries such as the Netherlands, Italy, and Switzerland.[82] In the preface, Etienne Hirsch, the former president of Euratom, confidently declared: "*There is one question which is well and truly solved, about which public opinion, identifying nuclear energy and bombs, remains highly sensitive: the risk of nuclear incident. Experience of operating a number of nuclear power stations for nearly twenty years has demonstrated that the precautions taken and the disciplines followed are perfectly effective.*"[83]

By the late 1960s, the foundations of EC and Euratom information policies had begun to change. In the wake of the "Empty Chair Crisis"—with de Gaulle's France boycotting European institutions, and member countries at loggerheads over Europe's future—neofunctionalist beliefs in the inevitability of an elite-driven European integration faded, among scholars and officials alike.[84] The "permissive consensus" among the public began to erode. Practice was slower to change: in 1971, the EC Commission announced that information activities should focus primarily not on the general public but on opinion leaders—trade unionists; agricultural, university and consumer groups; and youth.[85] Furthermore, communicating on nuclear risk still seemed superfluous at a time when no Western reactors had faced major accidents.

While many of the key recommendations of the so-called Tindemans Report (1975/76), presented by (and named for) the prime minister of Belgium, went unnoticed, the report indeed accelerated the shift in the EC communication policy. Communication would now increasingly seek legitimacy not only for the policy outcomes but also for the policy process.[86] Tindemans recommended a number of measures under the "Citizens' Europe" initiative, including enhanced protection for the environment and the rights of consumers. In the nuclear sector, Tindemans expressed concern for "the psychological reactions throughout the whole of Europe against the setting up of nuclear power stations." To solve the problem, he recommended the creation of a joint European nuclear safety regulator, following the model of the US Regulatory Commission, with full powers to control all aspects of nuclear energy. The "strictness, openness and in particular independence" of a European safety regulator would be indispensable in order to render "the necessary development of nuclear energy in Europe acceptable to public opinion."[87] Supranational supervision was all the more necessary, given that many nuclear power stations were planned for border regions (see chapter 9).

A Europe-wide regulator did not exist—and still does not exist today—but the reference to "psychological reactions" echoed the nascent concern for public perception and relations with society: psychological or even psychoanalytical explanations were often evoked to explain and remedy the "irrational fears" seemingly underlying opposition to nuclear power.[88]

Pressure from an increasingly transnational citizen protest against the EC nuclear policies compelled the EC Commission to organize public hearings on nuclear energy projects—common practice in many countries at the time. The so-called Brunner hearings, held in Brussels between November 1977 and January 1978, constituted the first large-scale public engagement ever organized by the European Commission.[89] In terms of our typology, the hearings entailed consultation, and to some extent even participation. The hearings helped to reframe the boundaries of the debate at the supranational level, compelled the Commission to engage with critical views from civil society, and consolidated the status of the European Environmental Bureau as an interlocutor and critic not only on nature protection issues but also on energy and nuclear policy.[90]

Following Chernobyl, strengthening the regulation of safety and public health, as well as improving information and communication systems, became primary objectives of Euratom and the European Communities, which introduced several short-term emergency actions and more lasting long-term measures to this effect.[91] At the international level, Euratom was an active partner in negotiating all four post-Chernobyl "nuclear-safety family" conventions, all under the auspices of IAEA: Convention on Early Notification of a Nuclear Accident (1986), Convention on Assistance in the Case of a Nuclear Accident or a Radiological Emergency (1986), Convention on Nuclear Safety (1994), and Joint Convention on the Safety of Spent Fuel Management and on the Safety of Radioactive Management (1997). However, Euratom only became party to the conventions in the early 2000s.[92]

NEA: Enhancing Public Understanding and Building Acceptance

As the EC began to grow beyond its six founding countries in 1973, the NEA expanded beyond Europe. When Japan became a full member in 1972, the agency dropped the "European" from its name and went simply by NEA. Australia joined NEA in 1973, Canada in 1975, and the United States in 1976. Relations with the public gradually took on increasing importance in NEA's work, spurred by the new context of public skepticism of nuclear energy. Similar to the nuclear expansion plans within Comecon and the EC, NEA sanguinely predicted in 1973 that nuclear energy would account for half of all electric power by the end of the century.[93] The "new energy situation" triggered by the first oil crisis compelled the agency to revisit its program of work.

Public understanding of nuclear energy gained increasing attention, alongside issues of radiological protection, safety of nuclear installations, waste management, and regulatory frameworks.[94] The foundations for today's organizational structure were established with the creation of the Committee on the Safety of Nuclear Installations (CSNI), Committee on Radiation Protection and Public Health (CRPPH), Committee on Technical and Economic Studies on Nuclear Energy Development and the Fuel Cycle (NDC), and the Working Group on Radioactive Waste—the predecessor of today's Radioactive Waste Management Committee (RWMC). The academic community was now recognized as a key stakeholder group, and independent expert groups were established under various committees.[95]

The Three Mile Island accident in 1979 did relatively little to advance NEA's stakeholder engagement work. Contrasting views were expressed within the Radiation Protection and Public Health Committee on Three Mile Island: a special session of the committee first defined the event as an "incident," then a 1980 report requalified it as an "accident." The committee's scientific secretary, Osvaldo Ilari, suggested action on emergency plans and post-accident work, yet some country delegates rejected the idea as exaggerated and unjustified.[96]

The controversies over ocean dumping of waste gained further attention from governments and the public. In the early 1970s, several member countries called for a halt to sea disposal campaigns to allow time for lessons to be learned. Along with the Belgians and the Dutch, UK officials considered public apprehension unfounded and wanted to press ahead, while Norway led the opposition. The NEA helped to facilitate agreement, first by creating a joint Research and Environmental Surveillance Program in 1981 to coordinate national research programs on sea disposal, and then by coordinating work that led, in 1993, to a total ban on ocean dumping under the 1972 London Convention on the Prevention of Marine Pollution.

The disputes over sea disposal helped to bring about geological disposal as the solution for the waste problem. In 1977, a joint landmark report by the committees on Radiation Protection and Public Health and Radioactive Waste Management laid the basis for international cooperation and currently prevailing principles in high-level waste management.[97] However, the report did not yet identify stakeholder involvement as a major issue.[98] The agency nevertheless sought to support and supplement "public understanding" activities in its member countries. In 1982, a high-level joint NEA-IAEA workshop concluded that public opposition was now a major obstacle for the further development of nuclear energy. In response, the NEA began publishing a regular newsletter in 1983. In the first issue, NEA Director-General Howard K. Shapar argued that the key objective of "public acceptance" could only be achieved if governments provided basic information to the public.[99]

The Chernobyl disaster triggered further public information and communication work, and contributed to closer cooperation between national safety authorities. The Committee on Nuclear Regulatory Activities was established in October 1988 and would later become an NEA forerunner in promoting public engagement.[100] Informing the public about incidents and accidents at nuclear installations constituted another area of work. As early as 1980, the NEA science committee had set up an incident reporting system, which the IAEA adopted four years later. In November 1986, an NEA workshop on public understanding considered a safety scale to provide fast and simple information to the public. A few years later, joint work by the NEA and the IAEA would lead to the adoption in 1990 of the International Nuclear Event Scale (INES) for characterizing, reporting, and communicating on nuclear incidents.[101]

On the whole, the NEA approach remained largely confined to one-way provision of information, designed to justify and legitimize decisions that industry and governments had already taken. For example, a 1988 workshop on public understanding of radiological protection concepts stressed the need for "giving information" so the public would "understand."[102]

"Nuclear Renaissance" and True Public Participation (1991–2015)?

The collapse of the Soviet Union and the dissolution of Comecon led to the gradual integration of the former Soviet bloc countries into the Western nuclear community, and associated efforts at further engaging the public. Work on international nuclear law now sought to balance between promotion and constraint—the respective priorities of the two former periods.[103] Public engagement progressed, reflecting the so-called participatory turn under way especially in the radioactive waste management policies of many member countries.[104] As the memory of Chernobyl gradually subsided, the nuclear industry embarked upon communication efforts to back up its self-declared "nuclear renaissance."[105] Euratom and the NEA supported industry and government efforts to promote nuclear energy as a solution to problems of energy security, "peak oil," climate change, and sustainable development.

Comecon left a significant legacy of twenty-eight Soviet-built reactors—either pressurized water reactors (the VVER) or Chernobyl-type RMBK channel graphite reactors—in Bulgaria, the Czech Republic, Slovakia, Hungary, Lithuania, and the German Democratic Republic, four of which came on line after the collapse of the USSR.[106] The safety of these reactors raised considerable citizen concern in the East and the West, and became a major challenge for the IGOs. In the newly independent Central and Eastern European states, Soviet technology was widely viewed as a symbol of imperialism, and antinuclear and anti-Russian attitudes spread quickly. With the demise of Comecon, Euratom remained the only organization with supranational regulatory powers in a significant number of areas concerning the civil nuclear industry.[107]

"Dirigiste" Euratom and Participatory European Community?

Dealing with the Soviet heritage also entailed attempts at "Westernizing" Eastern European practices of public information, communication, and participation. Despite numerous reform attempts, the Euratom Treaty still has an uneasy relation with the broader European Union framework and its attempts to bring the EU closer to the citizens, especially since the rejection of constitutional treaty in 2005 by Dutch and French voters.[108] The pronuclear member states typically wish to preserve the treaty unchanged, while those phasing out or having rejected nuclear power call for changes that would give a more prominent role to the EU Parliament or national parliaments in nuclear policies[109] and to the civil society on radiation protection and nuclear safety issues.[110] The EU Parliament has sought to extend its powers, both via legal challenges and via political pressure through parliament reports and resolutions.[111] The case law practice of the European Court of Justice has in turn helped to gradually expand the remit of the treaty.[112]

The 1998 Aarhus Convention on citizen access to environmental information, participation, and justice has been elemental in enhanced public participation in environmental matters but has a somewhat unclear role in the nuclear sector.[113] Its provisions concerning access to information and public participation are included in Euratom directives on the safety of nuclear installations (2009) and on radioactive waste management (2011), while the third pillar—access to justice—remains outside of the Euratom framework. Legal avenues for safeguarding access to information and public participation rely on the Directives on public information and Environmental Impact Assessment, to the extent that these apply to nuclear activities.[114] Only the European Union and its member states—but not Euratom—are contracting parties to the convention.[115] Based as it is on the idea of risk optimization—weighing the costs against the benefits of reducing radiation risk—Euratom's "as low as reasonably achievable" principle is in tension with the subsidiarity and precautionary principles of EU environmental legislation.[116] Since the convention gives the member states ample leeway in implementation, the information and engagement procedures concerning nuclear energy also vary greatly from one country to another.[117]

Five challenges currently facing Euratom public engagement work deserve attention. First, the promotional nature of the founding treaty undermines Euratom's credibility as an independent actor. Second, the legacy from the formative period of European integration—an information approach based on the deficit model of citizen knowledge—still lingers, alongside efforts toward public participation.[118] Third, the role of the Aarhus Convention in Euratom activities needs clarification. Fourth, while Euratom research projects have helped to foster and disseminate participatory tools and approaches in radioactive waste management across Europe, including in Central and Eastern Europe, progress in the other areas of nuclear policy has been modest.[119] Remarkably, the latest Eurobarometer opinion survey on nuclear safety dates from 2010, before Fukushima. Fifth, public participation is hampered by tensions between Euratom and other EU bodies, notably Directorates-General for Environment and Climate, the European Council, the Court of Justice, and Parliament, but also with the European Court of Human Rights, a key non-EU body with competencies in the area of nuclear energy.[120]

NEA: Waste, Radiological Protection, and the Emergence of "Stakeholders"

After the end of the Cold War, both the OECD and the NEA strengthened their outreach to nonmember countries. Intensified collaboration followed, also in the area of public information, with countries of Central and Eastern Europe and Asia,[121] and new members joined the NEA, including significant nuclear states such as South Korea (1993) and Russia (2012).[122] With thirty-four countries, the Nuclear Energy Agency accounts for about 80 percent of the world's installed nuclear capacity[123]—a figure that is declining with the development of nuclear in China and India and stagnation in the West.

Public participation advanced, especially in the areas of radiation protection and radioactive waste management. Several international seminars on public participation sought "to increase understanding amongst decision makers" on the topic.[124] The Committee on Radiation Protection and Public Health took note of the "changing public attitudes toward risk" and called for the "development of better mechanisms" for engagement.[125] The radioactive waste management committee explored the philosophical and ethical aspects, concluding that geological disposal was the best option.[126]

The emergence of sustainable development on the OECD and NEA agendas boosted NEA's public engagement work in the late 1990s. At the time, the OECD was searching for a new identity in the post–Cold War world, in the face of tighter European integration, globalization, and competition from other IGOs, while new actors such as the Western European Nuclear Regulators Association and the World Association of Nuclear Operators emerged as rivals to the NEA. The OECD responded by trying to craft its identity as a major international organization promoting sustainable development.[127] Consequently, the NEA increasingly framed its activities in this light,[128] and included sustainable development in its slightly modified mission statement in 1999.[129] It now portrayed nuclear power as a low-carbon, sustainable energy source and a guarantor of energy security.[130] The committees dedicated to safety, radiation protection, and waste management were in the lead in promoting public participation.[131] The Nuclear Law Committee studied the potential impact of new international conventions relating to environmental assessment and public participation.[132] The appointment of new division heads and scientific secretaries spurred stakeholder involvement work, as did the involvement of radiation protection and public health committee delegates who had personal experience working with the affected Chernobyl-area populations.[133] The term "stakeholders" made its way into NEA vocabulary,[134] while the agency commissioned sociologists specializing in issues of trust to adapt the concept to radiation protection and raise awareness.[135] A series of "science and values" workshops from 2008 to 2015 epitomized the efforts to integrate "radioprotection science" and "social values."[136]

However, the traditional approach of "reassuring the public" remained in focus even among those who pushed for public participation. Serge Prêtre, the Swiss chairperson of the radiation protection and public health committee from 1993 to 1996, evoked the need to create a climate of trust by listening to the public's "ancient fear" of radiation and cancer.[137] The IAEA's longtime radiation protection and public health delegate, Abel Gonzales, was even more outspoken, dismissing stakeholder involvement as one of "those Anglo-Saxon terms that sound crispy, are untranslatable and usually mean nothing or everything."[138]

The establishment of the Forum on Stakeholder Confidence in 2000 marked a milestone in NEA stakeholder engagement. The forum defined "stakeholder" as including not only those with an interest to defend but "any actor—institution, group or individual—with an interest or with a role to play in the process."[139] The forum has met annually, holding national workshops including lectures and topical case studies to share experience.[140] Incarnating the "culture of compromise" characteristic of the OECD in general,[141] the forum has provided a fairly nonpolitical and neutral venue for discussion on topics too conflict-laden to be discussed in a purely national setting.[142] It has been influential in drawing attention to and advancing reflection on the role of trust and confidence in radioactive waste management policy.[143]

Public participation still remains a relatively sensitive topic even within NEA, as does the concept of peer pressure—the OECD "trademark."[144] The NEA does not conduct full-blown nuclear policy reviews along the lines of those carried out on many other OECD policy areas, yet since the 1980s it has conducted independent reviews on specific technical and safety-related waste management topics, tailor-made to the demands from the reviewed country. These constitute ad hoc expert reviews, conducted at the initiative of and with full financing from the commissioning organization in the reviewed country, unlike the regular OECD reviews carried out by member countries as peers.[145] In the early 2000s, NEA member countries rejected the secretariat's suggestion that peer reviews including laypeople be introduced. Secretariat officials interviewed in 2019 were keen to underline that "we are not there to embarrass anybody," and viewed the idea of systematic nuclear-policy peer reviews with suspicion. A recent NEA brochure only mentions exchange of experiences and sharing of approaches on stakeholder engagement—in contrast with terms evoked in the description of more technical NEA work areas: "harmonization," "identification of best practices," and "development of shared strategies."[146] In two recent international NEA stakeholder workshops, in 2017 and 2019, few stakeholders outside of the NEA's habitual reference groups—government officials, experts, industry, and nuclear communications specialists—were invited.

Ultimately, efforts at fostering public engagement suffer from the agency's historical legacy and reputation as a promoter of nuclear energy. In principle, the NEA does not have a promotional role—its mission is to help keep the "nuclear option" open to countries that choose to use it.[147] In practice, the agency needs to constantly engage in difficult arbitration between countries developing nuclear power and those seeking to phase it out—the latter notably keen to ensure that the NEA remains "neutral." The rather adverse conjuncture in which the nuclear sector finds itself—especially since the 2011 Fukushima accident—has accentuated the feelings that public engagement appears more as a threat than as an opportunity.

Conclusions: Irreconcilable Tensions Between Promotion and Participation?

In the early years of the atom, nuclear sector IGOs had a distinctly promotional role, with public communication designed to persuade especially the opinion-building elites but also the public at large of the virtues of nuclear power. Comecon's promotional activities constituted a case in point, but Euratom and the European Nuclear Energy Agency engaged in similar and widespread efforts of public communication, employing a range of tools including visual arts, documentary films, and exhibitions. Comecon, which disappeared with the fall of the Soviet Union in 1991, never had a chance to demonstrate its possible ability and willingness to implement two-way public consultation or participation. To some extent the promotional role has followed the two remaining IGOs until the present day, thereby slowing the adoption of more advanced forms stakeholder engagement.[148] Institutional inertia is most clearly illustrated by Euratom's founding treaty, which has come to increasing tension with the evolving European Union legislation on citizen engagement. Euratom's support for and financing of R&D on stakeholder engagement has not come without creating friction, as demonstrated by some of the experiences of the HoNESt project team. The project faced as its counterpart what appeared to be an internally divided Euratom–European Commission research bureaucracy, torn between an at times rather zealous commitment to the promotion of nuclear energy on one hand, and a willingness to learn from social science research and stakeholder engagement on the other.

The differences in the trajectories of citizen engagement in the three IGOs analyzed in this chapter stem largely from their distinct functions, remits, and operating principles. Its rhetoric of collaboration notwithstanding, Comecon was essentially a tool for the USSR to exercise control over the Soviet bloc countries. Unlike the exclusively nuclear-centered NEA and Euratom, Comecon had the peaceful use of the atom as only one of its many areas of collaboration. Euratom wields considerable regulatory power, in contrast with the NEA, which relies almost solely on what political scientists and OECD insiders have described as "idea games" and "soft persuasion" via "peer pressure."[149] It does so notably via its old and, according to many observers, rather "undemocratic" founding treaty, which does not allow for much parliamentary and public scrutiny.[150] The concrete choices of policy implementation in Euratom remain largely in the hands of member states, while the interplay between the various EU institutions and the Court of Justice has created a complex framework comprising legal, administrative, and political elements, with the EU Parliament in a clearly secondary role. Prevention of nuclear proliferation has been Euratom's core competence since the beginning but remains outside the NEA's remit. This, together with the lack of regulatory power, helps to explain NEA's relative progress in public engagement. National delegates participating in NEA meetings can rather freely discuss even sensitive topics and try out new ideas, in the absence of a shadow of hierarchy and strong pressure from national capitals.

Widening public participation frequently faces resistance from within the IGOs, which are uneasy about engaging with more critical stakeholders. Such stakeholders themselves may in turn be reluctant to engage with organizations they perceive as part of the "nuclear lobby" and their attempts to "manufacture acceptance."[151] Overcoming such mutual "fears" may, however, be a precondition for the very survival of Euratom and NEA. Otherwise, both might fall into oblivion, perceived as increasingly irrelevant and inward-looking clubs of the "nuclear community."

Future research could usefully explore a topic that we could only evoke anecdotally: the diffusion of participatory policy approaches, via imitation, learning from best practice, socialization, and peer pressure. As in many of their member countries, also within the NEA and Euratom radioactive waste management has led this work in the nuclear sector. The respective roles of the member countries and the NEA and Euratom bureaucracies in bringing about such a shift merits further analysis. Such research could also tell us more about the challenges in the attempts to "democratize" nuclear-societal relations in the face of institutional inertia and entrenched asymmetries of power.

The dynamics and challenges of transferring Western ideas of public engagement to East European contexts constitutes another area for future exploration. After the demise of the Soviet Union, Euratom and the NEA took on many of Comecon's earlier tasks. This exposed some of the key differences between the legal traditions. For example, in Eastern Europe, EU legislation is typically applicable without having to go through a specific process of transposition into the national legislation.[152] The strict control and even suppression of civil society activism during the Soviet era has also created a particular Eastern European setting that renders unviable any attempts at directly transferring Western public engagement practices.

Finally, the challenges of participation could be analyzed in a truly global—and therefore far more challenging—context, by examining the incipient IAEA work in this area.

Notes

1. Matthias Schmelzer, *The Hegemony of Growth: The OECD and the Making of the*

Economic Growth Paradigm (Cambridge: Cambridge University Press, 2016); Jan-Henrik Meyer, "From Nature to Environment: International Organizations and Environmental Protection before Stockholm," in *International Organizations and Environmental Protection: Conservation and Globalization in the Twentieth Century*, ed. Wolfram Kaiser and Jan-Henrik Meyer (New York: Berghahn, 2017), 31–73.

2. Elisabeth Roehrlich, "Negotiating Verification: International Diplomacy and the Evolution of Nuclear Safeguards, 1945–1972," *Diplomacy and Statecraft* 29, no. 1 (2018): 29–50.
3. Mark Elam and Göran Sundqvist, "Public Involvement Designed to Circumvent Public Concern? The 'Participatory Turn' in European Nuclear Activities," *Risk, Hazards and Crisis in Public Policy* 1, no. 4 (2010): 203–29, doi:10.2202/1944-4079.1046.
4. Sabine Saurugger, "The Social Construction of the Participatory Turn: The Emergence of a Norm in the European Union," *European Journal of Political Research* 49, no. 4 (2010): 471–95; Francis Chateauraynaud and Didier Torny, *Les sombres précurseurs: une sociologie pragmatique de l'alerte et du risque* (Paris: Editions EHESS, 1999).
5. On the history of the IAEA and its attempts at public engagement in the wake of Chernobyl, see Elisabeth Roehrlich, *Dual Mandate: A History of the International Atomic Energy Agency* (Baltimore: Johns Hopkins University Press, 2021); Elisabeth Roehrlich, "The Cold War, the Developing World, and the Creation of the International Atomic Energy Agency (IAEA), 1953–1957," *Cold War History* 16, no. 2 (2016): 195–212; Elisabeth Roehrlich, "An Attitude of Caution: The IAEA, the United Nations, and the 1958 Pugwash Conference in Austria," *Journal of Cold War Studies* 20, no. 1 (2018): 31–57.
6. Gene Rowe and Lynn J. Frewer, "A Typology of Public Engagement Mechanisms," *Science, Technology, and Human Values* 30, no. 2 (2005): 251–90, doi:10.1177/0162243904271724.
7. Wilfried Konrad et al., "Comparative Cross-Country Analysis on Preliminary Identification of Key Factors Underlying Public Perception and Societal Engagement with Nuclear Developments in Different National Contexts," in *History of Nuclear Energy and Society (HoNESt) Consortium Deliverable no. 4.2* (2018), https://perma.cc/LJ4Q-HT3F.
8. Jan-Henrik Meyer, "Challenging the Atomic Community: The European Environmental Bureau and the Europeanization of Anti-Nuclear Protest," in *Societal Actors in European Integration: Polity-Building and Policy-Making 1958–1992*, ed. Wolfram Kaiser and Jan-Henrik Meyer (Basingstoke: Palgrave, 2013), 197–220; Jan-Henrik Meyer, " 'Where Do We Go from Wyhl?' Transnational Anti-Nuclear Protest Targeting European and International Organisations in the 1970s," *Historical Social Research* 39, no. 1 (2014): 212–35.
9. Cristian Bențe, "The Marshall Plan and the Beginnings of Comecon," *Society and Politics* 9, no. 2 (2015): 5–10.
10. Kazimierz Grzybowski, "International Organizations from the Soviet Point of View," *Law and Contemporary Problems* 29, no. 4 (Autumn 1964): 889.
11. For example, Matthew Evangelista, *Unarmed Forces: The Transnational Movement to End the Cold War* (Ithaca, NY: Cornell University Press, 1999).
12. V. I. Veksler, "Novyi Metod Uskoreniia Reliativistskikh Chastits," *Doklady Akademii*

Nauk SSSR 43, no. 8 (1944): 346–48; V. I. Veksler, "O Novom Metode Uskoreniia Reliativistskikh Chastits," *Doklady Akademii Nauk SSSR* 44, no. 9 (1944): 393–96; and JINR, "Istoriia," http://jinr.info/history/, accessed November 7, 2017. See also V. G. Kadyshevskii, ed., *Dubna—Ostrov Stabil'nosti. Ocherki po Istorii Ob"edinennogo Instituta Iadernykh Issledovanii (1956–2006 gg.)* (Moscow: Akademkniga, 2006); L. F. Zhidkova and Istoriia Dubny, *1956–1986* (Kimry: Izdatel'stvo "Filial GUPTO TOT" Kimrskaia Tipografiia, 2006); M. G. Shafranova, *Ob"edinennyi Institut Iadernykh Issledovanii: Informatsionno-biograficheskii Spravochnik*, 2nd ed. (Moscow: Fizmatlit, 2002).

13. *Pravda* and *Izvestiia*, January 18, 1955.
14. Sonja Schmid, "Nuclear Colonization? Soviet Technopolitics in the Second World," in *Entangled Geographies: Empire and Technopolitics in the Global Cold War*, ed. Gabrielle Hecht (Cambridge, MA: MIT Press, 2011), 125–54; George Ginsburgs, "Soviet Atomic Energy Agreements," *International Organization* 15, no. 1 (Winter 1961): 47–53.
15. According to Krige, the US pushed Atoms for Peace also to secure the predominance of its nascent industry in the growing nuclear sector. See John Krige, "The Peaceful Atom as Political Weapon: Euratom as an Instrument of U.S. Foreign Policy in the 1950s," *Historical Studies in the Natural Sciences* 38, no. 1 (2008): 5–44.
16. John Krige, "Atoms for Peace, Scientific Internationalism, and Scientific Intelligence," in *Global Knowledge Power: Science and Technology in International Affairs*, ed. John Krige and Kai-Henrik Barth (Chicago: University of Chicago Press, 2006), 161–81; *Proceedings of the International Conference on the Peaceful Uses of Atomic Energy*, 17 vols. (Geneva: United Nation, 1955–1956). The second conference, held over two weeks in September 1958, produced over thirty volumes of papers.
17. M. G. Meshcheriakov, *K 90-Letiiu so Dnia Rozhdeniia* (Dubna: 2000), 52–57. See also http://www.larisa-zinovyeva.com/первые-ускорители-дубны-основа-межд-3/.
18. I. N. Golovin, "Mirnyi Atom: U Istokov Sotrudnichestva," *Priroda* no. 5 (1988): 122–28.
19. Joseph E. Pilat, Robert E. Pendley, and Charles K. Ebinger, eds., *Atoms for Peace: An Analysis After Thirty Years* (Boulder, CO: Westview Press, 1985).
20. Christian Pineau and Christiane Rimbau, *Le grand pari, l'aventure du traité de Rome* (Paris: Fayard 1991), 158; Jean Marie Palayret, "Les décideurs français et allemands face aux questions institutionnelles dans la négociation des traités de Rome," in *Le couple France–Allemagne et les institutions européennes*, ed. Marie-Thérèse Bitsch (Brussels: Bruylant, 2001), 105–50.
21. Pineau and Rimbau, *Le grand pari*; Palayret, "Les décideurs français et allemands face aux questions institutionnelles dans la négociation des traités de Rome"; Michel Dumoulin, *Spaak* (Bruxelles: Editions Racine, 1999).
22. Gérard Bossuat, *L'Europe des Français, 1943–1959: La IV République aux sources de l'Europe communautaire* (Paris: Publications de la Sorbonne, 1966), 286; *Intergovernmental Committee on European Integration: The Brussels Report on Euratom* (abridged, English translation of document commonly called the Spaak Report), May 1956, http://aei.pitt.edu/42941/. According to Article 1 of the Euratom Treaty, the organization was to create the "conditions for the speedy establishment and growth of nuclear industries."

23. Ilina Cenevska, *The European Atomic Energy Community in the European Union Context: The "Outsider" Within* (Leiden and Boston: Brill Nijhoff, 2016), 22–25.
24. Ben Rosamond, *Theories of European Integration* (Basingstoke and New York: Palgrave, 2000); Dirk Spierenburg and Raimond Poidevin, *The History of High Authority of the European Coal and Steel Community: Supranationality in Operation* (London: Weinfeld and Nicholson, 1994); Kevin Ruane, *The Rise and Fall of the European Defence Community: Anglo-American Relations and the Crisis of European Defence (1950–1955)* (New York: St. Martin's Press, 2000); Cenevska, *The European Atomic Energy Community*, 20.
25. Mervyn O'Driscoll, *The European Parliament and the Euratom Treaty: Past, Present and Future*, EU Parliament Working Paper, Energy and Research Series ENER 114 EN, 2-2002 (Luxembourg: European Parliament, 2002), ix.
26. John Krige, "The Peaceful Atom as Political Weapon: Euratom and American Foreign Policy in the Late 1950s," *Historical Studies of Natural Sciences* 38, no. 1 (2008): 5–44. See also John Krige, *Sharing Knowledge, Shaping Europe: US Technological Collaboration and Nonproliferation* (Cambridge, MA: MIT Press, 2016); Gunnar Skogmar, *The United States and the Nuclear Dimension of European Integration* (Basingstoke and New York: Palgrave, 2004).
27. O'Driscoll, *The European Parliament and the Euratom Treaty*; Anna Södersten, *Euratom at the Crossroads* (Cheltenham, UK: Edward Elgar, 2014), 41–43; Cenevska, *The European Atomic Energy Community.*
28. O'Driscoll, *The European Parliament*, i.
29. Olivier Pirotte, in collaboration with Pascal Girerd, Pierre Marsal, and Sylviane Morson, *Trente ans d'expérience Euratom: La naissance d'une Europe nucléaire* (Brussels: Bruylant, 1988), 265.
30. O'Driscoll, *The European Parliament*, xi.
31. Richard T. Griffiths, ed., *Explorations in OEEC History* (Paris: OECD, 1997), 27.
32. Louis Armand, *Some Aspects of the European Energy Problem: Suggestions for Collective Action, Report prepared for the O.E.E.C.* (Paris: Organization for European Co-operation, June 1955), 7, 25–51; NEA, *NEA 50th Anniversary* (Paris: OECD Nuclear Energy Agency, 2008), 9.
33. Germany, Austria, Belgium, Denmark, France, Greece, Ireland, Iceland, Italy, Luxembourg, Norway, the Netherlands, Portugal, the United Kingdom, Sweden, Switzerland, and Turkey. Spain became a full member of the OEEC on July 20, 1959. NEA, *NEA 50th Anniversary.*
34. Gail M. Marcus, "The OECD Nuclear Energy Agency at 50," *Nuclear News* 51, no. 2 (2008), 28.
35. Henri Métivier, *Fifty Years of Radiological Protection: The CRPPH 50th Anniversary Commemorative Review*, Publication no. 6280 (Paris: OECD Nuclear Energy Agency, 2007), 20.
36. Achille Albonetti, *Europe and Nuclear Energy: The Atlantic Papers 2* (Paris: Atlantic Institute for International Affairs, 1972), 63.
37. "About Us—How We Work," Nuclear Energy Agency, accessed March 2021, https://www.oecd-nea.org/jcms/rni_6682.
38. Dick van Lente, ed., *The Nuclear Age in Popular Media: A Transnational History, 1945–1965* (New York: Palgrave Macmillan, 2012), 1–2.
39. Peter D. Cameron, "The Revival of Nuclear Power: An Analysis of the Legal Implications," *Journal of Environmental Law* 19, no. 1 (2007): 72–73.

40. Rowland Maddock, "Energy and Integration: The Logic of Interdependence in the Soviet Union and Eastern Europe," *Journal of Common Market Studies* xix, no. 1 (September 1980).
41. "Novosti," *Atomnaia Energiia* 27, no. 3 (1969): 244–45.
42. Epistemic communities are defined as "small networks of policy specialists congregate to discuss specific issues, set agendas, and formulate policy alternatives outside the formal bureaucratic channels, and they also serve as brokers for admitting new ideas into decision-making circles of bureaucrats and elected officials." Peter M. Haas, "Introduction: Epistemic Communities and International Policy Coordination," *International Organization* 46, no. 1 (1992): 31.
43. "Short Reports," *Atomnaia Energiia* 12, no. 1 (1962): 85.
44. "Short Reports," *Atomnaia Energiia* 16, no. 3 (1964): 279–81; "Short Reports," *Atomnaia Energiia* 28, no. 6 (1970): 536.
45. Sonja Schmid, "Celebrating Tomorrow Today: The Peaceful Atom on Display in the Soviet Union," *Social Studies of Science* 36, no. 3 (2006): 331–65; Paul Josephson, "Rockets, Reactors and Soviet Culture," in *Science and the Soviet Social Order*, ed. Loren Graham (Cambridge, MA: Harvard University Press, 1990), 168–91.
46. Sonja Schmid, "Shaping the Soviet Experience of the Atomic Age: Nuclear Topics in *Ogonyok*, 1945–1965," in *The Nuclear Age in Popular Media: A Transnational History, 1945–1965*, ed. Dick van Lente (New York: Palgrave Macmillan, 2012), 19–52.
47. E. M. Malitikov, " 'Znanie' za 70 Let," January 2008, http://malitikov.ru/znanie-za-70-let/.
48. *Ekonomicheskoe Sotrudnichestvo Stran-Chlenov SEV*, 16 vols., 1975–1990.
49. Jan-Henrik Meyer, *The European Public Sphere: Media and Transnational Communication in European Integration 1969–1991* (Stuttgart: Franz Steiner, 2010).
50. Maria Fritsche, *The American Marshall Plan Film Campaign and the Europeans: A Captivated Audience?* (London: Bloomsbury Academic, 2018).
51. Jackie L. Harrison and Stefanie Pukallus, "The European Community's Public Communication Policy 1951–1967," *Contemporary European History* 24, no. 2 (2015): 233–51.
52. Anne Bruch and Eugen Pfister, " 'What Europeans Saw of Europe': Medial Construction of European Identity in Information Films and Newsreels in the 1950s," *Journal of Contemporary European Research* 10, no. 1 (2014): 26–43.
53. Harrison and Pukallus, "The European Community's Public Communication Policy," 8.
54. Bruch and Pfister, " 'What Europeans Saw' "; Alexander Reinfeldt, "Communicating European Integration: Information vs. Integration?" *Journal of Contemporary European Research* 10, no. 1 (2014): 47–49. An influential stimulus for this approach was the book *The People's Choice*, by Paul F. Lazarsfeld, Bernard Berelson, and Hazel Gaudet, first published in 1944.
55. Reinfeldt, "Communicating European Integration," 45–46; see also Jürgen Elvert, *Die europäische Integration* (Darmstadt: Wissenschaftliche Buchgesellschaft, 2006), 1–2.
56. Reinfeldt, "Communicating European Integration," 46–47; Ernst B. Haas, *The Uniting of Europe: Political, Social, and Economic Forces, 1950–1957* (Stanford, CA: Stanford University Press, 1958).
57. Reinfeldt, "Communicating European Integration." The concept of permissive consensus goes back to Leon N. Lindberg and Stuart A. Scheingold, *Europe's*

Would-Be Polity: Patterns of Change in the European Community (Englewood Cliffs, NJ: Prentice Hall, 1970).

58. Thomas F. Cusack, "A Tale of Two Treaties," *Common Market Law Review* 40, no. 117 (2003): 126; Cenevska, *The European Atomic Energy Community*, 26.
59. O'Driscoll, "The European Parliament," viii–ix.
60. Albonetti, *Europe and Nuclear Energy*, 33.
61. O'Driscoll, "The European Parliament," 37; Andrew Barry and William Walters, "From EURATOM to 'Complex Systems': Technology and European Government," *Alternatives* 28, no. 3 (2003): 311.
62. Luis Echávarri, "Opening Remarks at the 50th Anniversary of the Nuclear Law Committee Colloquium, February 6, 2007," NEA/SEN/NLC(2007)2 (Paris: OECD Nuclear Energy Agency, 2007).
63. Marcus, "The OECD Nuclear Energy Agency at 50," 28.
64. The OEEC European Nuclear Energy Agency, https://www.iaea.org/sites/default/files/publications/magazines/bulletin/bull3-3/03305302326.pdf; Gail H. Marcus, *NEA and Its Committees: Historical Review of the First Fifty Years (1958–2008)*, review draft, accessed March 23, 2021, http://marcus-spectrum.com/documents/NEA Timeline.pdf.
65. Métivier, *Fifty Years of Radiological Protection*, 30. More generally on ocean dumping: Jacob Darwin Hamblin, *Poison in the Well: Radioactive Waste in the Oceans at the Dawn of the Nuclear Age* (New Brunswick, NJ: Rutgers University Press, 2008).
66. E.g., Jan-Henrik Meyer, " 'Atomkraft—Nej tak': How Denmark Did not Introduce Commercial Nuclear Power Plants," in *Pathways into and out of Nuclear Power in Western Europe: Austria, Denmark, Federal Republic of Germany, Italy, and Sweden*, ed. Astrid Mignon Kirchhof (Munich: Deutsches Museum, 2019), 74–123.
67. Albonetti, *Europe and Nuclear Energy*, 31.
68. Albonetti, *Europe and Nuclear Energy*, 35, 64–65.
69. Cameron, "The Revival of Nuclear Power," 73–77.
70. See for example, V.T. Tsurkov, "Some Results of the Work of the Comecon Scientific-Technical Coordinating Council on Fast Reactors," *Soviet Atomic Energy* 44, no. 6 (1978): 624–26.
71. E. V. Kulov, "Socialist Integration of Nuclear Science and Technology: Peaceful Atom in the Countries of Socialism. Collaboration Between Member-Nations of Comecon," *Soviet Atomic Energy* 47, no. 2 (1979): 673–74.
72. A. F. Panasenkov, "Sotrudnichestvo Stran-Chlenov SEV v Razvitii Atomnoi Energetiki i ego Rol' v Osushchestvlenii Polozhenii Dogovora o Nerasprostranenii Iadernogo Oruzhiia," *Biulleten MAGATE*, 22, no. 3/4 (1980): 176–17.
73. A. A. Sarkisov, *Atomnye stantsii Maloi Moshchnosti* (Moscow: Nauka, 2011), 3; V. A. Sidorenko, ed., *Istoriia Atomnoi Energetiki Sovetskogo Soiuza i Rossiii* 2 (Moscow: Kurchatov Institute, 2002), 101.
74. A. M. Petros'iants (ed.), *Iadernaia Industriia Rossii* (Moscow: Energoatomizdat, 2000), 933, and Petros'iants, *Atomnaia Nauka i Tekhnika SSSR* (Moscow: Atomizdat, 1987), 292–97.
75. See Yu Yurasov, "Agreement on Setting Up the Ineratominstrument Society," *Soviet Atomic Energy* 32, no. 4 (1972): 409–11; W. Dietzsch, "Interatominstrument-Ausstellung zur Leipziger Herbstmesse 1974," *Isotopenpraxis* 11, no. 3 (1974): 111–114; Isotope Rosatom, "History," accessed March 6, 2021, at http://www

.isotop.ru/en/about/history.htm; J. Jurasow, "Die internationale Wirtschaftsvereinigung Für kerntechnischen Gerätebau 'Interatominstrument,' " *Isotopenpraxis Isotopes in Environmental and Health Studies* 8, no. 11–12 (1972): 405–6.

76. B. Korda and I. Moravcik, "The Energy Problem in Eastern Europe and the Soviet Union," *Canadian Slavonic Papers/Revue Canadienne des Slavistes* 18, no. 1 (1976): 1–3.
77. A. M. Petros'iants, *Atomnaia Energiia v Nauke i Promyshlennosti* (Moscow: Energoatomizdat, 1984).
78. Tetiana Perga, "The Fallout of Chernobyl: The Emergence of an Environmental Movement in the Ukrainian Soviet Socialist Republic," in *Nature and the Iron Curtain: Environmental Policy and Social Movements in Communist and Capitalist Countries, 1945–1990*, ed. Astrid Mignon Kirchhof and John R. Mc Neill (Pittsburgh: University of Pittsburgh Press, 2019), 55–72; Susanne Bauer, Karena Kalmbach, and Tatiana Kasperski, "From Pripyat to Paris, from Grassroots Memories to Globalized Knowledge Production: The Politics of Chernobyl Fallout," in *Nuclear Portraits: Communities, the Environment, and Public Policy*, ed. Laurel Sefton MacDowell (Toronto: University of Toronto Press, 2017), 149–89.
79. Alexei Sissakian, "Joint Institute for Nuclear Research—Yesterday, Today, Tomorrow," *Bulletin of the Georgian National Academy of Sciences* 175, no. 2 (2007): 39–46.
80. Albonetti, *Europe and Nuclear Energy*, 19.
81. Albonetti, *Europe and Nuclear Energy*, 19.
82. Albonetti, *Europe and Nuclear Energy*, 16, 77.
83. Albonetti, *Europe and Nuclear Energy*, 5.
84. Reinfeldt, "Communicating European Integration," 53.
85. Reinfeldt, "Communicating European Integration," 49.
86. Reinfeldt, "Communicating European Integration," 54–55.
87. Leo Tindemans, "*European Union: Report by Mr. Leo Tindemans, Prime Minister of Belgium, to the European Council,*" *Bulletin of the European Communities Suppl.* 1, no. 76 (1976): 27.
88. For instance, Louis Timbal-Duclaux, "L'opposition à l'énergie nucléaire: Psychologie, sociologie, ethnologie, psychanalyse: quatre approches convergentes," *Revue générale nucléaire* 5 (1979): 501–5; Colette Guedeney and Mendel Gérard, *L'angoisse atomique et les centrales nucléaires* (Paris: Payot, 1973).
89. European Commission, *Open Discussions on Nuclear Energy. Held by the European Commission, Brussels, 29/11–1/12/1977 and 24–26/1/1978* (Luxembourg: Office for Official Publications of the European Communities, 1978).
90. Meyer, "Challenging the Atomic Community"; Meyer, " 'Where Do We Go from Wyhl?' "
91. Commission of the European Communities, *Nuclear Safety: The European Community following the Chernobyl Accident*, 12/89 (Luxembourg: Commission of the European Communities, August–September 1989), accessed March 23, 2021, http://aei.pitt.edu/4627/1/4627.pdf.
92. Södersten, *Euratom at the Crossroads*.
93. Métivier, *Fifty Years of Radiological Protection*, 32.
94. NEA, *NEA 50th Anniversary*, 17.

95. Métivier, *Fifty Years of Radiological Protection*, 33.
96. Métivier *Fifty Years of Radiological Protection*, 37.
97. NEA, *Objectives, Concepts and Strategies for the Management of Radioactive Waste Arising from Nuclear Power Programmes (Paris: OECD Nuclear Energy Agency, 1977).*
98. NEA, *Objectives, Concepts and Strategies*; Métivier, *Fifty Years of Radiological Protection*, 34–35.
99. Howard K. Shapar, "The Role of Governments in Promoting a Realistic Public Understanding of the Potentialities of Nuclear Power," *NEA News* 1 (December 1983): 15–17.
100. "History of the OECD Nuclear Agency: Timeline," NEA, last modified March 1, 2018, https://www.oecd-nea.org/general/history/timeline.html.
101. The International Nuclear Event Scale was partly inspired by a similar classification and information system that the French EDF had adopted in 1988 on an experimental basis. Boris Dänzer-Kantof and Félix Torres, *L'énergie de la France: De Zoé aux EPR, l'histoire du programme nucléaire français* (Paris: Editions François Bourin, 2013), 461–62; Sezin Topçu, *La France Nucléaire: L'art de gouverner une technologie* (Paris: Seuil, 2013), 232.
102. Mike Boyd, "Stakeholder Involvement and the CRPPH: A Learning Process From Chernobyl to Fukushima," presentation by CPPRH chair Mike Boyd at the NEA Workshop on Stakeholder Involvement in Nuclear Decision Making (Paris: NEA, January 17, 2017).
103. Cameron, *The Revival of Nuclear Power.*
104. Göran Sundqvist and Mark Elam, "Public Involvement Designed to Circumvent Public Concern? The 'Participatory Turn' in European Nuclear Activities," *Risk, Hazards, and Crisis in Public Policy* 1, no. 4 (2010): 203–29.
105. William J. Nuttall, *Nuclear Renaissance: Technologies and Policies for the Future of Nuclear Power* (New York: Taylor & Francis, 2005).
106. Finland has two VVER-440 at Loviisa. Romania was active in Comecon research and training programs but chose Canadian CANDU PHWRs, two of which at 650 MW are operational, the other three suspended or canceled. "Country Nuclear Power Profiles. Finland," IAEA, accessed March 21, 2021, https://cnpp.iaea.org/countryprofiles/Finland/Finland.htm; "Country Nuclear Power Profiles. Romania," IAEA, accessed March 21, 2021, https://cnpp.iaea.org/countryprofiles/Romania/Romania.htm.
107. Cenevska, *The European Atomic Energy Community*, 3.
108. Elizabeth Monaghan, " 'Communicating Europe': The Role of Organised Civil Society," *Journal of Contemporary European Research* 4, no. 1 (2008), 18–31; Södersten, *Euratom at the Crossroads*; Cenevska, *The European Atomic Energy Community.*
109. E.g., Cameron, "The Revival of Nuclear Power," 76.
110. Cenevska, *The European Atomic Energy Community*, 3.
111. Cenevska, *The European Atomic Energy Community*, 42.
112. Cenevska, *The European Atomic Energy Community*, 50–66.
113. *Convention on Access to Information, Public Participation in Decision-making and Access to Justice in Environmental Matters, done at Aarhus, Denmark, on 25 June 1998* (UN Economic Commission for Europe [UNECE]), accessed March 21, 2021, https://unece.org/DAM/env/pp/documents/cep43e.pdf.

114. Södersten, *Euratom at the Crossroads*, 152.
115. Cenevska, *The European Atomic Energy Community*, 11.
116. Södersten, *Euratom at the Crossroads*, 152.
117. Cenevska, *The European Atomic Energy Community*, 201.
118. Reinfeld, "Communicating European Integration," 55.
119. E.g., Gianluca Ferraro and Meritxell Martell, *Euratom Projects, Radioactive Waste Management and Public Participation: What Have We Learnt So Far? A Synthesis of Principles* (European Commission, Joint Research Centre, 2015), 6.
120. Cenevska, *The European Atomic Energy Community*, 204–21.
121. NEA, *NEA 50th Anniversary*, 27.
122. Marcus, *NEA and Its Committees*.
123. NEA Annual Report 2019, https://www.oecd-nea.org/upload/docs/application/pdf/2020-06/ar2019.pdf.
124. See, e.g., NEA, *Public Participation in Nuclear Decision Making—Participation du Public aux Décisions Nucléaires: Proceedings of an International Workshop, Paris, France, 4–6 March 1992* (Paris: OECD, 1993).
125. Métivier, *Fifty Years of Radiological Protection*, 21.
126. NEA, *Learning and Adapting to Societal Requirements for Radioactive Waste Management: Key Findings and Experience of the Forum on Stakeholder Confidence*, NEA Publication No. 5296 (Paris: OECD-NEA, 2004), 14; NEA, *The Environmental and Ethical Basis of Geological Disposal of Long-lived Radioactive Waste* (Paris: OECD-NEA, 1995), https://www.oecd-nea.org/rwm/reports/1995/geodisp-entire-report.html.
127. OECD High-Level Advisory Group on the Environment, *Guiding the Transition to Sustainable Development: A Critical Role for the OECD* (Paris: OECD, 1997).
128. NEA, *NEA 50th Anniversary*, 33; *NEA News* 19, no. 1 (2001), special issue on sustainable development.
129. The second part of the mission statement reads: "It strives to provide authoritative assessments and to forge common understandings on key issues, as input to government decisions on nuclear energy policy and to broader OECD policy analyses in areas such as energy and sustainable development." *The Strategic Plan of the Nuclear Energy Agency 2017–2022* (Paris: OECD Nuclear Energy Agency, 2016), 15, https://www.oecd-nea.org/general/about/strategic-plan2017–2022.pdf.
130. NEA, *NEA 50th Anniversary*, 33; NEA, *Nuclear Energy in a Sustainable Development Perspective* (Paris: OECD-NEA, 2000).
131. E.g., CNRA, *Future Nuclear Regulatory Challenges* (Paris: OECD-NEA, 2008) highlighted the interface between regulatory authorities and the public as a major future challenge.
132. NEA, *NEA 50th Anniversary*, 31.
133. Métivier, *Fifty Years of Radiological Protection*.
134. The term first appeared in the minutes of the 57th session of the CRPPH. Métivier, *Fifty Years of Radiological Protection*, 44.
135. Personal communication with an expert consultant with extensive experience of NEA stakeholder work, December 17, 2017.
136. Boyd, *Stakeholder Involvement and the CRPPH*.
137. Métivier, *Fifty Years of Radiological Protection*, 70.
138. Métivier, *Fifty Years of Radiological Protection*, 71.

139. Javier Lezaun, "Surfing the FSC," NEA, accessed March 23, 2021, https://www.oecd-nea.org/rwm/fsc/oxford.html.
140. "Forum on Stakeholder Confidence," NEA, accessed March 23, 2021, https://www.oecd-nea.org/rwm/fsc/workshops/.
141. Aynsley Kellow and Peter Carroll, "Exploring the Impact of International Civil Servants: The Case of the OECD," *International Journal of Public Administration* 36, no. 7 (2013): 482–91.
142. Rianne Mahon and Stephen McBride, "Standardizing and Disseminating Knowledge: The Role of the OECD in Global Governance," *European Political Science Review* 1, no. 1 (2009): 83–101.
143. *Implementing Geological Disposal of Radioactive Technology Platform: Vision Report* (European Commission Publications Office, European Research Area, 2009).
144. OECD, *Peer Review: An OECD Tool for Co-operation and Change* (Paris: OECD, 2003).
145. NEA, *International Peer Reviews for Radioactive Waste Management: General Information and Guidelines*, NEA publication no. 6082 (Paris: OECD-NEA, 2005), https://www.oecd-nea.org/rwm/reports/2005/nea6082-peer-review.pdf.
146. NEA, "Nuclear Energy Agency," brochure, April 2019, accessed March 23, 2021, http://www.oecd-nea.org/pub/nea-brochure.pdf, p. 8.
147. According to its current strategic plan, the NEA's mission is to facilitate the "safe, environmentally sound and economical use of nuclear energy for peaceful purposes," by helping the member countries to maintain and develop their scientific, technological, and legal capacities through international cooperation. *The Strategic Plan*, 15.
148. See, e.g., remarks by the former director of the German safety authority Wolfgang Renneberg at a witness seminar organized by the HoNESt project on October 16, 2018, Barcelona. (Mimeo, Barcelona: Pompeu Fabra University, 2019).
149. Martha Finnemore and Kathryn Sikkink, "International Norm Dynamics and Political Change," *International Organization* 52, no. 4 (1998): 887–917; Martin Marcussen, "Multilateral Surveillance and the OECD: Playing the Idea Game," in *The OECD and European Welfare States*, ed. Klaus Armingeon and Michelle Beyeler (Cheltenham: Edward Elgar, 2004), 13–31; OECD, *Peer Review*.
150. Cenevska, *The European Atomic Energy Community,* 9; Driscoll, *The European Parliament*.
151. The HoNESt project team encountered such concerns when organizing stakeholder workshops in 2017 and 2018.
152. Cenevska, *The European Atomic Energy Community,* 221–24.

PART 3

Perspectives

CHAPTER 5

Risky or Beneficial? Exploring Perceptions of Nuclear Energy over Time in a Cross-Country Perspective

Josep Espluga, Wilfried Konrad, Ann Enander, Beatriz Medina, Ana Prades, and Pieter Cools

Introduction

Nuclear power is a contested technology. Like certain other technoscientific developments, using fission energy to generate electricity has been the subject of long-lasting and still-ongoing debates on whether this source of energy production is the boon or bane of modern societies. Particularly in the aftermath of reactor accidents that have attracted global attention, advocates and opponents of nuclear energy have sharpened their arguments for and against the atom sector. For instance, as a consequence of the Fukushima accident in March 2011, Germany declared that it would phase out nuclear energy by the end of 2022; by contrast, the UK is pursuing the opposite path by pushing forward with new reactor projects. Even before the Fukushima accident, Germany and the UK were characterized by differing public views on nuclear energy. Eurobarometer data from a 2010 survey provide support for this claim. On the question "Should the current level of nuclear energy as a proportion of all energy sources be reduced, maintained the same or be increased?" UK and German participants responded, respectively, "Reduced," 25 percent vs. 52 percent; "Maintained the same," 39 percent vs. 37 percent; "Increased," 27 percent vs. 7 percent; and "Don't know," 9 percent vs. 4 percent. Looking at the average across all European Union member states, a high proportion of European citizens at this time seemed to have reservations about nuclear energy, with 34 percent in favor of reducing, 17 percent of increasing, and 39 percent of keeping the existing situation.[1] It is worth noting that these data only include the period up to 2009, thus predating the Fukushima event.

While some countries do show perception patterns in favor of nuclear energy, this finding of an overall tendency of reservations about nuclear power is also reflected in the balance of perceived risks and benefits. The results at the European level were:

1. The benefits of nuclear power as an energy source outweigh its risks: 35 percent.
2. The risks of nuclear power as an energy source outweigh its benefits: 51 percent.
3. Neither: 7 percent.
4. Don't know: 7 percent.[2]

A look at the Eurobarometer data from the 1980s to 2009 in six European countries (figure 5.1) further illustrates, across time and space, the wide range of profiles when it comes to perceptions of nuclear energy and its related risk-benefit trade-off..

What are the underlying reasons for favoring or rejecting nuclear energy or for evaluating nuclear power plants as risky or beneficial? How can we explain such a variety of perceptions? What is the specific role of perceived risks and benefits? How much weight do these factors have in shaping public opinions? We suggest answers to these questions based on the analysis of citizen perceptions of nuclear energy in seven European countries and the United States. First, we synthesize the lessons from the existing literature on perceived risks and benefits of nuclear energy, then we briefly outline our methodology. Next we present the results of the analysis of the eight historical country reports, and follow up with concluding remarks.

Background

There is a considerable literature on public perceptions of nuclear energy from the 1970s onward. This work has revealed that public acceptance of and opposition to nuclear technologies are the result of an interplay among numerous complex factors influencing and shaping perceptions and values,[3] and relate to a broad spectrum of interactions between people and institutions within local communities and within wider society.[4] Various approaches in the field of social theory of risk have been applied in order to understand the complex interplay among these factors. These include the psychometric paradigm,[5] the affective approach,[6] the cultural theory of risk,[7] the interpretative risk perception research,[8] and various governance approaches.[9]

Based on our analysis of the literature on nuclear perceptions, we propose a particular mode of interaction between the dimensions of risk:[10]

Fig. 5.1. Eurobarometer data on the perception of nuclear energy 1982–2009 in %

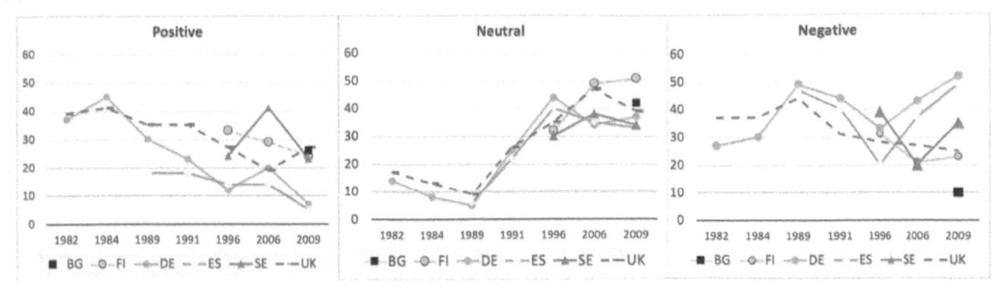

Sources and notes: Author's depiction based on Eurobarometer (1982–2009)

BG: Bulgaria; FI: Finland; DE: Germany (1982–1991 figures do not include data from the German Democratic Republic or East Germany; 1996–2009 means unified Germany); ES: Spain; SE: Sweden; UK: United Kingdom; Missing to 100% = don't know

Positive attitude combines answers to "worthwhile to develop" (1982–1996), "in favor" (2006), "increased" (2009)

Neutral combines answers "no particular interest" (1982), "no particular advantage" (1984, 1989), "neither develop nor abandon" (1991, 1996), "balanced views" (2006), "maintained the same" (2009)

Negative combines answers to "unacceptable risk" (1982–1996), "opposed" (2006), "reduced" (2009)

1. The degree of public acceptance (or toleration) of nuclear energy is mainly related to the perception of certain types of benefits and risks.
2. The perception of risks and benefits is strongly influenced by the degree of trust that people have in the institutions and companies promoting and regulating nuclear power and sites.
3. Furthermore, both risk and benefit perceptions and social trust are influenced by a set of antecedent variables, including affective feelings, "affective imagery," values (salient) or beliefs, and ideological and political orientations (e.g., pro-environmental ideological orientation). In addition, the persistence of already held attitudes toward nuclear energy tends to condition the possibility of changing these in the future, since a certain psychosocial inertia is a familiar phenomenon.

We now turn to the current debate on each of these three main assumptions underlying public acceptance of nuclear technology.

Perception of Benefits and Risks

The first theoretical assumption is that the degree of public acceptance (or toleration) of nuclear energy technology is mainly related to the perception of certain types of benefits and risks. Notably, the very same factor can be seen as a source of both risks and benefits. According to the literature review, the perceived risks and benefits of nuclear energy include two general domains: the perception of health, environment, and safety impacts; and economic issues.

PERCEPTION OF HEALTH, ENVIRONMENT, AND SAFETY IMPACTS

The perceived environmental risks of nuclear energy have included potential radioactive releases that could harm the environment and wildlife.[11] However, the perceived potential contribution of nuclear energy to mitigate climate change is also prominent in recent risk perception literature.[12] Acceptance is in some ways influenced by framing nuclear energy as a low-carbon technology, but there is also empirical evidence that people see both climate change and nuclear energy as problematic in terms of risks, in that they express only a "reluctant acceptance" of nuclear energy as a solution to climate change.[13] Thus, citizens seem to be prepared to accept nuclear energy when it is framed as a means of tackling climate change, while acceptance substantially diminishes when it comes to evaluating nuclear energy for itself ("unconditional acceptance").[14]

Concerns about short-term and chronic health impacts among citizens living near nuclear installations have been shown to influence perceptions on nuclear energy.[15]

On the one hand, safety is a factor underlying positive attitudes toward nuclear energy, in the sense that the perception of adequate security measures generates greater acceptance.[16] On the other hand, perceptions of security failures as linked to incidents and accidents are reported as a key factor in changing attitudes in a negative direction.[17] In fact, a considerable amount of data shows that acceptance of nuclear energy decreased after accidents such as those of Chernobyl and Fukushima.[18] Apart from security failures, other causes of incidents may be from intentional human acts. In this sense, the perception of risks emerging from terrorist attacks plays a role in the greater or lesser degree of acceptance of nuclear energy.[19]

PERCEPTION OF ECONOMIC ISSUES RELATED TO NUCLEAR ENERGY

We also examine the advantages and disadvantages of nuclear energy compared to other energy options, as well as the debate about energy consumption.[20] For example, a scenario of decreasing energy consumption could reduce the need to invest in nuclear energy, an issue discussed further in chapter 2 of this volume.

Alleged benefits of nuclear energy include its ability to foster economic growth and to improve the performance of industry sectors or companies.

Argumentation along these lines emphasizes economic priorities[21] and the prosperity of nations.[22]

Other issues relate to concerns about security of supply—that is, whether the nuclear power plants can reliably cover a major share of a country's energy needs[23] and guarantee a continuous and sufficient electricity supply.[24] The current debates also cover potential energy shortages, the risk of not having a sufficient energy supply to meet national demand, and the need for nuclear sources in order to avoid such problems.[25]

Trust

The second theoretical assumption is that perception of risks and benefits is strongly influenced by the degree of trust that people have in the institutions and companies promoting and regulating nuclear power and sites.

Trust in the nuclear governance institutions affects the perceived risk of nuclear power; higher trust and lower risk perceptions would thus predict positive attitudes toward nuclear power.[26] It is assumed that social trust can significantly influence local acceptance, and the degree of trust earned by the different actors involved in nuclear processes is an important underlying key factor.[27] For instance, trust in inspection authorities is crucial for the distinction between opposition and reluctant acceptance, while trust and the perceived honesty of industry and scientists—and their "competence" (confidence)—are key factors in nuclear acceptance.[28] The credibility and status of nongovernmental organizations has also proved to be a key factor.[29]

An important part of the work on social trust in risk issues is based on the "salient values similarity theory," which assumes that people who perceive that they share similar views with an actor (e.g., the managing agency) tend to trust this actor more than those who do not.[30]

NIMBY (Not In My Backyard) effects are mentioned in various studies related to nuclear waste disposal,[31] pointing to siting decisions as being one of the most critical issues in nuclear development in many countries.[32] The siting of nuclear facilities is sometimes linked to perceptions of an unjustified distribution of risks and benefits, which relates to a lack of social trust in institutions and companies involved in nuclear developments.[33] A factor strongly related to social trust is "fairness,"[34] in the sense that those people who believe a procedure is fair are more willing to accept a decision,[35] and that the influence of procedural fairness is even stronger for persons who hold high moral convictions.[36] One study found that the inability of the actors responsible for nuclear waste policy to link the technical, political, and procedural issues into an integrated approach explained part of the public opposition to nuclear decisions (e.g., power plants and/or waste siting).[37]

Political strategies are also relevant in explaining people's attitudes and behaviors. For instance, proponents and opponents alike use nuclear power and radioactive waste management issues as a strategic battleground to promote their respective positions, leading to social and political polarization.[38] People's perceptions of these political strategies can influence their views on nuclear energy in a concrete social and political context.

Sociocultural and Psychological Factors

The third theoretical assumption is that both benefit-risk perceptions and social trust are influenced by a set of antecedent variables, including affective feelings, values, beliefs, and ideological and political orientations.

Public perception of nuclear power is related to general beliefs and values, which configure personal ideological systems, such as emphasis on economic versus social priorities, attitudes toward technology and environmental concerns, the social meaning of economic growth, or beliefs about the centralization of decision-making.[39] Individuals express greater or lesser support for nuclear power depending on their adherence to certain social values (such as traditional, altruistic, etc.).[40] People are also more likely to act in favor of or against nuclear energy when personal norms are strong.[41] For instance, in one study people who expressed greater concern about climate change and energy security and more strongly adhered to environmental values were less likely to favor nuclear power.[42] People's trust in information sources is also influenced by political party support and other ideological background variables.[43]

In this light, changes in public views following the Fukushima accident can be seen as moderated by political ideology (e.g., environmental views) over time,[44] and/or could also be explained by prior support for nuclear power.[45]

Public conceptions of nuclear energy are also shaped by the possible military applications of the technology, notably by concerns over the potential proliferation of nuclear weapon technologies, for example, via uranium enrichment facilities.[46]

Concern about the local community is also an important determinant, a fact that may be related to nuclear issues being conceived as general, rather than personal, matters.[47]

Affective feelings about nuclear power appear to form a key factor,[48] and people who are opposed to nuclear power plants have also been found to mainly associate nuclear power plants with negative feelings.[49]

In sum, we cannot separate nuclear energy from questions about the kind of society in which people wish to live.[50] From this perspective, Lars Löfquist, professor at Uppsala University (Sweden), argues that closing down nuclear power plants cannot be done without major disturbances in ordinary people's lives and lifestyles, a fact that can have an effect on their perceptions.[51]

Methods

The empirical basis of our analysis consists of the short country reports (SCRs) produced for twenty countries by the team of HoNESt historians.[52] One objective of these SCRs is to provide historical narratives covering sixty years, a basis upon which HoNESt social scientists could construct their studies on public perceptions and engagement. The SCRs hence provided specific evidence that allowed social scientists to identify key events, actors, arguments, behaviors, and types of public engagement. We have chosen a sample of eight country reports for a more in-depth analysis of risk and benefit perceptions. The following criteria were applied to select the countries (see table 5.1): (1) different levels and types of public acceptance and social movements of opposition; (2) different political systems and cultural and democratic norms; and (3) a balance of different geographical locations across Europe and beyond (US)..

The conditions under which nuclear energy has been developed vary greatly over time, since each historical phase is characterized by a specific political, social, and economic context. In our analysis we distinguish between three main historical phases:

1. Phase 1: 1950–1970: Postwar developments and Atoms for Peace program.
2. Phase 2: 1970–1990: Economic growth and public mobilization. Three Mile Island and Chernobyl accidents affected public opinion.
3. Phase 3: 1990–2015: Fall of the Iron Curtain. Globalization. Climate change, peak oil, and the role of renewables. Fukushima accident.

Thus, the empirical evidence for our analysis (as available in the SCRs) comprises the interaction between nuclear energy and society in seven European countries and the US, from the 1950s to 2015. The evidence illustrates the arguments raised by the key actors concerning nuclear risks and benefits. We analyzed three actor types: proponents (companies, nuclear industry associations), affected people (civil society organizations, general public), public authorities and regulators (policymakers at different levels). Other closely related actors, such as scientific bodies and advisory committees, have also been taken into account.

Table 5.1. Country characteristics

Country	Geography (location)	Political system	Public acceptance
Bulgaria	East Europe	Soviet regime + Democracy	High
Finland	Nordic Countries	Democracy	High
F.R. Germany	Central Europe	Democracy	Low
Spain	Mediterranean Europe	Dictatorship + Democracy	Low
Sweden	Nordic Countries	Democracy	Medium
Ukraine	East Europe	Soviet regime + Democracy	Medium
United Kingdom	West Europe	Democracy	High
United States	North America	Democracy	Medium

Since historical country narratives have provided the empirical source of our analysis, the nature of these specific "primary data" needs some consideration. To further underline the interdisciplinary character of HoNESt, social scientists and historians together developed a "Guiding Framework" of questions to take into account when compiling the historical country reports, to enable social scientists to understand how different societies have reacted to nuclear developments. All of the SCRs therefore follow the same structure. In addition to a "Facts and Figures" section, they provide roughly five "Events" crucial for the country's nuclear history, a "Showcase" with a more in-depth depiction, and a comprehensive overview (narrative) of the country's historical context. A thematic analysis of all this written material was carried out to identify, analyze, and report patterns within data.[53]

While our results allow a general macro-level overview of the eight considered countries, readers should bear in mind that what is true in general might not be true in a given specific case (e.g., acceptance of nuclear energy can be broad in some countries, despite resistance in local communities, and vice versa).

Results

The risk- and benefit-related perceptions of nuclear energy described in this section are the results of an in-depth analysis of all the relevant data from the eight short country reports examined. We carefully read and coded each SCR, highlighting all the potential text fragments related to risks and benefits. After analyzing the general history and the respective five key events of the eight countries, we identified a total of ninety text fragments as directly related to risks or benefits of nuclear energy. We interpreted the results to clarify the underlying patterns in the use of arguments and discourses. For each identified pattern, we selected part of the most representative excerpt to illustrate the relevant section on results.

Risk-Related Perceptions of Nuclear Energy

Our analysis shows that the most frequently mentioned risks are those related to the possibility of accidents and radioactive contamination, including both damages and losses that may affect human health or the environment (especially aquatic, fluvial, or marine environments). Concerns also relate to the safety of nuclear facilities, as well as to episodes of stress and anxiety experienced by some people when confronted by the possibility of such risks materializing. The perception of risks resulting from the economic consequences that nuclear projects entail is also relevant, as are concerns related to the sustainability of nuclear projects.

HEALTH

Although a few references can also be found in other periods, most *health* concerns related to nuclear energy were reported during the period 1970–1990. This coincides with the period of greater social mobilization against nuclear projects around the world. Health and environmental arguments were fundamental in most of these protests, although other dimensions could also be relevant but were less evident or explicit.

Regarding actors, the affected people expressed most of the concerns about health effects. Proponents and public authorities (regulators, etc.) focused very little on this dimension, and when they did so they tended to stress the low probability of this potential harm, or to highlight possible positive applications of nuclear technologies in the health or agriculture sectors. One example of the first argument is found in the US country report: "According to several studies, the radiation doses of the approximately 2 million people in the affected region were very small and there would be no long-term health impacts."[54] An example of the second type of argument can be found in the Finland country report: "In addition, isotopes and medical use of radiation were going to cure cancer and other sicknesses and help to cultivate more productive plants for agriculture."[55]

ENVIRONMENT

In similar terms, although mainly absent during the first period of nuclear development, from the 1970s onward the *environmental* risk became a dominant argument among all involved actors. For this period and thereafter, environmentally related perceptions, either positive or negative, such as water, soil, and air pollution or climate change impacts, are described in the country reports. Affected people highlighted the potential (or actual, in cases like Chernobyl) environmental impacts of nuclear facilities. In contrast, we identified only a few references to these impacts on the part of proponents and public authorities, usually hinting at positive impacts such as mitigating climate change.

Regarding local affected people, the country reports show several cases of local actors arguing about the potential environmental dangers of hosting nuclear installations and advocating other technological options they perceived to be less risky. For instance, the Finnish report says that "[the fishing] community in near Hästholmen feared that thermal pollution would damage the fragile marine ecology of the Gulf of Finland."[56]

However, in some cases, affected people seemed to agree with a "reluctant acceptance" of nuclear energy because it could help in coping with the low-carbon energy and climate change challenges.[57] Probably the clearest case of this argument can be found in the UK report:

> Overall, public responses highlighted the impact of climate change on their willingness to accept the need for nuclear power. The privatised industry's efforts to portray nuclear as a low carbon technology seem to have worked, and most UK citizens believe that nuclear will have a significant part to play in the generation of electricity in the future. . . . A number of high-profile environmental writers and campaigners have changed their minds and now support nuclear power as part of the answer to the challenges posed by climate change. . . . As climate change continues to rate as a matter of concern for the public, nuclear power is perhaps seen as a "necessary evil."[58]

SAFETY

Safety debates are also identified in the country reports. In general terms, most of the safety concerns about nuclear energy arose after the 1970s. During the earlier period, only a few doubts about safety issues were found in the SCRs, namely the Finnish fears about Soviet designs and reported criticisms in Sweden from technical experts and politicians regarding reactor safety requirements. However, from the 1970s, conflicting attitudes appear among the public and specific experts about reactor safety, as affected people showed increased concerns about the location of the nuclear installations because of safety issues. Several cases are described in the SCRs, like this one from Germany:

> The pilot-scale project SNR 300 motivated promoters due to the limited uranium reserves and regulators hoped for an efficient utilization of the minerals by building this reactor. However, very soon the search for a site raised concerns among receptors [affected people] who demonstrated against the project, some of the demonstrators even came from the Netherlands as the chosen site was close to the country.[59]

After major nuclear accidents such as Three Mile Island and Chernobyl, demands for safety were reinforced. Nevertheless, it must be said that proponents and regulators tended to focus on high technological expertise and innovations as arguments for the guaranteed safety of the nuclear power plants.

ECONOMICS

Finally, the *cost* of nuclear energy projects has been cited as a justification for reluctance to invest in these projects and for canceling ongoing projects, but also for continuing with a project once initiated (in this case to avoid potentially greater economic losses from cancellation of projects in which resources had already been invested). In cases such as those of the Federal Republic of Germany and Bulgaria, promoters and regulators were critical of nuclear energy because of its high cost. For instance, the Bulgarian report says:

> Regarding nuclear power, former vice-minister of electrification Oved Tadzher remembers that his Ministry of Electrification officials were not convinced Bulgaria was ready to operate a nuclear station. According to Tadzher, these officials considered the nuclear plant too expensive and too sophisticated for Bulgaria's existing technological capabilities.[60]

The perceived high costs of nuclear projects resulted in the need for the state to play an active role in promoting nuclear development in all investigated countries (both democracies and dictatorships). In some cases, such as in Sweden, although affected people voted for a halt to the nuclear program (following the nonbinding referendum in 1980), this was not carried out because public authorities argued that it would involve economic losses of resources already invested. Arguments about the high costs of nuclear energy were mainly put forward by affected people and included not only the economic costs caused by major accidents but also the costs resulting from further regulations that were established following these accidents. Thus, in some ways, accidents can be construed as a factor *increasing the resources needed* to implement projects in the nuclear sector (e.g., to cope with the design/procedures for accident mitigation).

BRIEF COMPARISON

Interestingly, there are very few differences between countries: almost the same concerns (the same perceived risks) emerge in all studied countries. However, some differences between the historical periods can be detected. In the first period (1950–1970), there are few mentions of perceived risks, during the second period (1970–1990) references to risks multiply, and in the third period (1990–2015) mentions of perceived risks decrease again (but not to the low level found during the first period). However, since our analyses are primarily based on a qualitative approach, these quantitative assessments should be regarded only as general indications. We must also bear in mind that these findings on the perceived risks of nuclear energy are based on the account made by the respective reports of the HoNESt historian team.

Benefit-Related Perceptions of Nuclear Energy

As with perceived risks, perceived benefits are also relatively similar among all the countries in the sample. As there are nuclear proponents in all countries of our sample, for each SCR we found a range of actors arguing that nuclear energy will bring various benefits, especially in economic terms (jobs, socio-economic development, cheap electricity, or a guarantee of energy supply) but also to a lesser extent environmental and human health benefits.

ENVIRONMENT

Environmental impacts were referred to in various ways along the different historical phases. References to positive environmental effects of nuclear energy appeared in the first period (1950–1970) as a response by proponents and public authorities to early concerns occasionally raised by citizens, who were not generally greatly worried at this time. This is exemplified by a quotation from the Swedish report: "In the 1950s and 1960s, the largest and oldest environmental organization Svenska Naturskyddsförening had even demanded a faster introduction of nuclear power to save the remaining wild rivers."[61]

During the second period (1970–1990), when debates emerged about assumed negative impacts, promoters and public authorities tried to relativize or minimize the importance of these and, in some cases, to highlight the potential positive environmental impacts. An illustrative example appears in the Spanish report:

> Even the environmental impact of the NPP was suggested as an unquestionable advantage, as "heat emitted by the NPP—around 30 degrees in winter—will bring a tropical climate to the touristic destination of the Guadiana reservoirs. This change in the climate will be to the advantage of the farmers." . . . A report from the Ministry of Agriculture to substantiate this argument was commissioned.[62]

These arguments increased in the 1990s, when promoters and public authorities argued that without nuclear power stations, the commitments adopted in the international climate agreement could not be fulfilled. For instance, the Finnish report states: "Without nuclear energy Finland was forced to invest in conventional energy, and this decision defied the international agreements against the climate change."[63] According to this framing, nuclear energy would help to mitigate environmental risks.

Minor changes can be observed over time regarding benefits, specifically with respect to favorable environmental effects. In the first two phases (1950–1970 and 1970–1990), some actors talked about positive environmental impacts of nuclear energy production, such as temperature increases that could be beneficial for certain ecosystems, and the fact that nuclear technologies would pollute less than other industries. Since the 1990s, the discussion has increasingly concerned the benefits of nuclear energy in the fight against climate change.

HEALTH

There are few mentions of health benefits in the first period (1950–1970), but some have been identified in the second phase (1970–1990). The Finnish report mentions how the proponents of the national nuclear program stressed that radiation would be useful for medical health purposes.[64] Similar findings are documented in the US report, which references arguments aimed at diminishing the importance of radiation impacts on human health.[65] During the third period (1990–2015), there are even fewer references to health effects in the SCRs. This implies that health benefits were mainly argued during the 1970–1990 period, just when the social conflicts related to nuclear energy reached their highest levels in most of the Western countries.

SAFETY

Another benefit-oriented argument is that some people (mainly promoters/regulators) have a high degree of confidence in the technical safety of nuclear facilities. This argument is illustrated by the UK report:

> Supporters of nuclear energy emphasize the facts that nuclear power will help secure energy independence; does not produce greenhouse gases that contribute to global warming or air pollution; and is a proven, developed technology with a sixty-year history of safe operation in the UK. The safety of UK reactors is heavily emphasized and linked to the strict and well-established regulation of the UK nuclear industry.[66]

In general, the arguments varied across the actor groups. Proponents almost always emphasized a high degree of safety measures and standards, although in the Federal Republic of Germany public authorities tried to convince energy companies of nuclear safety (in the early years of the technology, when the companies were not very convinced of its viability). Regulators and public authorities were traditionally allied with the promoters, probably because nuclear technology needs the full support of the state, at least in its early stages. Over time they tended to act in a more autonomous way, sometimes publicly criticizing the promoters' actions (after Three Mile Island the independence of regulators was strengthened, and they became autonomous bodies). Certain groups directly involved in running nuclear facilities, such as local governments of municipalities hosting the nuclear power plants and plant workers, considered themselves to be informed about safety measures and viewed these positively.

ECONOMICS

Regarding the *economic* dimension, a few arguments found in the country reports merit mention. In their public communication, promoters and public authorities stressed *job creation* as a major benefit to people affected by nuclear energy infrastructures. Implicitly arguing on the basis of a risk-benefit model, business actors and regulators assumed that economic advantages would outweigh any possible concerns about risks (a calculation that, however, failed in several cases).

In countries such as the Federal Republic of Germany, Finland, Spain, and the US, public authorities and promoters presented nuclear energy as a trigger for socioeconomic development in the form of *technological and industrial modernization*, and even in some cases (such as Spain or Finland) as a path to joining the highly developed countries. However, there are also examples (the Federal Republic of Germany) where a shift in this idea was observed from the 1970s and 1980s.

The few references that we found in the country reports to the *guarantee of energy supply* appear predominantly in the early stages (the Finnish and Spanish reports). In the last period (1990—2015), this issue only rarely appeared.

Some country reports (such as Finland, Sweden, and the UK) take into account nuclear energy's impact on *energy prices*. This factor seems to be more present in the first and second considered periods than in the third one. Different attitudes can be detected among different actors, confronting the wish to maintain cheap tariffs with the need to move toward more sustainable energy systems and growth.

Relational and Contextual Factors Shaping Perceptions on Nuclear Energy

Although the perceived risks and perceived benefits are similar in all the countries studied, the social and institutional responses are very different. In order to explore this disparity, according to our theoretical framework we first need to understand how people perceive their relationships with institutions (trust), as well as what kind of sociocultural factors are part of the context in which the nuclear technology is perceived.

These are the main political-institutional factors identified in the SCRs shaping social trust:

- Low institutional trustworthiness, which draws attention to the fact that certain social sectors do not trust the institutions that manage or regulate nuclear energy. Distrust is related to the perception that these institutions have carried out some kind of incorrect or unethical behavior, for example by favoring private interests above the public, or by acting against the law or by keeping secrets (which at some point were revealed to the public). There are many examples of these behaviors in the SCRs. In countries such as the UK and Finland, a higher perception of trust in institutions has been found regarding nuclear development, but not so much regarding waste management.

- Political strategies or "games," such as the cases of political elections affecting decision-making on nuclear issues (Federal Republic of Germany), or political parties changing their opinion about nuclear developments when governing (Spain), or fighting between pro- and anti-European parties in some countries (mainly in Eastern Europe).
- Dependency on other countries as a factor influencing decision-making, leading national governments to adopt certain behaviors to gain energy autonomy or to avoid dependency. These concerns appear more in the Eastern European countries (Bulgaria and Ukraine), but also in Finland.

The analysis also found several sociocultural factors shaping public perceptions:

1. Sociocultural factors related to structural interactions:
   - Territorial identity conflicts (territorial comparative grievances; conflicts between economic activities and land uses, etc.).
   - National scientific pride (and also national military pride).
2. Sociocultural factors related to individual lifestyles:
   - Subjective attributes of risk: perception of difficulty of calculating risks, perception of low controllability of risk, unwillingness to be exposed, familiarity with the technology (and coping with similar risks in the past).
   - Conflicts of values: social conflicts related to preferences for different lifestyles, different economic and social development models, different attitudes toward pacifism/warmongering that nuclear development may entail, concerns about how future generations will judge current ones because of their management of nuclear energy, etc. These elements correspond to different ideologies or ways of understanding how society should ideally be and evolve.

These factors are also unevenly represented among the different countries, and correspondingly they can help to explain at least some of the different social responses to nuclear energy.

The overview allows us to see how the political-institutional factors are distributed: First, institutional trustworthiness is a factor widely represented among all the countries (perceived as low in all the countries, except the UK and Finland). The political *games* are relevant in all the countries but with a lower profile in the UK and the US. The *dependency* factor is present overall in the nuclear debates of Bulgaria, Ukraine, and Finland.

Regarding sociocultural factors, the *conflict of values* and *subjective attributes of risk* can be found in all the countries; *national pride* seems to be relevant in all the countries except Spain and the Federal Republic of Germany; and the *territorial identity conflicts* are present overall in Germany, Spain, and Sweden.

Conclusions

Our analysis has identified several key findings in terms of patterns of perceived risks and benefits, actors, and periods.

First, in the SCRs, arguments on perceived risks are relatively few in the first period (1950–1970), increase during the second period (1970–1990), and then decrease again somewhat in the third period (1990–2015). Generally, the types of risks mentioned are similar over the time intervals considered (basically health and safety concerns, environmental and economic risks). The trend is similar in all of the countries that we examined. Moreover, concerns about economic risks increased during the second period, when the costs of nuclear projects and economic uncertainties were used as arguments in nuclear power debates, a type of reasoning that continued in the third period (1990–2015). As expected, the affected people expressed plenty of concerns about all kinds of risks (health, safety, environmental, economic), while neither promoters nor public authorities/regulators focused greatly on these issues. These latter groups did not express concerns about risks, precisely because they considered nuclear safe—in all dimensions of risk/safety.

In terms of benefits, health and safety references were almost absent during the first period (1950–1970), but after the 1970s they increased in the form of promises and guarantees of safety, generally to counterbalance public concerns. Proponents always emphasized a high degree of safety measures and standards, as did regulators, although over time the latter tended to become more critical and demanding of the proponents, probably as a result of becoming autonomous bodies in most of the countries (a process discernible after the Three Mile Island event). Perception of economic benefits is quite constant but decreasing; safety guarantees to avoid health and environmental harm appear in the 1970s; environmental issues increase over time, especially when climate protection enters national and international policy agendas.

The most significant change appears in the environmental benefits. In the first and second periods, these were conceptualized in terms of less pollution of water resources, or as beneficial for new agricultural or touristic activities, but during the third period the arguments changed radically toward benefits for fighting climate change. This change is observable in the arguments of all actors, especially among proponents and public authorities, but also among the affected people in countries such as the UK and Finland.

Second, we note that the evolution of perceived risks and benefits across time is similar across countries. Nevertheless, public acceptance of nuclear energy is quite different in each country. This evidence clearly suggests that additional variables need to be addressed in order to understand public acceptance of nuclear issues.

In line with recent trends in social research on risk,[67] we suggest that the social acceptability of nuclear, besides risk-benefit perceptions, ultimately depends on public trust in the institutions that manage and/or regulate this technology (political-institutional dimension). This trust, in turn, is a function of sociocultural factors generated by the social climate over time (sociocultural dimension)—that is, a set of antecedent variables including affective feelings, values or beliefs, and ideological and political orientations.

This has clear implications for the study of public responses to and engagement with nuclear projects. To understand why the public accepts or rejects a particular nuclear project, it is not enough to analyze the health, environmental, or economic dimensions of risk perception. The political-institutional dimension (trust) and the sociocultural dimension (ideologies, identity conflicts, etc.) also need to be considered. This includes understanding how people perceive their relationship with institutions and companies, as well as the sociocultural factors shaping the social context in which the risks and benefits are perceived.

It can be hypothesized that opinions on the desirability of nuclear energy rely more on these sociocultural and political dimensions than strictly on the perceived risks and benefits. There is a long-standing debate about the gap between different social groups' assessments of risk, especially—but not solely—between experts and laypeople, a gap that can reduce the acceptance of certain technologies such as nuclear. Our results clearly support the hypothesis that there are factors underlying public perception other than the health, environment, and economic benefits and risks, and that public reactions depend on the context in which the nuclear technology is deployed. As suggested by risk society theorists such as Ulrich Beck and Anthony Giddens, reflexivity is at the very heart of the social transformation toward a risk society: individuals become aware of the contradictions and limitations of the institutions governing society. One of the most important contested institutions is that of scientific expertise and expert assessments.[68] How can trust be reestablished? How can collectively binding decisions be made when experts lacks credibility and long-term implications of their decisions are unknown or seem highly complex?

Summing up, according to our research, the key to understanding consensus and conflicts on nuclear energy lies in the sphere of trust in institutions on one hand and sociocultural contextual factors on the other. These contextual factors refer to issues of responsibility, moral judgments, principles of justice and equity, and accountability and legitimacy. These concerns have to some extent been addressed by governments and nuclear companies through a broad set of public relations strategies, risk communication tools, and public participatory processes, thus pushing engagement issues to the forefront of the debate. However, the findings from the historical data and sociological analyses show the complexity of nuclear-society interactions in the differing political, economic, and social contexts of the countries studied, pointing to a need to further develop and critically evaluate effective and justifiable methods for such engagements.

Notes

1. Eurobarometer, *Europeans and Nuclear Safety: Special Eurobarometer 324* (Brussels: Eurobarometer, 2010), 26.
2. Eurobarometer, *Europeans and Nuclear Safety*, 41.
3. For instance, John C. Besley, "Does Fairness Matter in the Context of Anger About Nuclear Energy Decision Making?," *Risk Analysis* 32 (2012): 25–38; Karen A. Parkhill et al., "From the Familiar to the Extraordinary: Local Residents' Perceptions of Risk When Living with Nuclear Power in the UK," *Transactions of the Institute of British Geographers* 35 (2010): 39–58; Nick F. Pidgeon, Irene Lorenzoni, and Wouter Poortinga, "Climate Change or Nuclear Power—No Thanks! A Quantitative Study of Public Perceptions and Risk Framing in Britain," *Global Environmental Change* 18, no. 1 (2008): 69–85; Wouter Poortinga, Nick F. Pidgeon, and Irene Lorenzoni, "Public Perceptions of Nuclear Power, Climate Change and Energy Options in Britain: Summary Findings of a Survey Conducted During October and November 2005," *Understanding Risk Working Paper 06-02* (Norwich: Centre for Environmental Risk, 2006); Vivianne H. Visschers and Michael Siegrist, "Fair Play in Energy Policy Decisions: Procedural Fairness, Outcome Fairness and Acceptance of the Decision to Rebuild Nuclear Power Plants," *Energy Policy* 46 (2012): 292–300.
4. John Whitton et al., "Theoretical, Methodological and Epistemological Challenges of the Multi-disciplinary History of Nuclear Energy and Society (HoNESt) Research Project" (paper presented at ESSHC 2016, Valencia, Spain, 2016).
5. Paul Slovic, "Perceived Risk, Trust, and Democracy," *Risk Analysis*, 13, no 6 (1993): 675–82; Paul Slovic, *The Perception of Risk* (London: Earthscan, 2000).
6. Paul Slovic and Ellen Peters, "Risk Perception and Affect," *Current Directions in Psychological Science* 15, no. 6 (2006): 322–25.
7. Mary Douglas and Aaron Wildavsky, *Risk and Culture* (Berkeley: University of California Press, 1982).

8. Tom Horlick-Jones and Ana Prades, "On Interpretative Risk Perception Research: Some Reflections on Its Origins; Its Nature; and Its Possible Applications in Risk Communication Practice," *Health, Risk, and Society* 11, no. 5 (October 2009): 409–30; Tom Horlick-Jones, Ana Prades, and Josep Espluga, "Investigating the Degree of 'Stigma' Associated with Nuclear Energy Technologies: A Cross-Cultural Examination of the Case of Fusion Power," *Public Understanding of Science* 21, no. 5 (2012): 514–33; Brian Wynne, "May the Sheep Safely Graze? A Reflexive View of the Expert-Lay Knowledge Divide," in *Risk, Environment and Modernity*, ed. Scott Lash, Bronislaw Szerszynski, and Brian Wynne (London: Sage, 1996), 44–83.
9. Ortwin Renn, *Risk Governance: Coping with Uncertainty in a Complex World* (London: Earthscan, 2008).
10. Josep Espluga, Beatriz Medina, and Wilfried Konrad, coords., "Case Studies Reports; In-Depth Understanding of the Mechanisms for Effective Interaction with Civil Society: Selected Case Studies," in *History of Nuclear Energy and Society (HoNESt) Consortium Deliverable no. 4.3* (2018), https://perma.cc/5SDY-LKCK.
11. Carmen Keller, Vivianne H. Visschers, and Michael Siegrist, "Affective Imagery and Acceptance of Replacing Nuclear Power Plants," *Risk Analysis* 32, no. 3 (2012): 464–77; Nan Li et al., "How Do Policymakers and Think Tank Stakeholders Prioritize the Risks of the Nuclear Fuel Cycle? A Semantic Network Analysis," *Journal of Risk Research* 21, no. 5 (2018): 599–621, doi:10.1080/13669877.2016.1223164; Michael Siegrist, George Cvetkovich, and Claudia Roth, "Salient Value Similarity, Social Trust, and Risk/Benefit Perception," *Risk Analysis* 20, no. 3 (2000): 353–62.
12. Vivianne H. Visschers, Carmen Keller, and Michael Siegrist, "Climate Change Benefits and Energy Supply Benefits As Determinants of Acceptance of Nuclear Power Stations: Investigating an Explanatory Model," *Energy Policy* 39, no. 6 (2011): 3621–29; Adam Corner et al., "Nuclear Power, Climate Change and Energy Security: Exploring British Public Attitudes," *Energy Policy* 39, no. 9 (2011): 4823–33; Pidgeon, Lorenzoni, and Poortinga, "Climate Change or Nuclear Power," 69–85.
13. Pidgeon, Lorenzoni, and Poortinga, "Climate Change or Nuclear Power," 69–85.
14. Corner et al., "Nuclear Power, Climate Change and Energy Security," 4823–33.
15. Keller, Visschers, and Siegrist, "Affective Imagery," 464–477; Ahmet Kılınç, Edward Boyes, and Martin Stanisstreet, "Exploring Students' Ideas About Risks and Benefits of Nuclear Power Using Risk Perception Theories," *Journal of Science Education and Technology* 22, no. 3 (2013): 252–66.
16. Mark Elam and Göran Sundqvist, "The Swedish KBS Project: A Last Word in Nuclear Fuel Safety Prepares to Conquer the World?," *Journal of Risk Research* 12, no. 7–8 (2009): 969–88; Marja Ylönen et al., "The (De) Politicisation of Nuclear Power: The Finnish Discussion After Fukushima," *Public Understanding of Science* 26, no. 3 (2017): 260–74.
17. Keller, Visschers, and Siegrist, "Affective Imagery," 464–77.
18. Lei Huang et al., "Effect of the Fukushima Nuclear Accident on the Risk Perception of Residents Near a Nuclear Power Plant in China," *Proceedings of the National Academy of Sciences* 110, no. 49 (2013): 19742–47; Michael Siegrist and Vivianne H. Visschers, "Acceptance of Nuclear Power: The Fukushima Effect," *Energy Policy* 59 (2013): 112–19; Vivianne H. Visschers and Lasse Wallquist, "Nuclear Power Before and After Fukushima: The Relations Between Acceptance,

Ambivalence and Knowledge," *Journal of Environmental Psychology* 36 (2013): 77–86.

19. Younghwam Kim, Wonjoon Kim, and Minki Kim, "An International Comparative Analysis of Public Acceptance of Nuclear Energy," *Energy Policy* 66 (2014): 475–83.
20. Li et al., "How Do Policymakers and Think Tank Stakeholders Prioritize," 599–621; Lars Löfquist, "After Fukushima: Nuclear Power and Societal Choice," *Journal of Risk Research* 18, no. 3 (2015): 291–303.
21. Joop van der Pligt, J. Richard Eiser, and Russell Spears, "Public Attitudes to Nuclear Energy," *Energy Policy* 12, no. 3 (1984): 302–5.
22. Paul Slovic et al., "Nuclear Power and the Public: A Comparative Study of Risk Perception in France and the United States," in *Cross-Cultural Risk Perception: A Survey of Empirical Studies*, ed. Ortwin Renn and Bernd Rohrmann (Boston: Springer, 2000), 55–102.
23. Keller, Visschers, and Siegrist, "Affective Imagery," 464–77.
24. Kilinç, Boyes, and Stanisstreet, "Exploring Students' Ideas," 252–66; Visschers, Keller, and Siegrist, "Climate Change Benefits and Energy Supply Benefits," 3621–29; Corner et al., "Nuclear Power, Climate Change and Energy Security," 4823–33.
25. James W. Stoutenborough, Shelbi G. Sturgess, and Arnold Vedlitz, "Knowledge, Risk, and Policy Support: Public Perceptions of Nuclear Power," *Energy Policy* 62 (2013): 176–84.
26. Stephen C. Whitfield et al., "The Future of Nuclear Power: Value Orientations and Risk Perception," *Risk Analysis* 29, no. 3 (2009): 425–37.
27. Visschers, Keller, and Siegrist, "Climate Change Benefits and Energy Supply Benefits," 3621–29; Yue Guo and Tao Ren, "When It Is Unfamiliar to Me: Local Acceptance of Planned Nuclear Power Plants in China in the Post-Fukushima Era," *Energy Policy* 100 (2017): 113–25.
28. Kim, Kim, and Kim, "An International Comparative Analysis of Public Acceptance of Nuclear Energy," 475–83; Siegrist, Cvetkovich, and Roth, "Salient Value Similarity," 353–62.
29. Markku Lehtonen, "Opening Up or Closing Down Radioactive Waste Management Policy? Debates on Reversibility and Retrievability in Finland, France, and the United Kingdom," *Risk, Hazards, and Crisis in Public Policy* 1, no. 4 (2010): 139–79.
30. Siegrist, Cvetkovich, and Roth, "Salient Value Similarity," 353–62; George T. Cvetkovich and Patricia L. Winter, "Trust and Social Representations of the Management of Threatened and Endangered Species," *Environment and Behavior* 35, no. 2 (2003): 286–307; Wouter Poortinga and Nick F. Pidgeon, "Exploring the Dimensionality of Trust in Risk Perception," *Risk Analysis* 23, no. 5 (2003): 961–73; John Walls et al., "Critical Trust: Understanding Lay Perceptions of Health and Safety Risk Regulation," *Health, Risk, and Society* 6 (2004): 133–50.
31. Keller, Visschers, and Siegrist, "Affective Imagery," 464–77; Hank Jenkins-Smith et al., "Reversing Nuclear Opposition: Evolving Public Acceptance of a Permanent Nuclear Waste Disposal Facility," *Risk Analysis* 31, no. 4 (2011): 629–44.
32. Matthew Cotton, *Nuclear Waste Politics: An Incrementalist Perspective* (London: Routledge, 2017).

33. Löfquist, "After Fukushima," 291–303.
34. Visschers and Siegrist, "Fair Play in Energy Policy Decisions," 292–300; Jon D. Miller, "The Measurement of Civic Scientific Literacy," *Public Understanding of Science* 7 (1998): 203–23.
35. John C. Besley, "Public Engagement and the Impact of Fairness Perceptions on Decision Favorability and Acceptance," *Science Communication* 32, no. 2 (2010): 256–80.
36. Siegrist, Cvetkovich, and Roth, "Salient Value Similarity," 353–62.
37. Peter Hocke and Ortwin Renn, "Concerned Public and the Paralysis of Decision-Making: Nuclear Waste Management Policy in Germany," *Journal of Risk Research* 12, no. 7–8 (2009): 921–40.
38. Hocke and Renn, "Concerned Public and the Paralysis of Decision-Making," 921–40.
39. Van der Pligt, Eiser, and Spears, "Public Attitudes to Nuclear Energy," 302–5.
40. Whitfield, Rosa, Dan, and Dietz, "The Future of Nuclear Power," 425–37.
41. Judith I. De Groot and Linda Steg, "Morality and Nuclear Energy: Perceptions of Risks and Benefits, Personal Norms, and Willingness to Take Action Related to Nuclear Energy," *Risk Analysis* 30, no. 9 (2010): 1363–73.
42. Corner et al., "Nuclear Power, Climate Change and Energy Security," 4823–33.
43. Annukka Vainio, Riikka Paloniemi, and Vilja Varho, "Weighing the Risks of Nuclear Energy and Climate Change: Trust in Different Information Sources, Perceived Risks, and Willingness to Pay for Alternatives to Nuclear Power," *Risk Analysis* 37, no. 3 (2017): 557–69.
44. John C. Besley and Sang-Hwa Oh, "The Impact of Accident Attention, Ideology, and Environmentalism on American Attitudes Toward Nuclear Energy," *Risk Analysis* 34, no. 5 (2014): 949–64; Marc Poumadère, "Before and After Fukushima: The Many Fronts of Managing the Nuclear Power Option," in *India's Risks: Democratizing the Management of Threats to Environment, Health, and Values*, ed. R. Moor and R. Gowda (Oxford University Press India, 2014), 250–74; Marc Poumadère, "Que faut-il savoir d'un accident nucléaire pour une meilleure prise en compte des conséquences sur la société?" in *Risques majeurs, incertitudes et décisions*, ed. Myriam Merad, Nicolas Dechy, Laurent Dehouck, and Marc Lassagne (Paris: MA Editions—ESKA, 2016), 265–87.
45. Siegrist and Visschers, "Acceptance of Nuclear Power," 112–19; Stoutenborough, Sturgess, and Vedlitz, "Knowledge, Risk, and Policy Support," 176–84; Visschers and Wallquist, "Nuclear Power Before and After Fukushima," 77–86.
46. Keller, Visschers, and Siegrist, "Affective Imagery," 464–77; Mariliis Lehtveer and Fredrik Hedenus, "Nuclear Power as a Climate Mitigation Strategy—Technology and Proliferation Risk," *Journal of Risk Research* 18, no. 3 (2015): 273–90.
47. Lennart Sjöberg and Britt-Marie Drottz-Sjöberg, "Fairness, Risk and Risk Tolerance in the Siting of a Nuclear Waste Repository," *Journal of Risk Research* 4, no. 1 (2001): 75–101.
48. Guo and Ren, "When It Is Unfamiliar to Me," 113–25; Visschers, Keller, and Siegrist, "Climate Change Benefits and Energy Supply Benefits," 3621–29.
49. Keller, Visschers, and Siegrist, "Affective Imagery," 464–77.
50. Van der Pligt, Eiser, and Spears, "Public Attitudes to Nuclear Energy," 302–5; Poumadère, "Before and After Fukushima," 250–74; Martin J. Goodfellow, Hugo R.

Williams, and Adisa Azapagic, "Nuclear Renaissance, Public Perception and Design Criteria," *Energy Policy* 39, no. 10 (October 2011): 6199–6210.
51. Löfquist, "After Fukushima," 291–303.
52. Available at http://www.honest2020.eu/d36-short-country-reports. A unified document with all the short country reports can also be found at https://hdl.handle.net/2454/38269.
53. Virginia Braun and Victoria Clarke, "Using Thematic Analysis in Psychology," *Qualitative Research in Psychology* 3 (2006): 77–101; Gareth Terry et al., "Thematic Analysis," in *The SAGE Handbook of Qualitative Research in Psychology*, ed. Carla Willig and Wendy Stainton-Rogers (Los Angeles: Sage Publications, 2017), 17–37.
54. Paul R. Josephson, "United States Short Country Report (version 2018)," in *History of Nuclear Energy and Society (HoNESt) Consortium Deliverable no. 3.6* (2019), https://hdl.handle.net/2454/38269.
55. Karl Eric Michelsen and Aisulu Harjula, "Finland Short Country Report (version 2017)," in *History of Nuclear Energy and Society (HoNESt) Consortium Deliverable no. 3.6* (2019), https://hdl.handle.net/2454/38269.
56. Michelsen and Harjula, "Finland Short Country Report," 50.
57. In the terms used by Karen Bickerstaff et al., "Reframing the Nuclear Debate in the UK: Radioactive Waste and Climate Change Mitigation," *Public Understanding of Science* 17, no. 2 (2008): 145–69.
58. Stuart Butler and Robert Bud, "United Kingdom Short Country Report (version 2018)," in *History of Nuclear Energy and Society (HoNESt) Consortium Deliverable no. 3.6* (2019), 51, https://hdl.handle.net/2454/38269.
59. Astrid M. Kirchhof and Helmuth Trischler, "Federal Republic of Germany Short Country Report (version 2018)," in *History of Nuclear Energy and Society (HoNESt) Consortium Deliverable no. 3.6* (2019), 19, https://hdl.handle.net/2454/38269.
60. Ivaylo Hristov and Ivan Tchalakov, "Bulgaria Short Country Report (version 2019)," in *History of Nuclear Energy and Society (HoNESt) Consortium Deliverable no. 3.6* (2019), 33, https://hdl.handle.net/2454/38269.
61. Arne Kaijser, "Sweden Short Country Report (version 2018)," in *History of Nuclear Energy and Society (HoNESt) Consortium Deliverable no. 3.6* (2019), 14, https://hdl.handle.net/2454/38269.
62. Mar Rubio-Varas et al., "Spain Short Country Report (version 2018)," in *History of Nuclear Energy and Society (HoNESt) Consortium Deliverable no. 3.6* (2019), 25, https://hdl.handle.net/2454/38269.
63. Michelsen and Harjula, "Finland Short Country Report," 25.
64. Michelsen and Harjula, "Finland Short Country Report," 34.
65. Josephson, "United States Short Country Report," 42.
66. Butler and Bud, "United Kingdom Short Country Report," 21.
67. For instance, Horlick-Jones and Prades, "On Interpretative Risk Perception Research," 409–30; Renn, *Risk Governance*; Espluga, Medina, and Konrad, *Case Studies Reports*.
68. Ulrich Beck, *Risk Society: Towards a New Modernity* (London: Sage, 1992); Ulrich Beck, Anthony Giddens, and Scott Lash, *Reflexive Modernization: Politics, Tradition and Aesthetics in the Modern Social Order* (Stanford, CA: Stanford University Press, 1994); Anthony Giddens, *Consequences of Modernity* (Cambridge, MA: Polity Press, 1990).

CHAPTER 6

Trust and Mistrust in Radioactive Waste Management: Historical Experience from High- and Low-Trust Contexts

Markku Lehtonen, Matthew Cotton, and Tatiana Kasperski

Introduction

Despite almost half a century of intensive efforts, the unresolved waste problem—what to do with high-level radioactive waste—remains the Achilles' heel of nuclear energy, a major impediment to nuclear new-build. Burial in deep geological repositories constitutes the reference option for international and national organizations responsible for nuclear waste management (NWM), yet no such repository is operational yet.[1] Apart from the formidable engineering task of constructing a safe repository, the key challenge facing these projects is that of identifying local communities with adequate geological conditions and a population willing to host such a facility. Since the late 1980s, site investigations in many countries stalled, one after another, in the face of vehement local opposition. Opposition was often underpinned by public mistrust of the safety and viability of the waste management technologies, the implementing organizations, and the involved institutions. To remedy the situation, industry and public authorities have since the early 1990s engaged in more participatory approaches in order to earn *social acceptance* or a "social license to operate" for their repository projects.[2] Building public trust and confidence in the repository concept, in the actors involved, and in the process, has become a key objective. Nuclear waste management has indeed become a test bed for participatory governance arrangements more generally, designed to address the long-term decline of trust in the governance of health and environmental risks related to science and technology.[3] The importance of trust is also recognized internationally, as illustrated by the creation, in 2000, of the Forum on Stakeholder Confidence (FSC) of the OECD Nuclear Energy Agency (NEA), to exchange information and best practice in the area of stakeholder engagement in nuclear waste management policy (see chapter 4 in this volume).

We explore trust-building efforts in five countries. Finland, France, and Sweden represent forerunners in repository development, whereas the United Kingdom (UK) provides a case with a history of failed efforts to generate consensus and to find a willing host community, despite extensive trust-building efforts. By comparing these four countries, we can also contrast the Nordic high-trust societies with the French and UK low-to-medium trust contexts. After analyzing these four countries, we present a very different case in Russia, which calls into question the assumption that trust would always necessarily be vital in nuclear waste management policy and illustrates the importance of the cultural and institutional context in trust-building. We caution against considering trust a silver bullet and illustrate the potential virtues of mistrust, notably in the form of "civic vigilance," which can help to strengthen governance of nuclear waste.[4] Trust cannot be simply *built*, for instance via participatory processes. Trust has to be *earned*, which requires not only allowing but nurturing constructive mistrust.

Three questions guide our comparison of nuclear waste management policies in Finland, France, Sweden, and the UK:

1. How have trust and mistrust evolved over the years?
2. Which measures have been implemented to build trust?
3. Which measures have worked, which have failed, and why?

For lack of sufficient information on the specific trust-building measures, we could not apply the same approach to the Russian case. This fifth case study therefore consists of a more general analysis of trust and mistrust in the nuclear sector.

We draw on research on the history of nuclear energy and society conducted as part of the HoNESt project; existing literature concerning nuclear waste management policies; semi-structured interviews conducted mainly between 2009 and 2016 with key actors in the five countries at the national, regional, and local levels (local and national-level politicians and authorities, civil society, the nuclear industry and waste management organizations, and academic researchers); and the authors' earlier work on the subject.[5] On all five countries, informal discussions conducted over the years with stakeholders and experts provided further inputs.

First, we briefly present the three dimensions of trust, mistrust, and trust-building. Second, we review the experience of trust-building in the nuclear waste management policies in our case study countries. We conclude with a discussion of the context-specific historical legacies of trust and nuclear waste management.

Definitions: Trust, Mistrust, and Trust-Building

In scholarly literature and in everyday usage, trust is almost invariably portrayed in positive terms. Research has shown its value for socioeconomic development, well-being, and a wide range of social processes, including environmental protection policy and the legitimacy of political power.[6]

We define trust[7] as a stance whereby individuals accept "believing without knowing," thereby placing themselves voluntarily in a position of vulnerability toward "the other," be it an individual or an institution.[8] A trusting individual takes the risk that the other might prove untrustworthy, yet since the choice is voluntary, it does not have to imply the feeling of losing power and control.[9]

Social trust is interpersonal. It can be further divided into generalized trust in other, unknown, members of society[10] and particularized (specific) trust in people we already know, and with whom we interact regularly, for example in our own social or demographic group.[11]

The main focus in this chapter is on institutional trust—the public trust in key institutions involved in nuclear waste management, such as nuclear safety authorities, the government, nuclear operators, government regulation, and environmental organizations. Institutional trust can entail *specific support*—in other words, individuals' judgment of what the institution *does* (its performance)—or *diffuse support*—that is, what the institution *is* and represents for the individual in question.[12] For example, one can trust specifically the present government coalition (specific support) or more generally the national government (diffuse support). A trusted institution is perceived as competent, sincere, transparent, and reliable in keeping its promises and possessing demonstrable ability to deal with mistrust and to avoid mismanagement or entanglement in political scandals.[13] In situations of long-standing institutional mistrust, attempts at trust-building via participation and openness can initially undermine trust.[14]

The concept of ideological trust relates to higher-level institutions, such as democracy, the state, market, and planning, and their legitimate roles in society.[15] More abstract than social and institutional trust, it is difficult to capture via quantitative surveys. Ideological trust concerns schemes of interpretation of reality, relating to means-ends relationships and strategies[16]—that is, to "wider abstract systems and ideas,"[17] such as economic growth models, government intervention in the economy, technological optimism, the precautionary principle, centralized or decentralized solutions,[18] or nuclear power as an electricity-generating option. Unlike social and institutional trust, ideological trust draws not on previous evidence or knowledge but "on an individual's or institution's place within wider social discursive structures." [19] Ideological trust entails high degrees of both emotionality and rationality[20]—a typical situation in controversies over nuclear power and nuclear waste management.

Downsides of Trust and Virtues of Mistrust

Excessive trust has its downsides, such as leniency in judging the trustee, delay in perceiving exploitation, and increased risk-taking.[21] High levels of the so-called bonding social capital can feed exclusion, homogeneous social networks, specific norms of reciprocity,[22] groupthink, the exclusion of different yet competent others,[23] and creation of sharp boundaries between "insiders" and "outsiders." [24] Trustful citizens may lack the motivation to participate in planning and decision-making, preferring instead to delegate power to trusted experts and institutions.[25]

Ultimately, as healthy suspicion, mistrust toward the powers that be constitutes a foundation for the vitality of a democratic system—a form of "civic vigilance," [26] responsibility, and countervailing power that helps citizens to hold political, economic, and cultural elites to account.[27] Organizations and procedures of regulation (e.g., auditing, evaluation, ranking, and benchmarking) represent an institutionalized form of mistrust and vigilance.[28]

Table 6.1. Summary of the key concepts relating to trust and mistrust

Type of trust/ mistrust	Social	Institutional	Ideological
Description	Generalized Particularized	Diffuse support Specific support	Legitimacy of and support to meta-level institutions
Sources of trust	Competence Sincerity		Worldviews, visions
	Normative predisposition in relation to an institution or an individual (trust)		
	Predictability, based on previous experience (confidence)		

Measures for Building Trust and Feeding Mistrust

Nuclear waste management actors (industry, governments, government experts) often stress the importance of public trust as an essential prerequisite of success. The technical and scientific analyses and design of the waste disposal solution constitute the most fundamental trust-building measures.[29] We focus on the nontechnological trust-building measures designed to strengthen trust in institutions. Four categories of measures can be highlighted: (1) voluntary opt-in and opt-out (i.e., voluntary engagement by the community and its ability to withdraw from the siting process); (2) participatory governance approaches; (3) economic support, including notably community benefit schemes; and (4) the creation of independent bodies of control and oversight.

Civil society actors critical of nuclear waste management policy, in turn, seek to build mistrust toward institutions via various campaign strategies, such as by disseminating critical information or by revealing cases of mismanagement or even corruption. They can also seek to undermine ideological trust by criticizing nuclear power as an electricity supply solution or, for example, condemning what they see as the private sector's undue involvement in nuclear policy (see table 6.2)..

Building Trust and Mistrust in Our Case Study Countries

Trust-building and mistrust-building efforts take place in greatly differing background conditions: surveys consistently show exceptionally high levels of interpersonal and institutional trust in Finland and Sweden, whereas France and the UK are at an average European level,[30] and Russia is at the opposite extreme from the Nordics, with particularly low levels of institutional trust.[31]

Finland

Internationally, Finland is often portrayed as an exemplar of democratic and consensual governance of nuclear waste,[32] whose disposal project has advanced while generating hardly any overt citizen opposition, largely thanks to careful long-term preparation and consistent implementation of the government strategy from 1983.[33] This consistency has helped to generate trust in the policy actors, not least among the nuclear power companies.

In 1978, nuclear operators were made legally responsible for managing their waste. At the time, the state company Imatran Voima (IVO, today, Fortum) exported its spent nuclear fuel to the USSR, whereas the privately owned Teollisuuden Voima (TVO) was searching for its own solution. TVO first aimed to survey the whole country to identify the geologically "best" site but soon

Table 6.2. Examples of trust-building and mistrust-building measures in nuclear waste management (NWM)

Trust-building (by actors responsible for NWM)	Independent bodies of control and oversight Participatory governance Stepwise decision-making Voluntary opt-in and opt-out Partnerships Community benefit schemes Broadening of debate to strategic questions (e.g., energy policy)	Communication and PR Dissemination of broader "schemes of interpretation of reality," e.g., via educational and scientific institutions
Mistrust-building (by civil society actors, counter-experts)	Dissemination of information and NGO campaigning Revelations of mismanagement and corruption, overlooked potential harmful impacts	Critique of nuclear power, and of private sector involvement in the nuclear sector

switched to a more pragmatic approach: in 1986, the company approached sixty-six potentially suitable municipalities to probe local acceptance.[34] Eurajoki, the host for two of Finland's four operating reactors, was the only municipality not selected on the basis of a systematic selection and elimination process but mainly to minimize waste transports. In 1987, TVO began preliminary site characterization in five municipalities. Local opposition pushed TVO to change strategy and gradually increase stakeholder involvement in the process.[35]

Two legislative decisions in 1994 triggered a participatory turn in nuclear waste policy: banning of nuclear waste exports as of 1996, and the Environmental Impact Assessment (EIA) Act, which made EIA mandatory. TVO and IVO established a joint company, Posiva, to develop a method for disposing of the waste in granitic host rock, following the Swedish KBS concept. Posiva engaged in an exceptionally long and thorough participatory EIA process in 1997–1999. Four municipalities were considered, but Posiva had already shifted its focus to the country's two nuclear municipalities—Loviisa and Eurajoki—where it expected little resistance. The government also undertook some trust-building measures: the candidate municipalities had their representatives in the social science working group within the government nuclear waste management research program,[36] and the Ministry of Trade and Industry granted a handful of non-governmental organizations a modest sum of €34,600 for their information activities.[37]

Although highly participatory, the EIA was undermined by a community benefit scheme—the so-called Vuojoki Agreement—that Posiva and TVO negotiated with municipal authorities.[38] The Vuojoki Agreement—vital for obtaining the municipality's acceptance—was negotiated on an *ad hoc* basis, behind the scenes, without the involvement of the local community and while the EIA process was still under way.[39] Eurajoki municipal leaders were also motivated by the generous tax bonuses from the planned new nuclear reactor, hoping that by accepting the repository project Eurajoki would beat Loviisa in rivalry for the reactor.[40]

Although the EIA had little impact on decisions,[41] critics describing it as "theatre play" designed to wear out the opponents,[42] the project itself faced little local or national opposition. In 2000, Eurajoki municipal council gave its approval. This paved the way for a 2001 Parliament ratification (votes 159–3) of a government Decision in Principle for the construction of a rock characterization facility (ONKALO), eventually to become a repository. A year later, Parliament approved the construction of a new nuclear reactor in Olkiluoto. Construction of ONKALO began in 2004. In 2015, the government granted a construction license for the repository. Host community approval is no longer needed for the operation license. Disposal of waste is to start in 2024, only slightly behind the schedule set out in a 1983 government decision.

Despite the near absence of overt public opposition, surveys show relatively low trust in repository safety among citizens in Eurajoki (41%) and nationally (36%).[43] Among Eurajoki citizens, 57 percent trust Posiva as a source of information.[44] By contrast, the nuclear safety regulator, STUK, enjoys overwhelming trust as an information source—82 percent both nationally and in Eurajoki.[45] However, Finns have far less trust less than most other Europeans in NGOs as source of information on nuclear matters.[46] Trust-building has benefited from the absence of reactor accidents and the excellent performance of the Finnish reactors.[47] The Chernobyl accident helped reinforce the perceptions of a sharp contrast between the supposedly safe and reliable Finnish technology management and the unreliable and reckless Soviet/Russian one.[48]

Earlier research has suggested that the smooth advancement of the siting process is due to a number of contextual factors other than local-level trust-building: public trust in the safety authority; high tolerance of nuclear power technology in the 9,000-inhabitant Eurajoki that owes its prosperity to nuclear industry;[49] and Posiva's success in portraying the project as yet another contribution by nuclear industry to local development.[50] In its partnerships with the nuclear industry, Eurajoki has been somewhat of a bystander, willing to fully delegate safety review to STUK and primarily tending to its economic interests.[51]

These features have been backed up by strong ideological trust in representative democracy, municipal autonomy, state bureaucracy, and legalism.[52] In such a context, anchoring the approval in a parliamentary decision was vital for the legitimacy of the project. Finns' strong trust in technology and science, rationality, and pragmatism have led some to portray Finland as an "engineering nation,"[53] with the "Finnish engineer" sometimes seen as almost a mythical figure.[54] The successful "Finlandization" of Russian reactor technology in Loviisa further buttressed this perception.[55] Skepticism toward participatory and deliberative approaches is still relatively widespread, among both civil servants and citizens.[56]

The Finnish case carries features of what could be called pragmatic trust: the repository project appears as inevitable, albeit an outcome of a legally correct process.

France

In line with the state-driven nuclear policy in France, in 1979 the government created the National Radioactive Waste Management Agency (Andra) to implement geological disposal. The relatively high trust in state institutions and engineering elites quickly deteriorated after the late 1980s, first in May 1986 with the discovery that the authorities had downplayed the true extent of the Chernobyl fallout,[57] and in 1987–1990, when Andra's site investigations triggered vehement local opposition. To reestablish trust, Prime Minister Michel Rocard declared, in February 1990, a moratorium on investigations. The landmark 1991 Waste Law[58] reopened the search in 1991 to include three different nuclear waste management options and opened the discussion to a wide range of actors.[59] It also introduced the concepts of reversible geological disposal and community benefit packages, and led to the establishment of external evaluating bodies, as well as local information and liaison committees.

Toward the late 1990s, local conflict arose again, in the context of declining public trust in the governance of risk[60] and following the government decision to locate an underground research laboratory in Bure, a small village in a remote, rural, sparsely populated, and socioeconomically declining region in the east of France. With the other sites eliminated one by one, Bure soon became the *de facto* candidate for hosting a repository.[61] This eroded local actors' (e.g., town and village mayors') trust in their own ability to influence the national state decisions. The fifteen-year period of opening up inaugurated in 1991 culminated in a mandatory public debate organized in 2005–2006 by the National Commission on Public Debate (CNDP). Even many opponents considered that the commission had succeeded in resisting pressures from vested interests, protecting its own integrity, and bringing new perspectives to the debate.[62] The chairperson of the committee responsible for the debate managed to earn (social) trust among the participants. However, trust in nuclear waste management institutions suffered as long-term subsurface storage—an option introduced by the public debate—was excluded from the subsequent parliamentary debate on the 2006 Waste Law.

The 2006 law confirmed reversible geological disposal as the reference option. In 2010, Andra submitted its proposal for the Cigéo[63] repository, designed to host the high- and medium-level waste from the country's fifty-eight operating reactors and situated in Bure but spanning the border between two departments (Meuse and Haute-Marne). Local actors widely mistrust Andra—the locally most visible representative of the state in the project. Some blame Andra's methods of land acquisition, whereas others feel betrayed by the absence of the promised socioeconomic development.[64] In 2013, the mandatory CNDP debate on Cigéo turned into a farce: obstructed by opponents, the public meetings were replaced by expert debates on the internet. An *ad hoc* consensus conference, held in early 2014, managed to repair some of the reputational damage to CNDP.[65] Andra plans to start the construction of its repository in 2022 and waste disposal in 2030, yet local protest against the repository project has intensified recently.

Legally-mandated economic support—€30 million per year for each of the two departments—has helped to win the trust of local politicians and citizens in the state's willingness to provide the promised socioeconomic benefits. However, critics have attacked the community benefit schemes for constituting illegitimate bribery, "buying a conscience," creating dependence, wasting public money on luxury projects instead of spurring endogenous socioeconomic development, and treating municipalities unequally.[66] The schemes have undermined trust in the sincerity of state nuclear waste management actors: critics widely suspect the state of implementing participatory governance for purely instrumental reasons, merely to gain acceptance.[67]

Mistrust has become entrenched throughout the long and conflict-ridden history of the French nuclear waste management policy, exemplified by local discourses of us against them (the state) and critique of the state, which has allegedly failed to deliver on its promises.[68] Mistrust relations are complex, reciprocal, and multidimensional, owing largely to the complex multilevel governance of the project. Mistrust extends to relations between the state representation (prefects) at the departmental level, the departmental authorities, Andra headquarters, and its local office in Bure.[69] The numerous evaluating and multi-stakeholder consultative bodies represent institutionalized mistrust, which, counterintuitively, helps to enhance trust in the process and in the waste management solution. However, this institutionalization also highlights the role of social trust: the presence of the same individuals in key positions within multiple organisms involved in nuclear waste management policy spurs suspicion of an "inner circle" driving the project.[70]

The mistrust is reciprocal: not only do locals mistrust national state actors, but the national-level project proponents often mistrust local mayors, labeling them egoists only interested in the community benefits.[71] Disputes over the distribution of benefits reflect mistrust between small communities near the site and larger municipalities farther afield. At the local level, mistrust coexists with relatively strong trust in the safety of the repository (78%),[72] in Andra as a source of information on the project (63%),[73] and in local mayors as the most trusted politicians, an overwhelming majority of whom supports the project.[74] At the national level, trust in the competence of safety authorities is high (76.5%),[75] but trust in their sincerity is relatively low (40% for the safety authority and 57% for its technical support organization).[76] The strong ideological trust in the state, seen as the only legitimate guardian of public interest,[77] buttresses institutional mistrust: Andra and the successive governments are accused of failing to (1) live up to the high standards expected from state actors, and (2) deliver the promised socioeconomic development. Arguably, without the strong ideological trust in the state, expectations would be lower, as would the mistrust resulting from the perception of "failed promises."

The French case could be described as one of resigned trust, characterized by ideological trust in the state; deep-seated reciprocal relations of institutional mistrust; resignation of local actors in the face of state decisions; and perception of the repository project as the only hope for an economically declining region.

Sweden

The Swedish approach to stakeholder engagement in nuclear waste management is held as a best-practice model throughout the world.[78] The 1977 Nuclear Stipulation Act obligated the nuclear operators to demonstrate an "absolutely safe" solution to the waste problem, as a precondition for new reactor licensing.

After a nine-month period of intensive research, the power companies came up with the Kärnbränslesäkerhet (KBS, or Nuclear Fuel Safety) method. In 1981, they created a joint company, today known as SKB, to develop and implement the solution.

In the early 1980s, SKB had to interrupt drillings in its fourteen test sites because of opposition mostly from local environmentalists. The local groups formed a national network called the Waste Chain, also involving academic geologists as counter-experts that contested the KBS design. In the early 1990s, SKB devised a new strategy based on voluntarism. It sent a letter to all 286 Swedish municipalities inquiring about their willingness to engage in site investigations and offering the option to withdraw at any time. Eight relatively poor northern municipalities, all without nuclear installations, volunteered. SKB chose two for pre-studies,[79] but both withdrew from the process following local referenda in 1995 and 1997.[80]

Left again without a willing host site, SKB changed strategy. Like its Finnish counterpart, it gave up the ambition of finding the geologically best site and sought instead sites with good enough host rock and favorable socioeconomic conditions (e.g., public attitudes, transport, and other infrastructure). It invited four nuclear municipalities for feasibility studies. Varberg rejected, Nyköping[81] did not make a formal decision, Östhammar said yes in just four months, whereas Oskarshamn gave its approval only after seventeen months of elaborate local deliberations. Already in 1992, when nominated by SKB as the preferred site for an encapsulation plant, Oskarshamn had dictated its own conditions for approval. The foundations for the "Oskarshamn model" of citizen involvement were thus laid.[82]

In 2002, the government approved SKB's choice of Oskarshamn and Östhammar. With the local populations largely in favor of the project, something of a beauty contest unfolded. Local politicians in both municipalities sought to convince SKB of their advantages as host sites but also initiated close cooperation. Oskarshamn further refined its model, setting up a highly elaborate municipal organization for participatory multistakeholder dialogue with SKB.[83] Further trust-building included participatory consultations as part of the EIA process; so-called dialogue projects initiated by the safety regulator;[84] substantial economic support to NGOs for communication and research (from the national waste management fund financed by the nuclear industry); and independent evaluation and societal debate organized by the National Council for Nuclear Waste.[85]

In 2009 SKB chose Östhammar as the repository host—citing geological criteria—while Oskarshamn would receive the encapsulation plant. Prior to the decision, the municipalities had negotiated with SKB a nearly €200 million benefit package ("value added program"), of which 25 percent would go to the repository host, and 75 percent to the "loser." From 2010 until the end of the repository construction, the program would finance projects in education, training, infrastructure, innovation, and business development.

In the licensing process, approval is required from both the Radiation Safety Authority and from the Land and Environment Court. The Swedish Environmental Code that underpins the court proceedings provides unique transparency, as demonstrated by the five-week court hearings in spring 2017 concerning SKB's construction license application. These included site visits and open dialogue among some twenty to thirty participants representing SKB, local and regional authorities, antinuclear NGOs, and academics critical of the disposal concept. On the final day of the hearing, all the participants thanked the court chair for a fair and impartial process.[86]

In January 2018, the safety authority approved the application while the Land and Environment Court required further specifications from SKB, notably concerning copper canisters. The court's decision was informed by criticism from corrosion researchers at the Stockholm Royal Institute of Technology. The municipalities postponed the planned local referenda on SKB's application. The government decision is still pending, in spring 2021. Unlike in Finland, the Swedish municipalities retain their right to withdraw until the end of the process. In principle, the government can override this veto in the name of national interest, but this is considered highly unlikely.[87]

Throughout the years, SKB has won the trust of local authorities and citizens, in respect to both the safety and the process, via negotiations and dialogue, in which the municipalities have been highly proactive.[88] Trust in the safety of the project is stronger than in Finland, both nationally (73% in short-term safety and 54% in long-term safety) and locally (86%).[89] Trust in SKB at the local level is high (82%), but trust in the regulator as source of information is lower (61%) in Östhammar than in the Finnish Eurajoki.[90]

The case illustrates the virtues of civic vigilance, via openness to external critique and counter-expertise, whose roots reach back to the 1970s.[91] Significant state support to environmental NGOs, municipalities, and academic research has laid the basis for critical analysis of the project and mistrust-building. NGOs and counter-experts have brought to the agenda deep boreholes disposal[92] as an alternative, while the National Council for Nuclear Waste has highlighted seismicity, financing, and long-term monitoring as weaknesses of SKB's concept.[93] Trust-building via dialogue and civic vigilance has been backed up by the Swedish context of solid ideological trust in representative democracy, municipal autonomy, and an independent court system.[94] The case could therefore be described as one of earning trust via constructive mistrust.

United Kingdom

In contrast with the three forerunner countries, and despite a series of attempts in 1982–1987 and 2008–2013, the UK has yet to find a willing host for its high-level waste. More often than reactor new-build, waste disposal projects in the UK have been recurrent targets for social contestation. Even in the immediate aftermath of Chernobyl, UK residents viewed radioactive waste as a greater risk than a Chernobyl-like reactor accident. The waste problem has for a long time held the UK public from supporting further construction of nuclear power.[95]

In 1976, the landmark "Flowers Report" by the Royal Commission on Environmental Pollution suggested making nuclear new-build conditional on a solution to the waste problem.[96] The report also called for the establishment of a nuclear waste planning and siting organization totally independent from industry.[97] In contrast with the latter recommendation, Nirex (the Nuclear Industry Radioactive Waste Executive), set up in 1982 to discuss and develop options, was made up of nuclear industry bodies. From 1987 onward, Nirex searched for a willing host for a repository, following a deficit-model approach, underpinned by the assumption that educating and persuading the ignorant public would suffice for winning acceptance. It attributed local opposition to NIMBYism and ignorance.[98] This led to a crisis of trust, which culminated in 1997 when, following a lengthy public inquiry, the government rejected Nirex's proposal for an underground rock characterization facility (RCF) at Sellafield.[99] Nirex's private-industry background further engendered mistrust and spurred uninvited forms of participation. The RCF debacle provided an opportunity for a fresh start and a participatory turn. The New Labour government emphasized community involvement[100] in a context of growing public mistrust of governance of science and technology, as people felt excluded from scientific and technical decisions significantly affecting their lives.[101] An Ipsos MORI poll from 2002 indeed illustrated the mistrust: only 22 percent of the citizens trusted government to be competent in nuclear waste management, while 51 percent thought it was excessively influenced by the industry.[102] As many as 75 percent trusted the environmental NGOs to tell the truth about nuclear waste management, while only 30 percent trusted the industry and 35 percent the government.[103]

A new independent advisory body, the Committee on Radioactive Waste Management (CoRWM), was entrusted with the task of inspiring public trust in the country's nuclear waste management policy.[104] It embarked in 2003–2006 on an unprecedented process of public and stakeholder engagement.[105] CoRWM was widely commended for its ability to earn trust via its plural composition, openness to public inputs, and analytical-deliberative approach.[106] None of the organizations involved or observing the process contested the final report's recommendations:[107] In light of current knowledge, geological disposal was preferred but should be supported by a robust program of research on interim storage, while volunteerism and broad participation should characterize the site selection.[108] A 2008 Eurobarometer survey revealed that only 8 percent of UK citizens would agree to let the government alone decide on nuclear waste management—the lowest figure in Europe, and well below the 21 percent in Finland and Sweden and 17 percent in France.[109] The positive CoRWM experience may in part explain this demand for direct participation in decision-making—rather than trusting in the government to make the right decisions, the citizens felt empowered to participate. However, the trust built via the CoRWM process was undermined in 2008, when the government used its conclusions to justify nuclear new-build—against CoRWM's explicit caveat that its conclusions concerned only legacy waste.[110]

A multistakeholder West Cumbria Managing Radioactive Waste Safely Partnership (2010–2013) put the volunteering approach to test. The partnership, set up by the three councils involved, was to organize stakeholder engagement to ensure that a wide range of community interests were involved in discussions. In January 2013, Cumbria County Council withdrew from the stepwise siting process. The lower-tier borough councils (Allerdale and Copeland) were willing to continue, but since an agreement from all three parties was required, the siting process stalled. Subsequent policy amendments, notably the introduction of the notion of Nationally Significant Infrastructure Projects (NSIPs), act to transfer power from local councils back toward the central government.[111]

The Cumbria area is neither nonnuclear like the French host region, nor is it a true nuclear community. The nearby Sellafield nuclear complex has over the years been subject to repeated health and environmental scandals, ever since the fire in 1957 at Windscale.[112] Considerable mistrust has built up over the years toward nuclear waste management policy institutions, as a result of broken promises, top-down technocratic decision-making, and perceived arrogance of the involved engineers and scientists. The historically established mistrust toward the government efforts also undermined the operation of the West Cumbria Partnership, pushing discussions to broader topics such as nuclear new-build, the types and scale of wastes being produced, and alternative waste management options.[113]

The UK case could be described as one of ambiguous mistrust. Citizens mistrusted Nirex, suspecting it of tending to its own rather than the public interest, yet the citizens did not trust the government to deliver on its promises either. Nirex's efforts to gain trust via better educating the presumably ignorant and incompetent public in fact engendered further mistrust. The case underscores the ambiguous relationships between a growing institutional mistrust of the Big Six energy companies,[114] the government's nuclear waste management policy,[115] as well as the long-standing ideological trust in both market-based energy policy solutions[116] and the community, for example, in the form of localism.[117] Despite this trust in the markets, as a private-industry-led organization Nirex was not seen as a legitimate and credible defender of the public interest, but instead triggered mistrust and uninvited forms of participation. The UK case also shows the heterogeneity of publics and ambiguities among the UK citizens torn between trust and mistrust of government scientists.[118] Table 6.3 summarizes the key forms of trust- and mistrust-building in the four countries described above..

Russia

The Russian nuclear waste management policy provides a useful contrast with the European cases, allowing us to highlight the importance of cultural-institutional context and to put into perspective the assumption of all-encompassing importance of trust.

A country that proudly claims to have been the first in the world to connect a nuclear reactor to the electricity grid (in Obninsk, in 1954), Russia has been late to start addressing nuclear waste management. Policies were first developed in the late Soviet and early post-Soviet period. In the mid-1990s, the Russian parliament attempted, in vain, to adopt a law on management of nuclear waste. The economic, political, and social consequences of the Soviet collapse also contributed to a protracted crisis in the nuclear sector. The turn of the century saw the end of political liberalization and a gradual reconsolidation of authoritarianism under Vladimir Putin. The simultaneous recovery of the economy and the nuclear industry translated into growing ambitions of nuclear development at home and abroad,[119] and brought nuclear waste issues onto the agenda.

An important case of protest against official policy followed the 2001 adoption of a law allowing spent nuclear fuel imports to Russia. This law generated exceptionally widespread opposition, not only from NGOs but also from politicians usually loyal to the power. Yet the state managed to neutralize opposition, and imports continue.[120] In 2011, Parliament adopted a law on the handling of radioactive waste and spent nuclear fuel, the outcome of over two decades of

Table 6.3. Summary of key forms of trust-building and mistrust-building in nuclear waste management (NWM)

	Finland	Sweden	France	UK
Municipal veto	Yes, until the Decision in Principle	Yes, until the end of the process	No	Uncertain
Participation, dialogue	EIA; public hearings; participation of social scientists in government NWM research program management	EIA hearings; multistakeholder dialogue projects; Land and Environment Court hearings	CNDP, CLIS	CoRWM; MRWS
Economic support	Tax benefits; modest "private" support agreement	No tax benefits; significant value-added program	Tax benefits; mandatory economic support; voluntary industry support	Promise of community benefit packages
Independent bodies of control and oversight	No	National Council for Nuclear Waste; support for counter-expertise	National Review Board; CLIS; HCTISN; counter-expertise organizations	CoRWM
Mistrust-building	Modest, limited public protest and criticism	MKG; other NGOs; government-supported counter-expertise	Vehement criticism and protest at the local level; active NGOs against the project	Local- and national-level NGO opposition; critique of policymaking processes

CNDP = Commission nationale du débat public
CLIS = Commission locale d'information et de surveillance
HCTISN = Haut Comité pour la transparence et l'information sur la sécurité nucléaire
MKG = Miljöorganisationernas kärnavfallsgranskning (the Swedish NGO Office for Nuclear Waste Review)
CoRWM = Committee on Radioactive Waste Management
MRWS = West Cumbria Managing Radioactive Waste Safely Partnership

debate among nuclear experts and environmental activists. The law established a new organization, the National Operator, which in October 2013 announced a list of thirty potential sites for geological repositories and interim storage facilities that were to become operational by 2025.

Russian society is characterized by low trust in political institutions: for example, in 2018 Russia had the lowest index of trust in institutions among twenty-eight countries included in the Edelman Trust Barometer.[121] In Russia, trust in Parliament and political parties has traditionally been low,[122] and trust in NGOs is particularly low.[123] This reflects the absence of civil society during the Soviet period as well as its weakness in the first post-Soviet decades. Since the late 2000s, Russian NGOs have faced an increasingly harsh climate: persecutions and prosecutions are frequent, and state actors recurrently portray NGOs as untrustworthy because of their foreign ties.[124] Trust in the institution of the presidency might constitute an exception to the generally low trust in political institutions throughout the entire post-Soviet period. However, this trust may be less institutional than interpersonal—in other words, people may trust their president but not the institution he represents.[125] If the trust toward the president of Russian Federation was low during Boris Yeltsin's presidency in the 1990s, it grew significantly during the Putin era in the 2000s and 2010s.

Weak institutional trust notwithstanding, polls indicate high trust in nuclear energy and nuclear safety. This seems surprising given the experience from the early 1990s, when the public learned about the severity of the nuclear accidents kept secret during Soviet time, most importantly the Kyshtym (1957) and Chernobyl (1986) disasters. In the early 1990s, only 14 percent of respondents favored the development of nuclear energy.[126] The mistrust was reciprocal: citizens mistrusted nuclear officials and experts, who in turn mistrusted the citizens, considering them irrational and overly emotional. Citizens' worries and discontent with the mismanagement of the Chernobyl disaster were dismissed as expressions of "radiophobia," an irrational fear of radiation.[127] More recent polls paint a very different picture of present-day Russia: in 2016, 58 percent of Russians viewed nuclear energy favorably, and 73 percent believed that modern nuclear power stations were safe.[128]

Little information is available on trust-building measures. In the mid-2000s, the reorganization of the Ministry of Atomic Energy into Rosatom, led by Sergei Kirienko, helped gain citizen trust in a Russian "nuclear renaissance."[129] Several changes that allow for some limited public input into nuclear waste management testify to a breakaway from the Soviet period. Public hearings have been introduced as part of the EIA procedures concerning repository projects. Several such hearings took place in the 2010s. These procedures nevertheless often appear as superficial adaptations to what remains a very closed and technocratic policymaking system. However, they do sometimes allow the public to express its dissent and safety concerns, and may eventually contribute to postponing or modifying new nuclear projects. This is what happened in December 2013, with a near-surface storage of low-level radioactive waste in Sosnovy Bor, the location of the Leningrad nuclear power plant, following rather contentious public hearings on the EIA report.[130] The project in its proposed form was abandoned, albeit officially for economic reasons.[131]

Most often, however, public hearings serve to stage local citizen support for nuclear projects. This results not only from deliberate attempts by the organizers to limit participation to loyal employees of local nuclear enterprises,[132] but it also follows from specificities of the territorial organization of nuclear infrastructure. For instance, civilian and military nuclear installations are often concentrated in the same area. Many of these areas are classified as the so-called closed military cities, or closed administrative-territorial formations (Zakrytoe Administrativno-Territorial'noe Obrazovanie, or ZATO), which were secret during the Soviet period and are still accessible only with special permission. Since the Soviet period, residents of these mono-industrial cities have enjoyed privileged access to social welfare and material goods, as well as professional and social prestige. These cities depend heavily on nuclear activities, whose further development the residents strongly support. These tightly knit and closed communities exhibit high levels of particularized social trust between the local residents, authorities, and nuclear experts. In addition to ZATOs, many nonclosed mono-industrial cities are connected with the nuclear industry and share similar features of strong local identity and high dependence on the nuclear industry.[133]

Among the thirty potential waste sites announced in 2013, several were in closed cities. This was the case of the ZATO Zheleznogorsk in Krasnoyarsk region in Siberia, foreseen as the country's only site for an underground research laboratory for deep geological disposal of high-level waste. Only residents from the closed city could participate in the July 2012 public hearings, and background documents and other materials were available only onsite.[134] Unsurprisingly, of the 276 participants, 179 voted to support the project and only 12 against it. In a second round of hearings, in July 2015, citizens and activists from outside the city and other parts of Russia were allowed to participate, and the documentation was made available on the internet. Yet supporters still constituted the vast majority: 254 voted for and 49 against.[135]

It is difficult to say whether these manifestations of support reflect trust in the industry's and state's nuclear waste management policy, or the very limited possibilities for public expression of mistrust, which prevails especially outside the cities hosting nuclear installations. The Russian case is an example of resigned trust originating partly from dependence on nuclear enterprises and the strong social and economic ties in the already nuclearized communities chosen as potential hosts for future waste sites. In a political climate that severely limits public expressions of dissent, this resignation also reflects the belief that viable or desirable alternatives do not exist and mistrust of NGOs and activists that defend such alternatives. What sets the Russian case apart is the seeming disconnection between the very low institutional trust on one hand, and the strong trust in the nuclear industry and safety of nuclear installations on the other. Nuclear-sector authorities are just as mistrusted as other government institutions in Russia, yet this does not seem to prevent citizens from strongly supporting nuclear power. We argue that the strong ideological trust in nuclear energy and nuclear industry as a representative and a manifestation of the greatness of the nation can help explain this apparent paradox. Institutional trust therefore appears as rather irrelevant for citizen support to specific nuclear technologies and projects.

Indeed, since the beginning of the atomic era, Soviet nuclear research and industry achievements were promoted as proof of Soviet technical, political, and economic grandeur.[136] After the dissolution of the Soviet Union, and recovery from the economic and political crises following the collapse, the Russian government proposed a series of nuclear energy development programs that foresaw rapid expansion of the industry. These programs conveyed the image of Russia's technological greatness, including its international influence through an aggressive policy of exporting nuclear technologies. Since the 2000s, the Kremlin has promoted this image as an aspect of state nationalism, which embraces a glorious Soviet past and a vision of the present and future—that of a modern and technologically advanced Russia.[137]

Conclusions

Our cases show the importance of historical legacies—negative in France and the UK, and positive in Finland and Sweden. In the former, the benefits of exemplary participatory processes were undermined by long-established institutional mistrust emerging from perceived repeated betrayals by the key nuclear waste management actors, past technocratic policy approaches, and events such as the "Chernobyl cloud affair"[138] in France. In the Nordic cases, the long-standing ideological and institutional trust in public- and private-sector actors, together with lack of accidents, form the basis for successful trust-building in nuclear waste policy from the 1990s and onwards. In Russia, the strong ideological trust in nuclear power can be traced back to even longer-term dynamics of ideological trust in the country's "greatness" and in nuclear energy as its concrete incarnation.

The cases also illustrate the reciprocity and self-reinforcing nature of trust and mistrust.[139] The French and UK cases illustrate the reciprocity of mistrust between locals and state actors, as the national-level project proponents labeled citizens and local mayors as either ignorant or as egoists only interested in the economic benefits. In the Finnish, Swedish, and Russian nuclear communities, the nuclear industry gained citizen trust as a reliable employer and safe operator of nuclear installations. In the two Nordic countries, a virtuous circle of increasing trust seemed to set in, spurred by multistakeholder dialogue, long-term planning, and economic prosperity that provided stability and enhanced trust. In Sweden, state support for counter-expertise illustrates that trust could be built by showing trust toward those critical of the project.

Our cases also confirm the notion that trust is hard to gain but easy to lose; negative events generate mistrust more easily than positive events strengthen trust.[140] The Chernobyl cloud affair in France and the use of CoRWM recommendations to justify nuclear new-build in the UK quickly fed mistrust, which the nuclear waste management actors have struggled with ever since. Obviously, the French Chernobyl example also shows the importance of events external to the specific project and nuclear waste actors in question. The Russian case again is an exception in that the profound mistrust triggered by the revelation of the Chernobyl events was either short-lived or simply failed to affect trust levels in the nuclear communities.

Success in building institutional trust crucially depends on how it links with social and ideological dimensions of trust. Strong particularized social trust and internal cohesion among the nuclear waste management actors fostered institutional trust in Finland and Sweden. Similar cohesion within the French nucleocracy decried by critics fed "us vs. them" perceptions and mistrust of the state among the local population. The role of ideological trust, in turn, is clearly visible in how strong trust in nuclear waste institutions, and the consensus-seeking regulatory styles in Finland and Sweden, are backed up by solid ideological trust in national- and local-level representative democracy and in the legitimacy of state bureaucracy and local authorities in collaboratively defining and defending the public interest.[141] The UK has a tradition of a more adversarial and mistrust-based style,[142] whereas in France, trust-based collaboration among an inner circle of experts coexists alongside adversarial relations between the state and the civil society.[143] In these two countries, mistrust toward the local authorities is widespread among state actors and partly institutionalized in legislation, as exemplified by the recent planning legislation reforms in the UK. The French example also reveals ambiguities between strong ideological trust in the state as the only legitimate defender of the public interest, and an equally strong and reciprocal institutional mistrust between the state and civil society actors. Local actors mistrust the state institutions (especially Andra) precisely because they have in the past failed to respect the norms of French republicanism. In Russia, the strong ideological trust seems to partly override the influence of institutional mistrust, although the seeming irrelevance of institutional trust partly follows from the planning and decision-making arrangements that provide few opportunities to citizens outside the closed nuclear communities to participate.

In the Nordic context of strong institutional and ideological trust, more direct forms of citizen engagement and civic vigilance can appear unnecessary and even doubtful to many.[144] However, the Finnish and Swedish cases are far from identical in this respect. The Swedish model carries traits of a mistrust-based regulatory style—which is arguably making inroads into Nordic administration more broadly[145] but does not yet seem to have affected the Finnish nuclear waste management policy. In Sweden, trust has been built via long processes of dialogue, counter-expertise, and open exploration of potential weaknesses of the technical solution—that is, via dynamic interaction between trust and mistrust. The Finnish example evokes the danger of institutional overtrust and reveals great deference to authorities (especially the safety authority), the rule of law, and project engineers, and a corresponding mistrust of environmental NGOs. These trust and mistrust features are institutionalized to the extent that administrative decisions alone suffice for construction and operation licenses, funding is not provided for NGOs, and counter-expertise tradition is absent.

Last, trust-building is inevitably shaped by contextual factors, notably the host community's familiarity with nuclear industry and spillover effects from nuclear energy policy. Earning trust has been difficult for the French nuclear waste management actors in a region lacking nuclear experience. This contrasts with the relative ease of trust-building in the Nordic nuclear communities, and with the redundancy of trust-building in the Russian nuclear communities under an authoritarian regime. The UK case shows that negative prior experience of nuclear installations can undermine trust. The UK nuclear waste management actors quickly lost the incipient trust gained via the participatory CoRWM process once the government used the committee's recommendations to justify nuclear new-build. In Finland, by contrast, the parliamentary approval for a repository in 2001 was widely interpreted as a valid justification for a new reactor. In France, mistrust was reinforced by uncertainties related to the waste inventory—which crucially depends on decisions concerning reprocessing and the future of the country's nuclear program. In Sweden, the long-term plans for nuclear phaseout have facilitated SKB's efforts to earn trust.

Our examples demonstrate how the outcomes of nuclear waste policies in our five case study countries have been shaped by enduring and context-specific historical legacies. These legacies are constituted over time, via dynamic interaction between varying dimensions of trust and mistrust. As such, they accentuate the old yet ever-relevant lesson of trust being hard to gain but easy to lose.

Notes

1. We employ the term *nuclear waste management* as a shorthand for the variation across countries in their respective choices in dealing with spent nuclear fuel and radioactive materials (e.g., reprocessing or direct disposal of spent fuel; disposal of high-level or medium-level radioactive waste). Our choice is also in line with the book's overall focus of on nuclear energy in exclusion of other (e.g., medical) uses of radioactive materials.
2. Mark Elam and Göran Sundqvist, "Public Involvement Designed to Circumvent Public Concern? The 'Participatory Turn' in European Nuclear Activities," *Risk, Hazards, and Crisis in Public Policy* 1, no. 4 (2010): 203–29, doi:10.2202/1944-4079.1046.
3. Francis Chateauraynaud and Didier Torny, *Les sombres précurseurs: une sociologie pragmatique de l'alerte et du risque* (Paris: Editions EHESS, 1999); K. Bickerstaff et al., "Reframing Nuclear Power in the UK Energy Debate: Nuclear Power, Climate Change Mitigation and Radioactive Waste," *Public Understanding of Science* 17, no. 2 (2008): 145–69.
4. Lucie Laurian, "Trust in Planning: Theoretical and Practical Considerations for Participatory and Deliberative Planning," *Planning Theory and Practice* 10, no. 3 (2009): 369–91.
5. E.g., Matthew Cotton, "Industry and Stakeholder Perspectives on the Social and Ethical Aspects of Radioactive Waste Management Options," *Journal of Transdisciplinary Environmental Studies* 11, no. 1 (2012): 17–35. Matthew Cotton, *Nuclear Waste Politics: An Incrementalist Perspective* (Abingdon: Routledge, 2017); Markku Lehtonen, "Deliberative Decision-Making on Radioactive Waste Management in Finland, France and the UK: Influence of Mixed Forms of Deliberation in the Macro Discursive Context," *Journal of Integrative Environmental Sciences* 7, no. 3 (2010): 175–96; Markku Lehtonen, "Opening Up or Closing Down Radioactive Waste Management Policy? Debates on Reversibility and Retrievability in Finland, France, and the United Kingdom," *Risk, Hazards, and Crisis in Public Policy* 1, no. 4 (2010): 139–79; Markku Lehtonen, "Megaproject Underway: Governance of Nuclear Waste Management in France," in *Governance of Nuclear Waste Management: An International Comparison*, ed. Achim Brunnengräber, Maria Rosaria Di Nucci, Ana María Isidoro Losada, Lutz Mez, and Miranda Schreurs (Wiesbaden: Springer, 2015), 117–38; Markku Lehtonen, "France Short Country

Report (version 2019)," in *History of Nuclear Energy and Society (HoNESt) Consortium Deliverable no. 3.6* (2019), https://hdl.handle.net/2454/38269; Markku Lehtonen, Matti Kojo, and Tapio Litmanen, "The Finnish Success Story in the Governance of a Megaproject: The (Minimal) Role of Socioeconomic Evaluation in the Final Disposal of Spent Nuclear Fuel," in *Socioeconomic Evaluation of Megaprojects: Dealing with Uncertainties*, ed. Markku Lehtonen, Pierre-Benoit Joly, and Luis Aparicio (London: Routledge, 2017), 83–110; Markku Lehtonen and Matti Kojo, "The Role and Functions of Community Benefit Schemes: A Comparison of the Finnish and French Nuclear Waste Disposal Projects," in *Conflicts, Participation and Acceptability in Nuclear Waste Governance: An International Comparison Volume III*, ed. Achim Brunnengräber and Maria Rosaria Di Nucci (New York: Springer VS, 2019), 175–205; Aarne Kaijser and Per Högselius, *Resurs eller avfall? Internationella beslutsprocesser kring använt kärnbränsle, SKB Rapport R-07-37* (Stockholm: SKB, 2007); Tatiana Kasperski, *Les politiques de la radioactivité: Tchernobyl et la mémoire nationale en Biélorussie contemporaine* (Paris: Pétra, 2020); Tatiana Kasperski, "Nuclear Dreams and Realities in Contemporary Russia and Ukraine," in *History and Technology* 31, no. 1 (2015): 55–80; Tatiana Kasperski, "Une transition vers plus de nucléaire? Analyse comparée des politiques énergétiques russe et ukrainienne," *Revue internationale de politique comparée* 24, no. 1 (2017): 101–25; Tatiana Kasperski, "From Legacy to Heritage: The Changing Political and Symbolic Status of Military Nuclear Waste in Russia," *Cahiers du Monde Russe* 60, no. 2–3 (2019): 517–38.

6. Malcolm Tait, "Trust and the Public Interest in the Micropolitics of Planning Practice," *Journal of Planning Education and Research* 31, no. 2 (2011): 157–71; Caterina Galluccio, "Trust in the Market: Institutions versus Social Capital," *Open Journal of Political Science* 8 (2018): 95–107, https://doi.org/10.4236/ojps.2018.82008; Paul J. Zak and Stephen Knack, "Trust and Growth," *Economic Journal* 111, no. 470 (2001): 295–321; Éloi Laurent, "Peut-on se fier à la confiance?," *Revue de l'OFCE* 1, no. 108 (2009): 14; Stefano Carattini, Andrea Baranzini, and Jordi Roca, "Unconventional Determinants of Greenhouse Gas Emissions: The Role of Trust," *Environmental Policy and Governance* 25, no. 4 (2015): 243–57.

7. For the sake of simplicity, we use the term *trust* to encompass both its traditional meaning as a normative judgment concerning an individual or entity, and the meaning of *confidence*, that is, a belief based on earlier experience that certain events will occur as predicted. Timothy E. Earle and Michael Siegrist, "Morality Information, Performance Information, and the Distinction Between Trust and Confidence," *Journal of Applied Social Psychology* 36, no. 2 (2006): 383–416; Niklas Luhmann, *La confiance: un mécanisme de réduction de la complexité sociale* (Paris: Economica, 2006); William J. Kinsella, "A Question of Confidence: Nuclear Waste and Public Trust in the United States After Fukushima," in *The Fukushima Effect: A New Geopolitical Terrain*, ed. Richard Hindmarsh and Rebecca Priestly (London: Routledge, 2016), 223–46.

8. Earle and Siegrist, "Morality Information, Performance Information, and the Distinction Between Trust and Confidence." Following Geoffrey M. Hodgson, we define institutions broadly as "systems of established and embedded social rules that structure social interactions." Organizations, in turn, are a specific type of institution. Geoffrey M. Hodgson, "What Are Institutions?," *Journal of Economic Issues* XL, no. 1 (2006): 18.

9. Josep Espluga, Ana Prades, Nuria Gamero, and Rosario Solà, "El papel de la 'confianza' en los conflictos socioambientales," *Política y Sociedad* 46, no. 1–2 (2009): 255–73.
10. Bo Rothstein and Dietlind Stolle, "The State and Social Capital: An Institutional Theory of Generalized Trust," *Comparative Politics* 40, no.4 (2008): 441–59.
11. Maria Bäck and Henrik S. Christensen, "When Trust Matters—A Multilevel Analysis of the Effect of Generalized Trust on Political Participation in 25 European Democracies," *Journal of Civil Society* 12, no. 2 (2016): 180.
12. Elina Kestilä-Kekkonen and Peter Söderlund, "Political Trust, Individual-level Characteristics and Institutional Performance: Evidence from Finland, 2004–13," *Scandinavian Political Studies* 39, no. 2 (2016): 141.
13. Sören Holmberg and Lennart Weibull, "Långsiktiga förändringar i svenskt institutionsförtroende," in *Larmar och gör sig till – SOM undersökningen 2016*, ed. Ulrika Andersson, Jonas Ohlsson, Henrik Oscarsson, and Maria Oskarson (Göteborg: SOM rapport 70, 2017), 39–57; Laurian, "Trust in Planning," 383–84; Seth P. Tuler and Roger E. Kasperson, *Social Distrust: Implications and Recommendation for Spent Nuclear Fuel and High Level Radioactive Waste Management: A Technical Report Prepared for the Blue Ribbon Commission on America's Nuclear Future* (Washington, DC: Blue Ribbon Commission on America's Nuclear Future, 29 January 2010).
14. Andrew Gouldson, Rolf Lidskog, and Misse Wester-Herber, "The Battle for Hearts and Minds? Evolutions in Corporate Approaches to Environmental Risk Communication," *Environment and Planning C: Government and Policy* 25, no. 1 (2007): 56–72. Laurian, "Trust in Planning."
15. Tait, "Trust and the Public Interest," 158.
16. Peter Söderbaum, "Values, Ideology and Politics in Ecological Economics," *Ecological Economics* 28, no. 2 (1999): 161–70; Peter Söderbaum, "Ecological Economics in Relation to Democracy, Ideology and Politics," *Ecological Economics* 95 (2013): 223.
17. Tait, "Trust and the Public Interest," 160; Niklas Luhmann, *Trust and Power* (New York: Wiley, 1979).
18. Söderbaum, "Values," 163.
19. Tait, "Trust and the Public Interest," 160.
20. J. David Lewis and Andrew Weigert, "Trust as Social Reality," *Social Forces* 63, no. 4 (1985): 973.
21. Sanjay Goel, Geoffrey G. Bell, and Jon L. Pierce, "The Perils of Pollyanna: Development of the Over-Trust Construct," *Journal of Business Ethics* 58, no. 1–3 (2005): 203.
22. Minna Santaoja, Markus Laine, and Helena Leino, " 'Joku palikka siitä puuttuu'—luottamuksen rakentuminen täydennysrakentamisen suunnittelussa," *Yhdyskuntasuunnittelu* 54, no. 1 (2016), http://www.yss.fi/journal/joku-palikka-siita-puuttuu-luottamuksen-rakentuminen-taydennysrakentamisen-suunnittelussa/; Robert D. Putnam, *Bowling Alone: The Collapse and Revival of American Community* (New York: Simon and Schuster, 2000), 19–21.
23. Johanna Kujala, Hanna Lehtimäki, and Raminta Pučėtaitė, "Trust and Distrust Constructing Unity and Fragmentation of Organizational Culture," *Journal of Business Ethics* 139, no. 4 (2016): 702.
24. Jan W. van Deth and Sonja Zmerli, "Introduction: Civicness, Equality, and

Democracy—A 'Dark Side' of Social Capital?" *American Behavioral Scientist* 53, no. 5 (2010): 631–39.
25. John R. Parkins and Ross E. Mitchell, "Public Participation as Public Debate: A Deliberative Turn in Natural Resource Management," *Society and Natural Resources* 18, no. 6 (2005): 536.
26. Laurian, "Trust in Planning."
27. Mark E. Warren, "Democratic Theory and Trust," in *Democracy and Trust* (Cambridge: Cambridge University Press,1999), 310; Laurent, "Peut-on se fier," 27; Olivier Allard, Matthew Carey, and Rachel Renault, "De l'art de se méfier," *Tracés* 31, no. 2 (2016): 14.
28. van Deth and Zmerli, "Introduction," 665.
29. E.g., Mark Elam, Linda Soneryd, and Göran Sundqvist, "Demonstrating Safety—Validating New Build: The Enduring Template of Swedish Nuclear Waste Management," *Journal of Integrative Environmental Sciences* 3 (2010): 197–210.
30. OECD, *Government at a Glance* (Paris: OECD, 2013), 30; Jan Delhey, Kenneth Newton, and Christian Welzel, "How General Is Trust in 'Most People'? Solving the Radius of Trust Problem," *American Sociological Review* 76, no. 5 (2011): 786–807.
31. "Edelman Trust Barometer: 2018 Global Report," https://www.edelman.com/sites/g/files/aatuss191/files/2018-10/2018_Edelman_Trust_Barometer_Global_Report_FEB.pdf.
32. E.g., OECD-NEA, "Stepwise Decision Making for the Disposal of Spent Nuclear Fuel in Finland," Workshop Proceedings, Turku, Finland, 14–16 November 2001 (Paris: OECD/NEA, 2002).
33. E.g., Juhani Vira, "Winning Citizen Trust: The Siting of a Nuclear Waste Facility in Eurajoki, Finland," *Innovations* (Fall 2006): 65–80; Matti Kojo, "The Strategy of Site Selection for the Spent Nuclear Fuel Repository in Finland," in *The Renewal of Nuclear Power in Finland*, ed. Matti Kojo and Tapio Litmanen (London: Palgrave Macmillan, 2009), 161–91.
34. Kojo, "Strategy of Site Selection."
35. Kojo, "Strategy of Site Selection."
36. Seppo Vuori, "Research and Development Activities Related to Management and Disposal of Spent Nuclear Fuel in 2001–2013," *VTT Technology* 190 (2014), http://www.vtt.fi/inf/pdf/technology/2014/T190.pdf.
37. Matti Kojo, "Changing Approach: Local Participation as a Part of the Site Selection Process of the Final Disposal Facility for High-Level Nuclear Waste in Finland" (The 10th International Conference on Environmental Remediation and Radioactive Waste Management—ICEM'05, Glasgow, Scotland, ICEM05-1239, September 4–8, 2005).
38. E.g., Antti Leskinen and Markku Turtiainen, *Interactive Planning in the EIA of the Final Disposal Facility for Spent Nuclear Fuel in Finland*, Posiva Working Report 2002-45 (Olkiluoto: Posiva, 2002); Pekka Hokkanen, *Kansalaisosallistuminen ympäristövaikutusten arviointimenettelyssä*, Acta Electronica Universitatis Tamperensis 683 (Tampere: Tampere University Press, 2008), http://urn.fi/urn:isbn:978-951-44-7178-0.
39. E.g., Hokkanen, *Kansalaisosallistuminen*, 215.
40. Kojo, "Strategy of Site Selection," 179–85.
41. E.g., Hokkanen, *Kansalaisosallistuminen*; Hannah Strauss, "Involving the Finnish

Public in Nuclear Facility Licensing: Participatory Democracy and Industrial Bias," *Journal of Integrative Environmental Sciences* 7, no. 3 (2010): 211–28.

42. Thomas Rosenberg, "What Could Have Been Done? Reflections on the Radwaste Battle, as Seen from Below" (Presented at the European Nuclear Critical Conference 2007, Helsinki, November 9–11, 2007), accessed 21 March 2021, http://sydaby.eget.net/kil/rosenberg1.htm.
43. Mika Kari, Matti Kojo, and Tapio Litmanen, *Community Divided: Adaptation and Aversion towards the Spent Nuclear Fuel Repository in Eurajoki and Its Neighbouring Municipalities* (Jyväskylä and Tampere: University of Jyväskylä and University of Tampere, 2010). http://urn.fi/URN:ISBN:978-951-39-4149-9; Tuuli Vilhunen, Matti Kojo, Tapio Litmanen, and Behnam Taebi, "Perceptions of Justice Influencing Community Acceptance of Spent Nuclear Fuel Disposal: A Case Study in Two Finnish Nuclear Communities," *Journal of Risk Research* (2019), doi:10.1080/13669877.2019.1569094.
44. Vilhunen et al., "Perceptions of Justice Influencing Community Acceptance."
45. Vilhunen et al., "Perceptions of Justice Influencing Community Acceptance"; Energiateollisuus, *Suomalaisten Energia-asenteet 2018*, https://energia.fi/files/3278/Energia-asenne_2018_MATERIAALIPANKKIKUVAT.pdf.
46. In a 2006 Eurobarometer study, only 17% of Finns considered NGOs as the most trustworthy source of information on nuclear safety—in stark contrast with 42% in France and 36% in the UK. *Eurobarometer 2006: Europeans and Nuclear Safety. Special Eurobarometer 271* (European Commission: February 2006), http://ec.europa.eu/public_opinion/archives/ebs/ebs_271_en.pdf.
47. Measured by performance indicators such as Lifetime Energy Availability Factor, Lifetime Unit Capability Factor, and Lifetime Unplanned Capability Loss Factor, the Finnish reactors consistently rank as among the best in the world. IAEA, *Operating Experience With Nuclear Power Stations in Member States, 2017 Edition* (Vienna: IAEA, Vienna, 2017), https://www-pub.iaea.org/books/IAEABooks/12246/Operating-Experience-with-Nuclear-Power-Stations-in-Member-States-in-2016.
48. Tapio Litmanen and Matti Kojo, "Not Excluding Nuclear Power: The Dynamics and Stability of Nuclear Power Policy Arrangements in Finland," *Journal of Integrative Environmental Sciences* 8, no. 3 (2011): 181.
49. Tapio Litmanen, Mika Kari, Matti Kojo, and Barry D. Solomon, "Is There a Nordic Model of Final Disposal of Spent Nuclear Fuel? Governance Insights from Finland and Sweden," *Energy Research and Social Science* 25 (2017): 19–30; Kojo 2009, "Strategy of Site Selection"; Lehtonen, "Opening Up"; Strauss, "Involving the Finnish Public."
50. Matti Kojo, "Lahjomattomien haukansilmien valvonnassa. Ydinjätteen loppusijoitushankkeen hyväksyttävyyden rakentaminen Posiva Oy:n tiedotusmateriaalissa," in *Ydinjäteihme suomalaisittain*, ed. Pentti Raittila, Pekka Hokkanen, Matti Kojo, and Tapio Litmanen (Tampere: Tampere University Press, 2002), 36–66; Sari Yli-Kauhaluoma and Harri Hänninen, "Tale Taming Radioactive Fears: Linking Nuclear Waste Disposal to the 'Continuum of the Good,'" *Public Understanding of Science* 23, no. 3 (2014): 328.
51. Mika Kari, Matti Kojo, and Markku Lehtonen, "Role of the Host Communities in Final Disposal of Spent Nuclear Fuel in Finland and Sweden," *Progress in Nuclear Energy* (March 2021), https://doi.org/10.1016/j.pnucene.2021.103632.

52. E.g., S. Puustinen, *Suomalainen kaavoittajaprofessio ja suunnittelun kommunikatiivinen käänne. Vuorovaikutuksen liittyvät ongelmat ja mahdollisuudet suurten kaupunkien kaavoittajien näkökulmasta*, Yhdyskuntasuunnittelun tutkimusja koulutuskeskuksen julkaisuja A34 (Espoo: Teknillinen korkeakoulu, 2006); Sari Puustinen, Raine Mäntysalo, Jonne Hytönen, and Karoliina Jarenko, "The 'Deliberative Bureaucrat': Deliberative Democracy and Institutional Trust in the Jurisdiction of the Finnish Planner," *Planning Theory and Practice* 18, no. 1 (2017): 71–88; Tapio Litmanen, "The Temporary Nature of Societal Risk Evaluation: Understanding the Finnish Nuclear Decisions," in *The Renewal of Nuclear Power in Finland*, ed. Matti Kojo and Tapio Litmanen (London: Palgrave Macmillan, 2009), 192–217.
53. An expression used by an interviewed energy industry representative in June 2016.
54. Harri Lammi, "Social Dynamics Behind the Changes in the NGOs Anti-Nuclear Campaign, 1993–2002," in Kojo and Litmanen, *Renewal of Nuclear Power Policy*.
55. Karl-Erik Michelsen and Aisulu Harjula, "Finland Short Country Report (version 2017)," in *History of Nuclear Energy and Society (HoNESt) Consortium Deliverable no. 3.6* (2019), https://hdl.handle.net/2454/38269.
56. E.g., Mikko Rask, "The Problem of Citizens' Participation in Finnish Biotechnology Policy," *Science and Public Policy* 30, no. 6 (2003): 441–54; Markku Lehtonen and Laurence De Carlo, "Diffuse Institutional Trust and Specific Institutional Mistrust in Nordic Participatory Planning: Experience from Contested Urban Projects," *Planning Theory and Practice* 20, no. 2 (2019): 203–20.
57. Lehtonen, "France Short Country Report."
58. The law defines reversibility as the "ability of successive generations to either continue the construction and then exploitation of successive phases of the disposal project or reexamine the earlier choices and modify the management solution." Assemblée nationale, "Proposition de loi précisant les modalités de création d'une installation de stockage réversible en couche géologique profonde des déchets radioactifs de haute et moyenne activité à vie longue," texte adopté no. 789, "Petite loi" (Paris: National Assembly, July 11, 2016), http://www2.assemblee-nationale.fr/documents/notice/14/ta/ta0789/(index)/ta.
59. Yannick Barthe, *Le pouvoir d'indécision: La mise en politique des déchets nucléaires* (Paris: Economica, 2006); Yannick Barthe, "Les qualités politiques des technologies: Irréversibilité et réversibilité dans la gestion des déchets nucléaires," *Tracés—Revue de Sciences humaines* 16, no. 1 (2009): 119–37; Céline Parotte, "L'art de gouverner les déchets radioactifs: Analyse comparée de la Belgique, la France et le Canada" (PhD diss., Université de Liège, 2016).
60. Chateauraynaud and Torny, *Les sombres précurseurs*.
61. Andrew Blowers, *The Legacy of Nuclear Power* (London: Routledge 2016).
62. Global Chance, "Débattre publiquement du nucléaire? Un premier bilan des deux débats EPR et déchets organisés par la Commission nationale du débat public," *Les cahiers de Global Chance* 22 (Paris: Global Chance, 2006), 64, http://www.global-chance.org/IMG/pdf/GC22.pdf.
63. Cigéo is short for Centre industriel de stockage géologique.
64. E.g., Markku Lehtonen and Matti Kojo, "The Role and Function of Community Benefit Schemes: A Comparison of the Finnish and French Nuclear Waste Disposal Projects," in *Conflicts, Participation and Acceptability in Nuclear Waste Governance*,

ed. Achim Brunnengräber and Maria Rosaria Di Nucci (Wiedsbaden: Springer Nature, 2019), 175–205.

65. Julie Blanck, "Gouverner par le temps: la gestion des déchets radioactifs en France, entre changements organisationnels et construction de solutions techniques irréversibles (1950–2014)" (PhD diss., Institut d'études politiques & Centre de sociologie des organisations, Paris, 2017), 459–60, https://hal.archives-ouvertes.fr/tel-01917434/document.
66. E.g., Élise Descamps, "A Bure, la manne controversée du stockage nucléaire," *La Croix*, April 27, 2011, https://www.la-croix.com/Actualite/Economie-Entreprises/Economie/A-Bure-la-manne-controversee-du-stockage-nucleaire-_NP_-2011-04-27-595766; Olivier Descamps, "Cigéo: le gouvernement a-t-il voulu acheter les consciences?" *Journal de l'environnement,* December 7, 2017, http://www.journaldelenvironnement.net/article/cigeo-le-gouvernement-a-t-il-voulu-acheter-les-consciences,88675; Jade Lindgaard, "Déchets radioactifs contre argent frais: l'équation de Bure," *Mediapart*, May 23, 2013; Gaspard D'Allens and Andrea Fuori, *Bure, la bataille du nucléaire* (Paris: Seuil and Reporterre, 2017).
67. E.g., Blowers, *Legacy*; Blanck, "Gouverner par le temps," 457–58.
68. Lehtonen and Kojo, "The Role and Function."
69. Lehtonen and Kojo, "The Role and Function."
70. Lehtonen and Kojo, "The Role and Function"; Markku Lehtonen, Matti Kojo, Tuija Jartti, Tapio Litmanen, and Mika Kari, "The Roles of the State and the Social Licence to Operate? Lessons from Nuclear Waste Management in Finland, France, and Sweden," *Energy Research and Social Science* 61 (March 2013), https://doi.org/10.1016/j.erss.2019.101353.
71. Lehtonen and Kojo, "The Role and Function."
72. *Ifop, Enquêtes locales auprès des riverains Principaux enseignements : Ifop pour l'Andra* (Centre de Meuse/Haute-Marne, March 2016), 6.
73. Trust in Andra as source of information concerning the project: The degree of trust reaches 67% in the communes nearest (< 15 km) the planned repository. Ifop, *Enquêtes locales auprès des riverains Principaux enseignements*.
74. L. Rouban, "La démocratie représentative est-elle en crise?" (Paris: La Documentation française, 2018).
75. Trust in the competence of regulatory authorities. *Baromètre IRSN: La perception des risques et de la sécurité par les Français* (Fontenay-aux-Roses: IRSN—Institut de radioprotection et de sûreté nucléaire, 2017), 129.
76. Trust that these organizations tell the truth about the nuclear issues. *Baromètre IRSN*, 129.
77. Sabine Saurugger, "Democratic 'Misfit'? Conceptions of Civil Society Participation in France and the European Union," *Political Studies* 55 no. 2 (2007): 384–404.
78. Mark Elam and Göran Sundqvist, "Meddling in Swedish Success in Nuclear Waste Management," *Environmental Politics* 20, no. 2 (2011): 246–63; Cotton, *Nuclear Waste Politics*, 17.
79. The municipalities of Storuman and Malå.
80. Arne Kaijser, "Sweden Short Country Report (version 2018)," in *History of Nuclear Energy and Society (HoNESt) Consortium Deliverable no. 3.6* (2019), https://hdl.handle.net/2454/38269.
81. Host for an experimental research reactor shut down in 2005.
82. Mark Elam and Göran Sundqvist, *Stakeholder Involvement in Swedish Nuclear Waste*

Management (Report published by Swedish Nuclear Power Inspectorate/Statens Kärnkraftsinspektion, 2007).

83. Each group had its own topic area: safety and geoscience, technology, environment, social science, encapsulation, and local information. Each consisted of two council members of the municipal council, one civil servant, two local citizens, and one external expert. Harald Åhagen, Torsten Carlsson, Krister Hallberg, and Kjell Andersson, "The Oskarshamn Model for Public Involvement in the Siting of Nuclear Facilities" (*Proceedings of VALDOR: Values in Decisions on Risk* Symposium in the RISCOM program addressing transparency in risk assessment and decision making, Stockholm, Sweden, June 13–17, 1999).
84. Göran Sundqvist, *The Bedrock of Opinion: Science, Technology and Society in the Siting of High-Level Nuclear Waste* (Dordrecht: Kluwer Academic Publishers, 2002).
85. Litmanen et al., "Is There a Nordic Model."
86. This paragraph is based on observations by Arne Kaijser, who attended the court hearings.
87. OECD-NEA, *Partnering for Long-Term Management of Radioactive Waste: Evolution and Current Practice in Thirteen Countries* (Paris: OECD Nuclear Energy Agency, 2002), 84–85.
88. E.g., Kari, "Role of the Host Communities"; Elam and Sundqvist "Public Involvement." Anecdotal evidence suggests that SKB may have benefited from its acronym, which resembles that of many state agencies (personal communication from a Swedish NGO activist, August 27, 2018).
89. Per Hedberg and Sören Holmberg, "Åsikter om energi och kärnkraft. Forskningsprojektet Energiopinionen i Sverige. Resultat av SOM-undersökningen 2017" (Gothenburg: SOM-institutet, April 2018), Table 6, https://medarbetarportalen.gu.se/digitalAssets/1689/1689440_32.---sikter-om-energi-och-k--rnkraft-2017.pdf.
90. Demoskop, "*Attityder till kärnkraften: Forsmark*," *Rapport till Vattenfall*, December 6, 2017; Novus, "*SKB Opinionsmätning Östhammar våren 2019*," Report for SKB, May 3, 2019, https://www.skb.se/wp-content/uploads/2019/05/Novusrapport_Opinionsm%C3%A4tning-2019-%C3%96sthammar_SKB.pdf; Vilhunen et al., "Perceptions."
91. Kaijser, "Sweden Short Country Report."
92. This option entails burying waste in extremely deep boreholes (up to 5 km beneath the surface), and hence relies on the thickness of the natural geological barrier to guarantee safety.
93. Kärnavfallsrådet, *Kunskapsläget på kärnavfallsområdet 2016: Risker, osäkerheter och framtidsutmaningar* (Stockholm: Statens offentliga utredningar, SOU, 2016), 16; Cotton, *Nuclear Waste Politics*, 18.
94. E.g., Erika Säynässalo, "Nuclear Energy Policy Processes in Finland in a Comparative Perspective: Complex Mechanisms of a Strong Administrative State," in Kojo and Litmanen, *Renewal of Nuclear Power*; Nazem Tahvilzadeh, "Understanding Participatory Governance Arrangements in Urban Politics: Idealist and Cynical Perspectives on the Politics of Citizen Dialogues in Göteborg, Sweden," *Urban Research and Practice* 8, no. 2 (2015): 238–54.
95. Cotton, *Nuclear Waste Politics*.
96. Named after the chair of the committee in charge of the report, Sir Brian Flowers, a former UKAEA official and a respected nuclear physicist.

97. Cotton, *Nuclear Waste Politics*, 72.
98. Andrew Blowers, "Nuclear Waste and Landscapes of Risk," *Landscape Research* 24, no. 3 (1999): 241–64; Stuart Butler and Robert Bud, "United Kingdom Short Country Report (version 2018)," in *History of Nuclear Energy and Society (HoNESt) Consortium Deliverable no. 3.6* (2019), https://hdl.handle.net/2454/38269.
99. Blowers, "Nuclear Waste"; Gordon MacKerron and Frank Berkhout, "Learning to Listen: Institutional Change and Legitimation in UK Radioactive Waste Policy," *Journal of Risk Research* 12, no. 7 (2009): 989–1008.
100. Peter Simmons and Karen Bickerstaff, "The Participatory Turn in UK Radioactive Waste Management Policy," in *Proceedings of VALDOR-2006*, ed. K. Andersson (Stockholm: Congrex Sweden, AB, 2006), 529–36.
101. Bickerstaff et al., "Reframing Nuclear Power," 151.
102. Bickerstaff et al., "Reframing Nuclear Power," 157.
103. Bickerstaff et al., "Reframing Nuclear Power," 157.
104. MacKerron and Berkhout, "Learning to Listen."
105. CoRWM engaged over 5,000 people in eight discussion groups, four citizens' panels, an open access online discussion guide, a schools' project, a national stakeholder forum, stakeholder roundtables at fourteen nuclear sites, open meetings, consultation documents, and correspondence by letter or email. CoRWM, *Managing our Radioactive Waste Safely—CoRWM's Recommendations to Government* (London: Committee on Radioactive Waste Management, 2006), 6.
106. E.g., Jason Chilvers, "Towards Analytic-Deliberative Forms of Risk Governance in the UK? Reflecting on Learning in Radioactive Waste," *Journal of Risk Research* 10, no. 2 (2007): 197–222; Jason Chilvers, "Deliberating Competence: Theoretical and Practitioner Perspectives on Effective Participatory Appraisal Practice," *Science, Technology, and Human Values* 33, no. 2 (2008): 155–85; Paul Dorfman, ed., *Nuclear Consultation: Public Trust in Government* (London: Nuclear Consultation Working Group, 2008), http://www.ourfutureplanet.org/newsletters/resources/Nuclear consult.com%20-%20Nuclear%20Consultation%20Public%20Trust%20in%20 Government.pdf; Alec Morton, Mara Airoldi, and Laurence D. Phillips, "Nuclear Risk Management on Stage: A Decision Analysis Perspective on the UK's Committee on Radioactive Waste Management," *Risk Analysis* 29, no. 5 (2009): 764–79; MacKerron and Berkhout, "Learning to Listen."
107. Cotton, *Nuclear Waste Politics*, 198.
108. CoRWM, *Managing Our Radioactive Waste*.
109. Eurobarometer, *Attitudes towards Radioactive Waste* (Brussels: European Commission, 2008), 40–41.
110. MacKerron and Berkhout, "Learning to Listen."
111. Butler and Bud, "United Kingdom Short Country Report." The white paper stated that ministers would prefer to work with public support but reserved the right to take more aggressive action on planning if "at some point in the future such an approach does not look likely to work." DECC, *Implementing Geological Disposal: A Framework for the Long-Term Management of Higher Activity Radioactive Waste* (London: Department of Climate Change & Energy, July 31, 2014).
112. Butler and Bud, "United Kingdom Short Country Report"; Blowers, *Legacy*. In 1981, the government sought to improve Windscale's reputation by renaming it Sellafield.
113. Cotton, *Nuclear Waste Politics*, 212.

114. HOL, *Energy Prices, Profits and Poverty: Fifth Report of Session 2013–14, Volume I: Report, together with formal minutes, oral and written evidence* (London: House of Commons Energy and Climate Change Committee, 2013); Markku Lehtonen and de Laurence De Carlo, "Community Energy and the Virtues of Mistrust and Distrust: Lessons from Brighton and Hove Energy Cooperatives," *Ecological Economics* 164 (October 2019).
115. Bickerstaff et al., "Reframing Nuclear Power"; *Eurobarometer, Attitudes towards Radioactive Waste* (Brussels: European Commission, 2008).
116. Ian Rutledge and Philip Wright, *UK Energy Policy and the End of Market Fundamentalism* (Oxford: Oxford University Press, 2011); Florian Kern, Caroline Kuzemko, and Catherine Mitchell, "Measuring and Explaining Policy Paradigm Change: The Case of UK Energy Policy," *Policy and Politics* 42, no. 4 (2014): 513–30.
117. Paul Hildreth, "What Is Localism, and What Implications Do Different Models Have for Managing the Local Economy?" *Local Economy: The Journal of the Local Economy Policy Unit* 26, no. 8 (2011): 702–14.
118. Cotton, "Industry and Stakeholder Perspectives."
119. Kasperski, "Une transition."
120. Natalia Melnikova et al., "Russia Short Country Report (version 2018)," in *History of Nuclear Energy and Society (HoNESt) Consortium Deliverable no. 3.6* (2019), 50–54, https://hdl.handle.net/2454/38269.
121. "Edelman Trust Barometer 2018," 41–44.
122. Vladimir Shlapentokh, "Trust in Public Institutions in Russia: The Lowest in the World," *Communist and Post-Communist Studies* 39, no. 2 (2006): 153–74; Vadim O. Kiselev, "Doverie k politicheskim institutam v Rossii: opyt sotsiologicheskogo monitoringa," *Monitoring obshchestvennogo meniia* 6, no. 124 (2014): 51–64.
123. "Edelman Trust Barometer 2018," 37, 43.
124. Susan Stewart and Jan Matti Dollbaum, "Civil Society Development in Russia and Ukraine: Diverging Paths," *Communist and Post-Communist Studies* 50 (2017): 207–20.
125. Shlapentokh, "Trust"; Kiselev, "Doverie."
126. VtsIOM, "O Chernobyle: 30 let spustia," April 26, 2016, https://wciom.ru /analytical-reviews/analiticheskii-obzor/o-chernobyle-30-let-spustya.
127. Kasperski, "Les politiques," 264–70; Aliaksandr Novikau, "What Is 'Chernobyl Syndrome?' The Use of Radiophobia in Nuclear Communications," *Environmental Communication* 11, no. 6 (2017): 800–9, doi:10.1080/17524032.2016.1269823.
128. "VTsIOM."
129. Melnikova, "Russia Short Country Report," 15.
130. Liia Vandysheva, "Reportazh: Ne na zhizn,' a na smert': Spory vokrug stroitel'stva PZRO v Sosnovom Boru ne utikhali do pozdnego vechera," *Bellona*, December 30, 2013, https://bellona.ru/2013/12/30/reportazh-ne-na-zhizn-a-na-smert-spory/.
131. Vladimir Griaznevich, "Rosatom meniaet kontseptsiiu iadernogo mogil'nika v Sosnovom Boru," *RBK*, July 15, 2015, https://www.rbc.ru/spb_sz/15/07/2015/55a 6058e9a79475cb1318d71.
132. Andrei Ozharovskii, "Kak gramotno i effektivno uchastvovat' v obshchestvennykh slushaniiakh?" *Ekologiia i Pravo* 2, no. 66 (2017): 18–20.
133. Natalia Mel'nikova, *Fenomen zakrytogo atomnogo goroda* (Iekaterinburg: Bank kul'turnoi informatsii, 2006); Dmitrii Iu. Faikov, *Zakrytye administrativno-territorial'nye obrazovaniia: 'Atomnye' goroda* (Sarov, Russia: FGUP RFIaTs-VNIIEF,

2010); Kate Brown, *Plutopia: Nuclear Families, Atomic Cities, and the Great Soviet and American Plutonium Disasters* (Oxford: Oxford University Press, 2012).
134. Anna Kireeva, "Zheleznogorsk gotovitsia k stroitel'stvu mogil'nika radioaktivnykh otkhodov," *Bellona*, July 7, 2012, https://bellona.ru/2012/07/30/zheleznogorsk-gotovitsya-k-stroitels/.
135. Administration of ZATO Zheleznogorsk of the Krasnoyarsk region, August 10, 2015. Decision n°1228 Ob utverzhdenii protokola obshchestvennykh slushanii na temu: "Materialy obosnovaniia litsenzii na razmeshchenie i sooruzhenie ne otnosiashchegosia k iadernym ustanovkam punkta khraneniya RAO, sozdavaemogo v sootvetstvii s proektnoi dokumentatsiei na stroitel'stvo ob"ektov okonchatel'noi izoliatsii RAO (Krasnoiarskii krai, Nizhne-Kanskii massiv) v sostave podzemnoi issledovatel'skoi laboratorii (vkliuchaia materialy otsenki vozdeistviia na okruzhaiushchuiu sredu)," https://ipravo.info/krasnoyarsk1/legal71/511.htm.
136. Paul R. Josephson, "Nuclear Culture in the USSR," *Slavic Review 55*, no. 2 (1996): 297–324; Sonja Schmid, "Celebrating Tomorrow Today: The Peaceful Atom on Display in the Soviet Union," *Social Studies of Science 36*, no. 3 (2006): 331–65.
137. Kasperski, "Nuclear Dreams."
138. Lehtonen, "France Short Country Report."
139. Paul Slovic, "Perceived Risk, Trust and Democracy," *Risk Analysis* 13, no. 6 (1993): 677.
140. Cf. Slovic,"Perceived Risk."
141. Cf. Puustinen et al., "Deliberative Bureaucrat."
142. E.g. James M. Jasper, *Nuclear Politics: Energy and the State in the United States, Sweden, and France* (Princeton, NJ: Princeton University Press, 1990), 72.
143. Cyrille Foasso, *Atomes sous surveillance: Une histoire de la sûreté nucléaire en France* (Brussels: Peter Lang, 2012); Michael Mangeon and Frédérique Pallez, "Réguler les risques nucléaires par la souplesse: Genèse d'une singularité française (1960–1985)," *Gérer et Comprendre. Annales des Mines* (Les Annales des Mines, 2017).
144. E.g., Lehtonen and de Carlo, "Diffuse Institutional Trust."
145. Puustinen et al., "Delibrative Bureaucrat"; S. Montin, "Från tilltrobaserad till misstrobaserad styrning: Relationen mellan stat och kommun och styrning av äldreomsorg," *Nordisk Administrativt Tidsskrift* 92, no. 1 (2015): 58–75.

CHAPTER 7

Nuclear Power and Environmental Justice: The Case for Political Equality

Matthew Cotton

Introduction

This chapter considers differing environmental justice landscapes in relation to the historical development of nuclear technologies. There are four parts to the analysis. The first is devoted to a discussion of the environmental justice concept and associated principles, divided into distributive, procedural, and recognition components. The second outlines an overarching conceptual framework of political equality for the purposes of normatively evaluating the historical development of nuclear technology. The third part applies the conceptual framework in detail to the historical case of the United Kingdom across three descriptive periods: (1) economic, energy, and military securitization (1940–1997); (2) technocratic failures and the move to stakeholder engagement (1979–2007); and (3) the return to securitization in the face of anthropogenic climate change (2007–present). Periodization facilitates discussion of the implications of technology choices, governance structure, and nuclear facility management policies on prevailing justice outcomes. Finally, the chapter concludes with a brief discussion of the conceptual relevance of political equality to the nuclear fuel cycle within a European context under conditions of a return to environmental securitization around energy demand and climate change.

Environmental Justice

The concept of environmental justice is strongly associated with the civil rights movement in the United States. Within the movement was a recognition that the marginalization of African Americans and other people of color within civil society was reflected in concurrent environmental degradation. Case studies such as Adeline Gordon Levine's *Love Canal*[1] and Robert Bullard's highly influential *Dumping in Dixie*[2] revealed the extent to which the siting of toxic chemical and radioactive wastes, municipal waste dumps, and other locally unwanted land uses (LULUs) fell into communities predominantly populated by African American, Hispanic, and Asian American communities, and how local social movements empowered by the broader civil rights movement successfully linked environmentalism with social justice in political discourse.[3] Given the socioeconomic inequalities inherent to racially marginalized communities when compared to their more affluent (and white) neighbors, the environmental risks associated with the toxic environments become disproportionately borne by the poorer community. Éloi Laurent notes that when comparing environmental justice in the US and European contexts (specifically in relation to the UK, though with reference to the European Union), the differing underlying philosophies of public policy are significant. In the US, policy structures traditionally recognize the universality of the natural rights granted to individuals and aim to curb discrimination against individuals in exercising those rights, whereas in Europe, the emphasis is usually upon correcting the social processes that produce inequality.[4] Moreover, in Europe, analysis of distributional inequality is less overtly tied to institutionalized *environmental racism* than in the US.[5] As Ilaria Berreta notes, environmental justice in Europe is more likely to be perceived, analyzed, and framed through the lens of class/income than through racial and ethnic categories. Though there are notable examples of overt environmental racism experienced by Roma communities in Central and Eastern Europe or among migrant communities in France, there are significant disparities in cultural, legal, and public policy approaches between the US and the EU, owing to historical and institutional differences. Given that environmental justice in the US was born in the context of the broader civil rights movement, over a forty-year period, grassroots environmental justice mobilization and protest has focused on the outcomes and consequences of public policy measures. The length and intensity of political action in the US to matters of distributional inequality have resulted in the development, in the Environmental Protection Agency and other federal bodies, of specific methods of environmental justice or environmental equity appraisal, within which only racial minorities (and not low-income communities) are recognized as legally binding categories by US federal law. Race-based environmental injustice is thus a basis for court action, whereas income is not. In the EU, by contrast, a much broader array of policy effects, impacts, and distributional inequalities are considered across a range of legislation.[6] Yet the common factor is that environmental justice can be explained by social and economic geographies of marginalization—how environmental resource development, pollution, resource access, and well-being are tied to the social and political processes that produce poverty, marginalization, and social exclusion.[7]

Since the 1960s, the academic literature on environmental justice has centered on three main dimensions: *distribution*, *procedure*, and *recognition*. In cases where the geographic dispersal of environmental risks and the adverse socioeconomic outcomes of poor communities intersect, we can call this *distributive environmental justice*. Whether from a point source (such as, for example, a radioactive waste management facility with a leaking container) or a nonpoint source pollution (such as particulate emissions from increased traffic to move nuclear materials through a community), the locational relationship between those who benefit from the facility and those who suffer the impacts is potentially unfair, and so the question of how to redress that imbalance has both moral and political implications. The disparity between beneficiaries and those who bear the burdens of environmental risk is "intensely geographical."[8] Nuclear energy has been subject to ongoing evaluation as an environmental justice issue.[9] This is because the beneficiaries of nuclear energy primarily include electricity customers, financiers, and industry shareholders. Injustice defined in these terms is primarily economic. Nuclear facilities are megaprojects: large-scale, complex, high-stakes sociotechnical infrastructure projects commissioned by public authorities and delivered through partnerships between public and private organizations, with high capital costs and long lead times for construction.[10] These costs are borne often by taxpayers, so that the private investment can be secured by public authorities. The private sale of electricity within liberalized energy markets primarily benefits the owners and investors in the plant, and so they become concentrated among a small number of private individuals. However, nuclear power has several negative social and economic externalities. These include risks (including physical hazards and perceived risks) associated with catastrophic accidents, such as the Chernobyl or Fukushima Daiichi nuclear accidents[11] or potential terrorist activity,[12] as well as those associated with routine processes of the nuclear fuel cycle such as uranium mining,[13] spent-fuel reprocessing,[14] smaller-scale "normal" nuclear accidents during plant operations,[15] routine discharges of radioactive materials,[16] the disposal of radioactive waste,[17] and the transportation of radioactive and nonradioactive materials.[18] These risks are geographically situated, and so become concentrated within site-specific community locations. The communities that bear the risks do not directly benefit from the transactions between producer and consumer within nuclear-powered electricity markets, and so, as with any environmental externalities, the costs associated with such risks are not reflected in the market price.[19] Such market failure must be redressed to ensure that fairness is established with respect to the impact of nuclear energy technology development on affected communities.

When environmental injustices emerge through uneven distributional effects, the issue of how such benefits and burdens are apportioned becomes increasingly political. This means that the decisions about unwanted developments become subject to a second procedural dimension of environmental justice. Procedure can be interpreted in terms of isonomy (that all individuals are treated equally under the law) and access to due process—that legal rights (including property and access rights) are upheld, that communities are suitably informed about developments, that opportunities to object are adhered to, and access to legal redress is available when agreements cannot be reached. This (legalistic) interpretation is evident in European commitments to the UNECE Convention on Access to Information, Public Participation in Decision-Making and Access to Justice in Environmental Matters (the Aarhus Convention), which defines justice broadly in this way.[20] Yet broader interpretation of procedure includes access to decision makers, voting rights, communicative planning processes through public participation, and the implications that environmental decision-making has for social sustainability and community cohesion.[21] Related to this is the concept of recognition-related justice concerns—that is, the capacity of individuals (and of the broader communities that support them) to have a voice, and to have their cultural, social, and moral values respected throughout processes of environmental decision-making.[22] Key aspects of recognition injustice are those of humiliation, degradation, and disrespect. For example, a local protest group might halt the construction of a locally unwanted land use in their area, and so redress distributed environmental injustice. However, if they become subject to campaign smears or intimidation on the way to achieving their aim, or if their concerns are dismissed as irrational, NIMBYist, or rural, the subsequent emotional toll to achieve distributive justice is itself unjust. Though recognition-related elements are in some respects less tangible and more subjective than other quantifiable or otherwise verifiable justice claims (such as those around risk and benefit distribution or capacity to vote on decisions), they nevertheless get to the heart of individuals' experiences of environmental policies, plans, and developments. It is both the process and the way decision-making is conducted, and not just the outcome, that is important to achieving recognition justice.

Linking the Three Dimensions of Environmental Justice

The trinity of justice dimensions is mutually reinforcing and interdependent. Where environmental harms are borne by specific communities in particular places, their capacity to act to prevent such harm (or at least to ameliorate its worst effects) and to have compensation, representation in due process, judicial oversight, and protection from prejudice, racial, cultural, and economic marginalization will often appear simultaneously within any given environmental management problem. Procedural concerns link directly to distributive concerns because the central and local government institutional apparatus and political context have a powerful influence on the just allocation of environmental harms and economic benefits within society.[23] These in turn both link to recognition justice because opposition to unwanted developments involves psychosocial impacts such as threats to identity, personal safety and privacy, voice, and community cohesion. Fair outcomes therefore depend on establishing fairness, honesty, trustworthiness, accountability, and transparency in the processes that resolve disputes, distribute environmental risks, and allocate resources, and these must be examined in concert with one another.[24] Of concern in this chapter is a normative ethical framework for environmental justice evaluation that brings together these different components, and the novelty lies in the application of these principles across the historical development of nuclear technology and policy. Of interest in this regard is the work of Kristen Shrader-Frechette, as proposed in her book *Environmental Justice: Creating Equality, Reclaiming Democracy*.[25] Shrader-Frechette's overarching concern is with ensuring environmental justice as social equality through democratic means. Given the ongoing concern within the nuclear sector around opportunities for participatory decision-making and public engagement within policy and facility planning, this framework provides a useful evaluative yardstick against which to assess the ethical performance of different nuclear contexts within Europe.

Shrader-Frechette argues that threats to equality and informed consent are inherent to the understanding of environmental justice. She draws principally upon two theories: John Rawls's[26] philosophy of justice-as-fairness and Ronald Dworkin's[27] notion of political equality. Rawls argues that fair distributional outcomes can only be decided from an original position whereby decision makers are ignorant of the status and conditions of those whom the decisions affect; Dworkin argues that fairness can be ensured only when all citizens are given equal consideration and concern with respect to decisions over distributive outcomes. From this Shrader-Frechette defines an overarching Principle of

Prima Facie Political Equality (PPFPE) that brings together these two concepts in relation to environmental justice disputes. At its heart, the PPFPE defines a distributive justice claim—that benefits and burdens should be fairly apportioned within society on the basis of a participatory justice principle. This principle asserts that any claim to overturn environmental equality based on land rights, economic development outcomes, or military gains (for example) is normally an insufficient justification to violate environmental equality. In short, one will violate the PPFPE by treating people in different locales differently without having morally relevant grounds for such discrimination.[28] Shrader-Frechette also argues that such redistributive policies are defensible on utilitarian grounds, as equality of access to such services nearly always serves the greater economic good. Within the PPFPE Shrader-Frechette asserts that "equality is defensible, and that only different or unequal treatment requires justification," in the sense that the onus for justifying undue environmental risks lies with those who propose developments that raise environmental inequalities and not those who oppose them.[29] This is something of a departure from planning and policymaking processes in many democratic nations. Across Europe the most common models of infrastructure planning and development are those that (increasingly) emphasize "streamlined" policy development and site development application processes. Liberalization of utility markets leads to the formation of the regulatory state,[30] and public authorities consequently take less control over site selection and instead provide regulatory procedures that allow the public opportunity to object (such as through planning inquiries) or to actively participate through different sorts of consultative and engagement mechanisms.[31] Under the PPFPE it behooves public authorities to involve citizens in strategic planning governance across regions and nations, asserting a "strong democratic"[32] right to be involved in the processes of decision-making, design, and project implementation. Thus, the PPFPE is critical of the procedural way megaprojects such as those associated with nuclear power development are planned for, given the current onus on objectors to halt environmental injustices through protest and direct action.

"Equality of treatment under the law" is the next key component, and it is "proportional to the strength of one's claims to it"—that is, in practice it may vary according to individual circumstances, compensation due to one's individual needs, or society's general interest in providing incentives for certain kinds of actions.[33] This is an important concept of legal egalitarianism: individuals must be treated fairly and in a nondiscriminatory way as independent beings by legal and policy processes (sometimes referred to as the principle of isonomy), and all are subject to, and have access to, the same laws of justice (due process) as defined in human rights legislation.[34] Distributive justice is defined in the PPFPE as "morally proper apportionment of benefits and burdens" (if environmental harm occurs, equality is therefore ensured through economic redistribution or by providing equality of economic opportunity in return). This then relates to a concurrent need for participative justice (itself a form of procedural justice), which Shrader-Frechette argues involves "institutional and procedural norms that guarantee all people equal opportunity for consideration in decision-making." This second facet requires that "stakeholder and expert deliberation [be] given equal weight" to one another. In other words, there is a democratization of expertise beyond the bounded rationality of techno-scientific specialists within nuclear policy domains. Shrader-Frechette's PPFPE is thus consonant with the concept of postnormal science[35] that opens up the complex "wicked" environmental policy problem of nuclear power to broader civic engagement and the input of lay experts[36] to ensure participative justice.

Further to the requirement for heterogeneous stakeholder participation, such stakeholders—including affected citizens—should be given "the same rights to consent, due process, and compensation that medical patients have." The argument is that it is unethical to expose people to environmental risks without first obtaining informed, competent, and autonomous consent, free of coercion, with access to relevant information concerning the risks/harms and capability to understand the relevant information and use it in individual decision-making.[37] Informed consent is given based upon a clear appreciation and understanding of the facts, implications, and consequences of development action. This aspect of participation is a form of recognition justice—that adequate informed consent is about respect for a person's dignity,[38] not simply gleaning information from them that is relevant for a policymaking process.

Conceptual Framework

From the foregoing discussion I break apart the PPFPE into four component elements (or subprinciples) that form a set of evaluative criteria for assessing nuclear power policy and planning processes:[39]

1. The justification principle: The onus for justifying the impositions of environmental/health burdens on individuals or affected communities rests with the proponent of the development (polluter, developer, policymaker, lobbyist, etc.), not with the opponent of development.
2. The compensation principle: Equal rights are asserted under that law, and unequal treatment must be compensated for (primarily through

economic means of wealth redistribution or in-kind benefits such as increased community economic opportunity or investment in services or infrastructures).

3. The engagement principle: Stakeholders, including heterogeneous publics, must have access to unbiased and complete information about environmental impacts and harms and access to decision-making control over issues that affect their local environment.
4. The autonomy principle: Affected communities and other stakeholder groups, including heterogeneous publics, must be included in decision-making through fair participatory processes over environmental decision-making and operate free from coercion, and affected individuals must give free, informed, and autonomous consent to environmental change given all of the aforementioned criteria.

To summarize: The PPFPE is valuable to the evaluation of the ethical dimensions of environmental justice, given the integrative nature of the principle—it allows the articulation of participative and distributive dimensions in concert with one another. It is possible therefore to use the four aforementioned elements of the principle as an "evaluative yardstick" from which to discuss the normative ethical implications of nuclear energy.[40] The novelty of the analysis presented in this chapter lies in how normative concepts of environmental justice evaluation through the Principle of Prima Facie Political Equality can be applied retrospectively to evaluate the technological and political history of nuclear energy development.

In the rest of this chapter, I apply the quadripartite model of political equality as a mechanism to understand environmental justice in UK nuclear energy policies, concluding with some broader lessons for European nuclear energy policy processes. I treat the UK as a critical case for nuclear energy policy development and environmental justice, as it features a complex interconnected sociotechnical system of military and civilian nuclear technologies, reprocessing, plutonium production, spent-fuel management, intermediate-high level radioactive waste management, and geological disposal mechanisms.

Nuclear Power and Environmental Justice in the UK

This chapter examines the history of the United Kingdom's nuclear programs as an environmental justice issue. The UK is a critical case for nuclear technology development in the sense that it contains numerous complex and often contradictory facets. The UK nuclear program is characterized by strong association between nuclear weapons and civilian power production, secretive governance practices during the Cold War, technocratic infrastructure planning, experimentation in deliberative decision-making, and public ambivalence toward the acceptability of nuclear energy programs.[41] I briefly summarize and periodize this complex and contentious policy history, in the sense that the political history of the technology is categorized into discrete blocks of time with relatively stable characteristics, using descriptive abstraction to discuss the similarities and differences between these characteristics.[42] I identify three overlapping periods pertinent to the environmental justice analysis of UK nuclear:

1. Economic, energy, and military securitization (1940–1997)
2. The deliberative turn (1976–2007)
3. Climate change securitization (2007–present)

In table 7.1, the principles are mapped against the periods, such that it is possible to see that early violation of the PPFPE on the grounds of military securitization, state enterprise, and public interest justification in the first period gives way to progress and innovation in egalitarian environmental decision-making that upholds all four principles. This is followed by an erosion of this progress in the period since 2007 as concern over rapid deployment of low-carbon technologies creates a new climate securitization discourse that supersedes principles of political equality in environmental decision-making. The recurrent feature of securitization is significant. Securitization is a process by which political authorities draw power back to the center under conditions of urgency and often poorly defined national interest. Securitization experts have often critiqued this policy framing as a means for states to curtail the rights of citizens.[43] For example, this phenomenon is seen in state responses to global terrorism, where rights to protest and rights to privacy are threatened under the guise of ensuring public safety. What we see, therefore, is a binary divide appearing between securitization and political equality where securitization characterizes the first and third periods of nuclear policy history in the UK and political equality the second period.

Economic, Energy, and Military Securitization (1940–1997)

Early UK nuclear technology policy is characterized by the simultaneous development of weapons and civilian electricity generation in a symbiotic relationship. In 1943 the secret Tube Alloys nuclear weapons project was led by the UK with Canadian involvement. This was later subsumed into the Manhattan Project through mutual agreement with the United States, though by the end

Table 7.1. Analysis principles and periods of nuclear policy

Phases/ Principles	Economic, energy, and military securitization (1940–1997)	The deliberative turn (1976–2007)	Climate change securitization (2007–present)
Justification	National interest (military secrecy); state enterprise (postwar energy security in the face of coal shortages); soft power (scientific prestige)	Democratic renewal; failed technocratic siting processes; push for stakeholder engagement in decision-making	Rapid low carbon transition; weakening of community and local authority decision-making control
Compensation	Employment to high-skilled nuclear workers	Community benefits to site volunteers	Privatized utilities; no compensation for new nuclear power stations; state sponsorship through strike pricing
Engagement	Public inquiries into nuclear facilities and limited site-based consultation	Public and stakeholder engagement, primarily on radioactive waste management options and site selection	Streamlining; infrastructure planning legislation to streamline decision-making; limited consultation detached from the need case for nuclear power
Autonomy	Peripheralization–(nuclear industry dependency)	Partnership working and local authority decision-making	Representative democracy; governance through ministerial oversight of nationally significant infrastructure projects

Note: The dark shaded sections represent a violation of the principle, the lighter shaded sections represent partial violation of the principle, and the unshaded sections uphold the principle.

of the Second World War the US withdrew its involvement, and the UK developed the High Explosive Research program in 1947. It is important to note that this was initially a civilian rather than solely military-led program—headed by the Ministry of Supply.[44] However, the development of the Atomic Energy Research Establishment in Harwell in Oxfordshire and the development of the first small research reactor GLEEP were important steps for the establishment of the UK's nuclear technology sector. Then, from early research reactors, the UK developed larger production facilities. The bomb program required facilities for uranium enrichment, nuclear fuel production, and fission reactors (and later fuel reprocessing facilities) to produce plutonium, and so there was a mutually reinforcing technology policy and development platform between civilian and military sectors.

Early prototype Magnox nuclear facilities were successfully constructed in the early 1950s.[45] The first Magnox reactor, Calder Hall at Windscale, went under construction in 1953 and was later connected to the newly formed national grid electricity transmission network in 1956, thus creating the UK's (and indeed the world's) first facility to provide commercially produced electricity.[46] In the wake of the international Atoms for Peace campaign, structures changed. After the start of the construction of Calder Hall, in 1954, the Atomic Energy Authority Act created the United Kingdom Atomic Energy Authority (UKAEA), an authority with the overall responsibility for the UK's nuclear energy program, which included responsibility for developing civilian nuclear technology. The primary focus was the development of the so-called fast breeder reactor (FBR)—a design that worked on the principle of creating more fissile material than it consumes,[47] with the former World War II wartime airfield at Dounreay in Caithness, in Northern Scotland, selected for this purpose (the Dounreay Fast Reactor program) in 1954. The first civilian program in the UK involved successive phases of construction throughout the 1950s and 1960s.

During this period, the nuclear deterrent, formed from a conjunction of military, technological, political, economic, and psychological currents, beginning in 1952, persuaded the Churchill government to adopt a full-scale nuclear deterrent after World War II.[48] The civilian applications and public benefits were stressed in the political rhetoric around the construction of the early Windscale site and proposed Dounreay sites—the primary goal in both cases was the production of plutonium from uranium for nuclear weapons production. The process of irradiating uranium to produce plutonium generates significant amounts of heat. Massive heat generation requires a disposal mechanism—so using this waste heat to produce steam within a turbine was a logical end use for the by-product of the weapon production process.[49] The civilian application of nuclear energy is an example of early industrial ecology: a process of designing a sociotechnical nuclear energy with a loop between waste (heat) and resource (electricity) cycles within the productive system for generating plutonium.

In political terms, this period is framed through the justificatory principles of state enterprise,[50] national interest, and state secrecy.[51] The nascent Cold War prompted a shrouding of nuclear technology development under the auspices of state secrecy, a decision-making process that put military strategic goals in primacy.[52] It is interesting to note, therefore, that in the early 1950s there was little consultation or parliamentary debate around the development of this first civil nuclear power program—informed consent of even elected representatives (let alone direct citizen consent) was truncated in the technology development process. It appears that the announcement in 1955 of the civilian nuclear power program seems to have taken many, including members of parliament, by surprise.[53] At its heart, the decision to pursue the twin goals of nuclear deterrence and civilian electricity generation is characteristic of technocratic decision-making, in the sense that deliberation on nuclear technology governance, and its potential impact, was largely excluded from democratic scrutiny through parliamentary deliberation—an issue across a range of European policy contexts.[54] Democratic institutions were co-opted into the technology decision-making processes by a politically insulated elite group. Although seven nuclear facilities were subject to public inquiries, all were granted planning permission. The public inquiries, though ostensibly vehicles to improve public scrutiny of nuclear technology, failed to open up the technology to broader scrutiny due to successful "boundary work" by the Atomic Energy Authority to demarcate scientific from nonscientific concerns.[55] Thus, autonomy for representative democratic authorities to represent the environmental justice interests of affected constituents within nuclear communities was traded off under a secretive policy framing of national interest that was nonetheless poorly defined. Indeed, direct government engagement with heterogeneous publics within affected communities on nuclear technology development was prohibited. It was left to other civil society actors, notably the Atomic Scientists' Association (ASA), to provide active community/public-level engagement on the technology platform and associated policy implications. The ASA's Atom Train traveling exhibition was the primary mechanism of this public engagement.[56] The British nuclear scientists' movement organized the exhibition in collaboration with government offices and private industry in 1947–1948, both providing public access to state nuclear knowledge and simultaneously clashing with the interests of the emerging British national security state in the early Cold War.[57]

The legitimization of postdemocratic nuclear decision-making through state enterprise and national justification has, in the period since the Second World War, rested on the proposition that the state's intervention in technology and land development is necessary to safeguard the public interest against private and sectional interests on the one hand,[58] and adverse military interests on the other.[59] What constitutes the public interest has always been contentious, in part because it is extremely difficult to define in public discourse, even by those who use it as a legitimizing concept.[60] In nuclear technology policy terms, the national interest could be broadly categorized in two ways. First, nuclear weapons were a strong sociopsychological driver for the postwar development of the British state. At a point of declining imperial power, the "hard" technological and military power of nuclear weapons assured continued geopolitical influence for the nation-state—a component of a British nationalist sentiment that persists to this day.[61] Second, from 1957 the government began to promote electricity generation by nuclear power as an alternative to coal-fired power stations, as a means both to reduce the political bargaining power of national coal miners' unions[62] and to establish an alternative energy technology pathway to domestic coal production. The upshot of this political justification is that the policy discourse combined military, economic, and energy securitization, and that urgent threats of Soviet military capability and fuel supply threats from dwindling domestic coal production repressed the emergence of counter-discourses concerning egalitarian, community-level autonomy and engagement justice principles.

In one sense, we can understand this nuclear technology development program under military securitization as emerging under conditions of governmental dishonesty that fundamentally violate the principles of the PPFPE. The link between civilian and military applications, and the technological secrecy that resulted through governance mechanism of the UKAEA served to mask the true scale and intent of nuclear military ambitions at local scales of nuclear site development. In this sense, the justification for nuclear technologies is predominantly top-down (specifically government-led) in nature. In essence, the government imagines the scale at which justification is needed, and then communicates this "downward"—to MPs and then to local site communities that accept nuclear facilities in their midst, thus violating the autonomy and justification principles—without a capacity to represent local interests due to the lack of transparency in the decision-making process. The concept of imagining also relates to how the technology is socially constructed—as one that is in some sense a public technology[63] rather than solely a private or state interest (see chapter 8 in this book);[64] but this is tempered with a more bottom-up civilian-led component of nuclear industry support, commonly understood now as a reluctant acceptance of nuclear power.[65] However, though energy security and the prestige derived from nuclear weapons were powerful initial drivers for government cooperation in the development of a civilian nuclear program, this began to shift toward the end of the 1950s. Roger Williams asserts that it was during this initial period of nuclear development and expansion in the late 1940s and early 1950s that government appeared to be rushing toward the development of a viable national nuclear technology platform due to military imperatives; but toward the end of the 1950s, this had given way to a political desire to establish prestige through world leadership in civilian nuclear technology.[66] Justification for nuclear moved from the hard power of military dominance to the soft power[67] of technology transfer and scientific dominance by the late 1950s—but, again, under the PPFPE this is insufficient justification for localized environmental harm.

In terms of nuclear technology governance, during this initial period of nuclear technological expansion, the governance strategy was to establish the industry and segregate political oversight from its production,[68] as well as to promote a commitment to government "boosterism" (i.e., a highly supportive attitude) and a sense of unchallenged technological optimism within UK society.[69] In terms of justification, the principle of compensation to the regions was a significant driver. This was argued in nuclear-supportive policy circles through local employment on plant construction and operation, and other indirect income and expenditure benefits filtering through into the local community as the industry becomes an established employer.[70] Promised economic benefits (both direct and indirect) are common push factors for policymakers to consent to high asset-value infrastructure megaprojects over other "softer" social value considerations.[71] The justification from a public governance perspective lies in this interpretation of a compensation principle—that development of large-scale sociotechnical systems brings jobs and production revenues that offset any associated environmental externalities (or at least dampen adverse reactions to such externalities). The underlying assumption is that nuclear is a long-term local employer, and that this will stimulate public buy-in and hence ameliorate environmental distributive injustices. However, the history of nuclear energy development in the UK during this period (as in many European country contexts) shows that the high capital intensity, long lead times for production, and the politically controversial nature of the megaproject is characterized not by generating net benefits but by incurring significant costs due to mistakes (technical, political, or ethical) being greatly amplified by the scale of the project.[72] Among the most significant of these socioenvironmental costs is the long-term management of nuclear waste (discussed in the next section).[73]

Though high-value nuclear employment is created, this in turn stimulates conditions of peripheralization[74]—where the mass influx of a highly skilled nuclear workforce creates opportunities for economic development for employees but the crowding out of other industries (including clean tech, lower-skilled labor, or tourism) due to the technological stigmatization that follows in the wake of nuclear development,[75] and a growing economic dependency on a single nuclear employer generates new socioeconomic threats to the community. Principally, it creates dependence upon a polluting industry—such that plant workers and their families will support nuclear, even when the land use is damaging to environmental and public health,[76] and unpopular among neighboring communities (the so-called doughnut effect).[77] I conclude, therefore, that the emphasis during this period on large-scale nuclear developments to meet the political goals of economic securitization through job creation does not sufficiently reach the threshold for the compensation principle.

The Deliberative Turn (1976–2007)

Throughout the early technologically optimistic period of nuclear expansion, the problem of radioactive waste remained peripheral to government nuclear energy policy, despite some prominent voices in the industry expressing doubts over the technical feasibility of the disposal of fission products.[78] In the 1950s and 1960, the primary factors that went into site selection and evaluation of nuclear facilities were access to cooling water, suitable geology for building foundations, proximity to the national grid, and proximity to areas of demand, all of which took priority over social, political, and justice dimensions of nuclear development.[79] Moreover, it is only comparatively recently that the myriad environmental, technological, and social implications of long-term radioactive waste management have been overtly addressed as a political issue in their own right.[80] Notable in this regard was the 1976 Flowers Report[81] that signaled that unbridled nuclear expansion was morally impermissible to future generations if nuclear waste was not managed by an independent committee to protect local communities and broader society from the associated radiological risks.[82]

It is notable that despite growing concerns over the management of nuclear facilities in terms of their environmental performance, safety, and public health, the culture of secrecy and national interest justifications persisted within the nuclear technology community up until the 1990s. The residual optimism from the development of the civilian nuclear program masked the problem from public scrutiny and controversy during this period. Policymakers in the 1960s and 1970s had confidence in the eventual development of a technological fix to the waste problem,[83] as nuclear science and engineering expertise had demonstrated such technical competence in the development of the nuclear program itself. Few in government doubted that eventually they "would be able to deal with the nuclear garbage."[84] It was clear, therefore, that although the need for the political shielding of nuclear technology secrets from espionage had diminished by the early 1990s, the governance of the industry remained hidden from public view and thus continued to violate the autonomy principle. A persistent culture of secrecy meant that radioactive waste management organizations like the Nuclear Industry Radioactive Waste Executive (Nirex Ltd.) failed to adopt a culture of transparency, openness, and community engagement despite persistent failures to "site" a repository for long-lived intermediate and high-level radioactive wastes in successive phases in the 1980s and 1990s. This culminated in the failure, in 1997, to secure planning permission for a rock characterization facility (RCF)—a test site for a deep geological repository—in Cumbria near Sellafield.[85]

The consequence of technological obduracy (favoring deep geological disposal without question) and institutional arrogance was the nuclear establishment's failure to build major nuclear waste infrastructure through purely technocratic means. Following the failed RCF proposal, in 1997 the Labour government instigated a root-and-branch review of nuclear waste governance practices through the Managing Radioactive Waste Safely policy program. MRWS set up the Committee on Radioactive Waste Management (CoRWM, pronounced "quorum"), composed of a variety of experts in nuclear physics, radiological safety, social sciences, and human rights. CoRWM differed from most scientific advisory committees in that it put public engagement at the heart of its decision-making process.

CoRWM oversaw the largest ever state-run public engagement program on a contentious policy issue in the UK. This created a rebalancing and rescaling of the relationship between state actors and local communities, taking seriously the expertise of nonspecialist "lay" citizens in the assessment of radioactive waste management options (from deep geological disposal to long-term surface storage and other esoteric options such as disposal in subduction zones).[86] At the time, this was heralded as participatory-deliberative turn in policymaking[87]—a much-lauded shift in the relationship between technological authority, the state, and the citizenry. CoRWM's work took place in the context of the nuclear industry's persistent failures to gain public support for any type of development. New nuclear power stations (such as Sizewell B) and spent-fuel reprocessing facilities (Thermal Oxide Reprocessing Plant, or THORP) were subject to lengthy public inquiries, and nuclear waste disposal sites remained elusive. In each case, "decide-announce-defend" politics—whereby experts could decide what was best for the country and then use the technical authority's justification for technology deployment—were deployed, violating the justification principle in each case. The MRWS program recognized that justification for radioactive waste policy required scientific and technical authority to be, to quote Churchill, "on tap, not on top."[88]

The CoRWM report to government recommended deep geological disposal, with safe interim storage, which was not radically different from the Nirex proposals in the 1980s and '90s. However, the proposals had legitimacy through political equality gained from upholding the principles of autonomy and engagement, and as such captured support from a broad array of stakeholders and political authorities. Moreover, CoRWM built upon international best practice derived from Canadian and Swedish cases in recommending a voluntary site selection process. This involves communities stepping forward to engage in the process of site selection without coercion from state interests—giving them a right to withdraw from proposals and empowering them to make decisions that affect their lives and livelihoods in a way that no previous nuclear technology decision had. I conclude therefore that the Flowers Report established a principle of environmental justice that influenced a shift toward successive governments taking seriously the issue of local environmental protection from nuclear waste and nuclear energy facilities. Yet it was only in the period following the 1997 failure of the RCF proposal by the corporation Nirex that the Labour government (elected in 1997) took seriously the issue of political equality in decision-making. What followed was a period of experimentation in achieving participatory justice, first in radioactive waste management options, and then in site selection for a geological disposal facility (GDF). We can therefore characterize this period as a paradigmatic change in the nature of environmental governance in the nuclear sector that upheld all four components of the PPFPE.

Climate Change Securitization (2007–Present)

Since 2007, the Labour government has changed its policy from pronuclear to nuclear as an instrument to combat anthropogenic climate change.[89] This created an inherent tension within the environmental justice outcomes of nuclear energy policy. What we see is that the Principle of Prima Facie Political Equality embodied in the emergent participatory-deliberative turn of the early 2000s began to give way in the mid-2000s to a new policy discourse of securitization. Whereas in the first period, the aforementioned securitization discourse concerned the development of nuclear weapons and the threat of coal shortages in the post–World War II era, now securitization relates to rapid decarbonization of electricity infrastructure and a perceived energy gap between rising demand and falling supply from the decommissioning of old nuclear sites (sometimes referred to as "keeping the lights on," which has been a staple of nuclear energy policy framing since the technology's inception).[90]

What we see, however, is that the change in policy became subject to challenges on the basis of participative justice. The aforementioned Aarhus Convention links legal rights (based on isonomy and due process) to the legal redress of environmental disputes. The Aarhus Convention is a core component of the European approach to ensuring legal access to an enforcement of environmental rights and is representative of the broader deliberative turn identified in the previous period. It is notable, therefore, that environmental nongovernmental organizations that campaign against nuclear power have brought legal proceedings against nuclear policy and planning developments, under the Aarhus Convention, in the UK and the Netherlands.[91] There are two notable examples. The first, in 2007, was the Greenpeace-initiated claim for judicial review seeking a quashing order under Aarhus terms in respect to the government's decision to support nuclear new-build as part of the UK's future energy mix. The review found that, given the huge importance of the nuclear new-build issue, in order to be Aarhus-compliant in these circumstances, only the promised "fullest public consultation" that upholds the justification and engagement principles would have been adequate.[92] The judicial review considered the justification of the policy change to have been made without sufficient access to public participation among key stakeholders, including the public. It was the process of consultation in advance of the policy change that was criticized, not nuclear power itself. This case is similar to that of the proposed Borssele nuclear power station in the Netherlands, where Greenpeace successfully appealed to the Aarhus Convention Compliance Committee (ACCC). The ACCC found that the power station proposal showed noncompliance with the convention because the operators did not carry out proper public participation as obliged under Article 6(10). What is notable in both cases is that the fullest possible consultation is now a powerful legal criterion for evaluating the justification for nuclear projects. Equality and environmental justice under the law are thus determined by the deliberative capacity of publics to substantively engage with nuclear energy policy and planning and strategic policymaking processes.[93] For the first time, the principles of engagement and autonomy have teeth in holding signatory governments to account for political equality established through participatory justice, and nuclear in the UK was an important test case in this regard.

Despite the setbacks for the government from Aarhus Convention challenges to nuclear policy, the power of the securitization framing continues to override claims for political equality during this post-2007 period. There is growing evidence that government support for new nuclear build is geared toward the maintenance of nuclear engineering skills necessary for the renewal of the UK's nuclear weapons program (the replacement of the current Trident weapons program).[94] Thus, the cycle of military technological development of nuclear weapons has had a powerful influence on the restriction of community rights under similar securitization discourse seen in the first period of the UK's nuclear history. Similarly, in the United Kingdom, and across Europe, energy security and climate mitigation policy frames have given way to changes in planning reform to speed up the rate at which major infrastructure projects such as nuclear power stations (and also major wind farms, biomass, waste incineration, and electricity transmission lines) are built, alongside electricity market reform to provide long-term sales contracts for power and a capacity market. All of these changes, however, violate the principles of autonomy and engagement: policy processes to streamline low-carbon infrastructure development curtail discussion of the local characteristics of the places in which infrastructure is built, and the threats to rural identity and amenity value and from radiological risk are downplayed in favor of reducing greenhouse gas emissions.

The securitization-led approach means that the justification for the construction of new infrastructure is decided in advance of the individual projects being proposed. In other words, the need case for nuclear energy alongside other local energy technologies becomes an established part of government energy policy, and cannot be overridden by local government, local planning authorities, or the input of community stakeholders. This is because the mechanism of public inquiry that had the power to halt or amend nuclear developments in the previous two periods (for example, the inquiries into the Windscale Piles fire,[95] construction of the THORP,[96] or the Sizewell B power station[97]) is curtailed in this third period. The new securitization framing fundamentally violates all of the principles outlined above. It also leads to growing public dissatisfaction with local planning authorities (deemed to be ineffective and powerless institutions), adverse reactions to scientific and technical authority, and the rise of coordinated collective action campaigns against major infrastructure projects. The violation of the core components of PPFPE will lead to growing institutional mistrust, vocal collective action through social movements of opposition, and greater legal challenges to the development of nuclear installations under the Aarhus Convention, as seen in the UK and the Netherlands.

Conclusions

We can treat the UK's nuclear history, in very broad terms, as showing a tremendous expansion and then rapid contraction in the application of environmental justice principles to policy and technology development across the three periods identified. The UK is showing an unusual trend toward the expansion of new-build nuclear energy at a time when the majority of other nuclear-power European nations are facing a declining nuclear sector. In Spain, the government plans to close the last of its nuclear reactors by 2035 and coal power plants by 2030. In Germany there is a moratorium on new nuclear build in the wake of the Fukushima Daiichi disaster, such that public opinion remains broadly opposed to nuclear power, and there is virtually no political support for building new nuclear plants. In Belgium, initial support in 2007 for the renewal of nuclear energy has since been reversed, and as of March 2018 the government reaffirmed a phaseout policy for existing nuclear stations. In France, though the government remains committed to the nuclear sector, current policies aim to reduce nuclear capacity from the current 70 percent to 50 percent by 2035. In Sweden, there is declining nuclear power under a capacity tax regime. Across Europe nuclear power is threatened by a number of factors. These include the falling cost of renewable energy, improvements in energy storage capacity, and low fossil fuel costs (particularly natural gas following the shale gas boom in the United States and Canada). Other contributory factors, such as the influence of the Fukushima Daiichi disaster, which demonstrated the global nature of radiological risks, are mixed across the continent.[98] Though they were highly influential in Germany's decision to phase out nuclear, in other countries public concerns over radiological risk following the nuclear disaster appear to have had little influence on government policy.

When looking across Europe as a whole, there is a common securitization policy frame that necessitates urgent action on energy supply and climate change mitigation goals. Yet carbon dioxide reduction goals under 2050 time limits are largely being achieved by the expansion of renewable energy, energy efficiency innovation and cost-saving measures in heating and transport, energy reduction across supply chains, community energy policy, microgeneration at the domestic scale, and regulatory measures to improve the energy performance of new housing stock. When it comes to nuclear energy, governments don't deem nuclear energy as essential to meeting securitization goals. This means that support is dwindling because construction of new nuclear plants will bring European governments into conflict with increasingly energized and globally connected (through social media) protest groups motivated by localized environmental justice concerns. Phasing out nuclear in favor of less risky (both environmentally and financially) renewable schemes is one way for governments to avoid difficult environmental justice challenges. We see a declining interest in new nuclear build in spite of growing post–Paris Agreement commitments to decarbonize electricity in Europe because public opinion is shifting away from nuclear in most cases.[99] However, though there is evidence of political ambivalence toward nuclear power among most existing nuclear-power-producing European nations, political decision-making over nuclear is not due to environmental injustice claims forcing governments to reject proposals. Rather, environmental justice and political equality are peripheral or secondary concerns. The high cost and capital intensity, long lead times, and unresolved environmental issues surrounding waste disposal make new nuclear build a far less attractive option under conditions of falling renewable prices than it was a decade ago, and so political hesitancy is growing. Environmental justice and political equality do remain important considerations, however. Across Europe and the Global North, the growing presence of the Skolstrejk för klimatet (School Strike for Climate) and Extinction Rebellion movements show a rapidly growing concern for global environmental justice. At present, these movements remain agnostic about nuclear energy as a solution to mitigating anthropogenic climate change. However, given the background of this growing environmental justice activism, it behooves governments engaging in nuclear energy renewal to consider the principles of justification, engagement, compensation, and autonomy with care, because when these are violated, this stimulates place-protective action, culminating in social movements opposing the technology. When local movements for environmental protection on the grounds of political equality gain traction within broader global environmental movements, they gain greater capacity to block developments. It is therefore necessary for pronuclear governments and developers to move past the assumption that justification from climate and energy security will be enough to convince local communities of the legitimacy of a development. When justification does not consider engagement and autonomy, then resultant community action to redress the environmental injustice slows development, builds mistrust of technical authorities and government institutions, and will stymie the so-called nuclear renaissance.

Notes

1. Adeline Gordon Levine, *Love Canal: Science, Politics, and People* (Lexington, MA: Lexington Books, 1982).

2. Robert Bullard, *Dumping in Dixie: Race Class and Environmental Quality* (Boulder, CO: Westview Press, 1994).
3. Troy W. Hartley, "Environmental Justice: An Environmental Civil Rights Value Acceptable to All World Views," *Environmental Ethics* 17, no. 3 (1995): 277–89; Susan L. Cutter, "Race, Class and Environmental Justice," *Progress in Human Geography* 19, no. 1 (1995): 111–22; David E. Camacho, ed., *Environmental Injustices: Political Struggles, Race, Class and the Environment* (North Carolina: Duke University Press, 1998).
4. Éloi Laurent, "Issues in Environmental Justice within the European Union," *Ecological Economics* 70 no. 11 (2011): 1846–53.
5. Liv Raddatz and Jeremy Mennis, "Environmental Justice in Hamburg, Germany," *The Professional Geographer* 65, no. 3 (2013) 495–511; Krista Harper, Tamara Steger, and Richard Filčák, "Environmental Justice and Roma communities in Central and Eastern Europe," *Environmental Policy and Governance* 19, no. 4 (2009): 251–68.
6. Ilaria Beretta, "Some Highlights on the Concept of Environmental Justice and Its Use," *e-cadernos CES* 17 (2012), http://journals.openedition.org/eces/1135.
7. Joan Martinez-Alier, *The Environmentalism of the Poor: A Study of Ecological Conflicts and Valuation* (Cheltenham: Edward Elgar Publishing, 2003); Isabelle Anguelovski and Joan Martinez-Alier, "The 'Environmentalism of the Poor' Revisited: Territory and Place in Disconnected Glocal Struggles," *Ecological Economics* 102 (2014): 167–76.
8. Gordon P. Walker and Harriet Bulkeley, "Geographies of Environmental Justice," *Geoforum* 37, no. 5 (2006): 655–59.
9. Andrew Blowers and David Pepper, "The Politics of Nuclear Power and Radioactive Waste Disposal: From State Coercion to Procedural Justice?," *Political Geography Quarterly* 7, no. 3 (July 1988): 291–98; Pius Krütli et al., "The Process Matters: Fairness in Repository Siting for Nuclear Waste," *Social Justice Research* 25, no. 1 (2012): 79–101; Dean Kyne and Bob Bolin, "Emerging Environmental Justice Issues in Nuclear Power and Radioactive Contamination," *International Journal of Environmental Research and Public Health* 13, no. 7 (2016), doi:10.3390/ijerph13070700; Matthew Cotton, "Environmental Justice as Scalar Parity: Lessons from Nuclear Waste Management," *Social Justice Research* 31, no. 3 (2018): 238–59.
10. Markku Lehtonen, Pierre-Benoit Joly, and Luis Aparicio, *Socioeconomic Evaluation of Megaprojects: Dealing with Uncertainties* (Abingdon: Routledge, 2016); Naomi J. Brookes and Giorgio Locatelli, "Power Plants As Megaprojects: Using Empirics to Shape Policy, Planning, and Construction Management," *Utilities Policy* 36 (2015): 57–66.
11. Frank N. von Hippel, "The Radiological and Psychological Consequences of the Fukushima Daiichi Accident," *Bulletin of the Atomic Scientists* 67, no. 5 (September/October 2011 2011), https://doi.org/10.1177/0096340211421588; Kristin Shrader-Frechette, "Nuclear Catastrophe, Disaster-Related Environmental Injustice, and Fukushima, Japan: Prima-Facie Evidence for a Japanese 'Katrina,'" *Environmental Justice* 5, no. 3 (2012): 133–39; Richard Hindmarsh, *Nuclear Disaster at Fukushima Daiichi: Social, Political and Environmental Issues* (Abingdon: Routledge, 2013); Brian Wynne, "Sheep Farming After Chernobyl: A Case Study in Communicating Scientific Information," *Environment* 31, no. 2 (1989): 10–39;

Ulrich Beck, "The Anthropological Shock: Chernobyl and the Contours of the Risk Society," *Berkeley Journal of Sociology* 32 (1987): 153–65.

12. Gavin Cameron, *Nuclear Terrorism: A Threat Assessment for the 21st Century* (London: Palgrave Macmillan, 1999); Mark E. Byrnes, David A. King, and Philip M. Tierno. Jr., *Nuclear, Chemical and Biological Terrorism: Emergency Response and Public Protection* (Boca Raton, FL: Lewis Publishers, 2003); Paul Slovic, "Terrorism as Hazard: A New Species of Trouble," *Risk Analysis* 22, no. 3 (2002): 425–26.
13. Howard Hu, Arjun Makhijani, and Katherine Yihand, eds., *Nuclear Wastelands: A Global Guide to Nuclear Weapons Production and Its Health and Environmental Effects* (Cambridge, MA: MIT Press, 1995).
14. Terence Lee, "Assessment of Safety Culture at a Nuclear Reprocessing Plant," *Work and Stress* 12, no. 3 (1998): 217–37; Dominique Pobel and Jean-Francois Viel, "Case-Control Study of Leukaemia Among Young People Near La Hague Nuclear Reprocessing Plant: The Environmental Hypothesis Revisited," *BMJ* 314, no. 7074 (1997): 97–119.
15. Charles Perrow, *Normal Accidents: Living with High Risk Technologies* (Princeton, NJ: Princeton University Press, 1999).
16. Britt-Marie Drottz-Sjoberg, *Medical and Psychological Aspects of Crisis Management During a Nuclear Accident* (Stockholm: RHIZIKON Risk Research Reports, 1993).
17. Kärnavfallsrådet, *Nuclear Waste State-of-the-Art Report 2016: Risks, Uncertainties and Future Challenges* (Kärnavfallsrådet—The Swedish Council for Nuclear Waste: Stockholm, 2016); Michael E. Kraft, "Policy Design and the Acceptability of Environmental Risks: Nuclear Waste Disposal in Canada and the United States," *Policy Studies Journal* 28, no. 1 (2000): 206–18; Paul Slovic, Mark Layman, and James H. Flynn, "Received Risk, Trust and Nuclear Waste: Lessons from Yucca Mountain," in *Public Reactions to Nuclear Waste*, ed. Riley E. Dunlap, Michael E. Kraft, and Eugene A. Rosa (Durham, NC: Duke University Press, 1993); Douglas Easterling and Howard Kunreuther, *The Dilemma of Siting a High-Level Nuclear Waste Repository*, Studies in Risk and Uncertainty (Boston: Kluwer Academic Publishers, 1995).
18. Mary Riddel and W. Douglass Shaw, "A Theoretically-Consistent Empirical Model of Non-Expected Utility: An Application to Nuclear-Waste Transport," *Journal of Risk and Uncertainty* 32, no. 2 (2006): 131–50.
19. Sherman Folland and Robbin Hough, "Externalities of Nuclear Power Plants: Further Evidence," *Journal of Regional Science* 40, no. 4 (2000): 735–53.
20. Matthew Cotton, "Public Participation in UK Infrastructure Planning: Democracy, Technology and Environmental Justice," in *Engaging with Environmental Justice: Governance, Education and Citizenship*, ed. Matthew Cotton and Bernado H. Motta (Oxford: Inter-Disciplinary Press, 2011); Juan R. Palerm, "Public Participation in Environmental Decision-Making: Examining the Aarhus Convention," *Journal of Environmental Assessment Policy and Management* 1, no. 2 (1999): 229–44; Raphael Heffron and Paul Haynes, "Challenges to the Aarhus Convention: Public Participation in the Energy Planning Process in the United Kingdom," *Journal of Contemporary European Research* 10, no. 2 (2014): 236–47.
21. John Whitton et al., "Conceptualizing A Social Sustainability Framework for Energy Infrastructure Decisions," *Energy Research and Social Science* 8, no. 7 (2015): 127–38, https://doi.org/10.1016/j.erss.2015.05.010; Katherine Witt, John Whitton, and Will Rifkin, "Is the Gas Industry a Good Neighbour? a Comparison of UK and Australia

Experiences in Terms of Procedural Fairness and Distributive Justice," *The Extractive Industries and Society* 5, no. 4 (2018): 547–56; Catherine Gross, "Community Perspectives of Wind Energy in Australia: The Application of a Justice and Community Fairness Framework to Increase Social Acceptance," *Energy Policy* 35, no. 5 (2007): 2727–36; Jean Hillier, "Beyond Confused Noise: Ideas Toward Communicative Procedural Justice," *Journal of Planning Education and Research* 18, no. 1 (1998): 14–24.

22. David Schlosberg, "The Justice of Environmental Justice: Reconciling Equity, Recognition, and Participation in a Political Movement," in *Moral and Political Reasoning in Environmental Practice*, ed. Andrew Light and Avner de-Shalit (Cambridge, MA: MIT Press, 2003); Darren A. McCauley et al., "Advancing Energy Justice: The Triumvirate of Tenets," *International Energy Law Review* 32, no. 3 (2013): 107–10.
23. David Schlosberg, *Defining Environmental Justice: Theories, Movements, and Nature* (Oxford: Oxford University Press, 2007); Alice Kaswan, "Distributive Justice and the Environment," *North Carolina Law Review* 81 (2002): 1031–48.
24. Rick L. Lawrence, Steven E. Daniels, and George H. Stankey, "Procedural Justice and Public Involvement in Natural Resource Decision Making," *Society and Natural Resources* 10, no. 6 (1997): 577–89; Susan L. Senecah, "The Trinity of Voice: The Role of Practical Theory in Planning and Evaluating the Effectiveness of Participatory Processes," in *Communication and Public Participation in Environmental Decision Making*, ed. Stephen P. Depoe, John W. Delicath, and Marie-France Aepli Elsenbeer (Albany: SUNY Press, 2004); Dorceta E. Taylor, "The Rise of the Environmental Justice Paradigm," *American Behavioural Scientist* 43, no. 4 (2000): 508–80; Susan Clayton, Amanda Koehn, and Emily Grover, "Making Sense of the Senseless: Identity, Justice, and the Framing of Environmental Crises," *Social Justice Research* 26, no. 3 (2013): 301–19.
25. Kristin Shrader-Frechette, *Environmental Justice: Creating Equality, Reclaiming Democracy* (New York: Oxford University Press, 2002).
26. John Rawls, *A Theory of Justice*, 2nd ed. (Oxford: Oxford University Press, 1999).
27. Ronald Dworkin, *Taking Rights Seriously* (Cambridge, MA: Harvard University Press, 1978); Ronald Dworkin, *The Theory and Practice of Autonomy* (Cambridge: Cambridge University Press, 1988).
28. Shrader-Frechette, *Environmental Justice*.
29. Shrader-Frechette, *Environmental Justice*.
30. Giandomenico Majone, "The Rise of the Regulatory State in Europe," *West European Politics* 17, no. 3 (1994): 77–101.
31. Matthew Cotton, "Planning, Infrastructure and Low Carbon Energy," in *The Routledge Companion to Environmental Planning and Sustainability*, ed. Simin Davoudi, Richard Cowell, Iain White,, and Hilda Blanco (Abingdon: Routledge, 2019): 248–56.
32. Benjamin Barber, *Strong Democracy: Participation Politics for a New Age* (Berkeley: University of California Press, 1984).
33. Shrader-Frechette, *Environmental Justice*.
34. Tsachi Keren-Paz, *Torts, Egalitarianism and Distributive Justice* (Abingdon: Routledge, 2018); Allen Buchanan, "The Egalitarianism of Human Rights," *Ethics* 120, no. 4 (2010): 679–710.
35. Silvio Funtowicz and Jerome Ravetz, "Science for the Post-Normal Age," *Futures* 25, no. 7 (1993): 739–55.

36. Judith Petts and Catherine Brooks, "Expert Conceptualisations of the Role of Lay Knowledge in Environmental Decisionmaking: Challenges for Deliberative Democracy," *Environment and Planning A* 38 (2005): 1045–59.
37. Shrader-Frechette, *Environmental Justice*, 24–29, 77; see also Matthew Cotton, "Fair Fracking? Ethics and Environmental Justice in United Kingdom Shale Gas Policy and Planning," *Local Environment* 22, no. 2 (2017): 185–202.
38. Lawrence O. Gostin, "Informed Consent, Cultural Sensitivity, and Respect for Persons," *JAMA* 274, no. 10 (1995): 844–45.
39. Derived from Cotton, "Fair Fracking?"
40. Thomas Webler, " 'Right' Discourse in Citizen Participation: An Evaluative Yardstick," in *Fairness and Competence in Citizen Participation*, ed. Ortwin Renn and Thomas Webler (Dordrecht: Springer, 1995): 35–86; Cotton, "Environmental Justice as Scalar Parity."
41. Ioan Charnley-Parry et al., "Principle for Effective Engagement," in *History of Nuclear Energy and Society (HoNESt) Consortium Deliverable no. 5.1* (2017), https://perma.cc/FXQ2-Y5M2; Josep Espluga, Beatriz Medina, and Wilfried Konrad, "Case Studies Reports: In-Depth Understanding of the Mechanisms for Effective Interaction with Civil Society: Selected Case Studies," in *History of Nuclear Energy and Society (HoNESt) Consortium Deliverable no. 4.3* (2018), https://perma.cc/5SDY-LKCK.
42. Timoteus Pokora, "A Theory of the Periodisation of World History," *Archiv orientálni* 34 (1966): 602–5; William A. Green, "Periodization in European and World History," *Journal of World History* 3, no. 1 (1992): 13–53:
43. Thierry Balzacq, *Understanding Securitisation Theory: How Security Problems Emerge and Dissolve* (Abingdon: Routledge, 2010).
44. Stuart Butler and Robert Bud, "United Kingdom Short Country Report (version 2016)," in *History of Nuclear Energy and Society (HoNESt) Consortium Deliverable no. 3.6* (2019), https://hdl.handle.net/2454/38269.
45. The term "Magnox" refers to an early design of pressurized, carbon dioxide–cooled and graphite-moderated nuclear reactors that use unenriched natural uranium fuel and a magnesium oxide alloy to clad the fuel as it enters the reactor.
46. NDA, *The Magnox Story* (Harwell: Nuclear Decommissioning Authority, 2008), http://www.nuclearsites.co.uk/resources/upload/Magnox%20Brochure2.pdf.
47. Alan Edward Waltar and Albert Barnett Reynolds, *Fast Breeder Reactors* (New York: Pergammon Press, 1981); Claire Le Renard, "The Superphénix Fast Breeder Nuclear Reactor: Cross-border Cooperation and Controversies," *Journal for the History of Environment and Society* 3 (2018): 107–44.
48. Butler and Bud, "United Kingdom Short Country Report."
49. "Nuclear Power Generation Development and the UK Industry," Department of Trade and Industry, accessed January 28, 2005, http://www.dti.gov.uk/energy/nuclear/technology/history.shtml; Brian Wynne, *Rationality and Ritual: The Windscale Inquiry and Nuclear Decisions in Britain* (Bucks: The British Society for the History of Science, 1982).
50. John Simpson, *The Independent Nuclear State: The United States, Britain and the Military Atom* (Dordrecht: Springer, 1983).
51. Andrew Blowers, *The Legacy of Nuclear Power* (Abingdon: Earthscan, 2016).
52. Tony Hall, *Nuclear Politics: The History of Nuclear Power in Britain* (London: Penguin 1986).

53. Ian Welsh, *Mobilising Modernity: The Nuclear Moment* (London: Routledge, 2000).
54. Dolores L. Augustine, *Taking on Technocracy: Nuclear Power in Germany, 1945 to the Present*, vol. 24 (Oxford: Berghahn Books, 2018); Wilfried Konrad and Josep Espluga, "Comparative Cross-Country Analysis on Preliminary Identification of Key Factors Underlying Public Perception and Societal Engagement with Nuclear Developments in Different National Contexts," in *History of Nuclear Energy and Society (HoNESt) Consortium Deliverable no. 4.2* (2018), https://perma.cc/LJ4Q-HT3F.
55. Elizabeth Rough, "Policy Learning through Public Inquiries? The Case of UK Nuclear Energy Policy 1955–61," *Environment and Planning C: Government and Policy* 29, no. 1 (2011): 24–45.
56. Butler and Bud, "United Kingdom Short Country Report."
57. Christoph Laucht, "Atoms for the People: The Atomic Scientists' Association, the British State and Nuclear Education in the Atom Train Exhibition, 1947–1948," *The British Journal for the History of Science* 45, no. 4 (2013): 591–608.
58. Heather Campbell and Robert Marshall, "Utilitarianism's Bad Breath? A Re-Evaluation of the Public Interest Justification for Planning," *Planning Theory* 1, no. 2 (2002): 163–87.
59. Paul A. Anderson, "Justifications and Precedents As Constraints in Foreign Policy Decision-Making," *American Journal of Political Science* 25, no. 4 (1981): 738–61.
60. Andrew Lister, "Public Justification and the Limits of State Action," *Politics, Philosophy & Economics* 9, no. 2 (2010); Campbell and Marshall, "Utilitarianism's Bad Breath?"
61. Anthony Heath, Bridget Taylor, Lindsey Brook, et al., "British National Sentiment," *British Journal of Political Science* 29, no. 1 (1999).
62. Margaret Gowing and Lorna Arnold, *Independence and Deterence: Britain and Atomic Energy, 1945–1952*, vol. 2 (London: MacMillan, 1974).
63. Helmuth Trischler and Robert Bud, "Public Technology: Nuclear Energy in Europe," *History and Technology* 34, no. 3–4 (2018): 187–212.
64. Butler and Bud, "United Kingdom Short Country Report"; Trischler and Bud, "Public Technology."
65. Karen Bickerstaff, Irene Lorenzoni, Nicholas Pidgeon, et al., "Reframing Nuclear Power in the UK Energy Debate: Nuclear Power, Climate Change Mitigation and Radioactive Waste," *Public Understanding of Science* 17, no. 2 (2008): 145–69.
66. Roger Williams, *The Nuclear Power Decisions* (London: Croom Helm, 1980).
67. Joseph S. Nye, "Soft Power," *Foreign Policy* 80 (1990): 153–71.
68. Gordon Mackerron and Frans Berkhout, "Learning to Listen: Institutional Change and Legitimation in UK Radioactive Waste Policy," *Journal of Risk Research* 12, no. 7–8 (2009): 989–1008.
69. Blowers and Pepper, "The Politics of Nuclear Power and Radioactive Waste Disposal."
70. John Glasson, Dominique van Der Wee, and Brendan Barrett, "A Local Income and Employment Multiplier Analysis of a Proposed Nuclear Power Station Development at Hinkley Point in Somerset," *Urban Studies* 25, no. 3 (1988): 248–61.
71. Markku Lehtonen, "Evaluation of 'the Social' in Megaprojects: Tensions, Dichotomies, and Ambiguities," *International Journal of Architecture, Engineering and Construction* 3, no. 2 (2014): 98–109.

72. David Collingridge, "Lessons of Nuclear Power US and UK History," *Energy Policy* 12, no. 1 (1984): 46–67.
73. Jason Chilvers and Jacquelin Burgess, "Power Relations: The Politics of Risk and Procedure in Nuclear Waste Governance," *Environment and Planning A* 40, no. 8 (2008): 1881–1900; Matthew Cotton, *Nuclear Waste Politics: An Incrementalist Perspective* (Abingdon: Routledge, 2017).
74. Andrew Blowers and Pieter Leroy, "Power, Politics and Environmental Inequality: A Theoretical and Empirical Analysis of the Process of Peripheralisation," *Environmental Politics* 3, no. 2 (1994): 197–228.
75. Tom Horlick-Jones, Ana Prades, and Josep Espluga, "Investigating The Degree of 'Stigma' Associated with Nuclear Energy Technologies: A Cross-Cultural Examination of the Case of Fusion Power," *Public Understanding of Science* 21, no. 5 (2012): 514–33; Spencer R. Weart, *Nuclear Fear: A History of Images* (Cambridge, MA: Harvard University Press, 1988).
76. Brian Wynne, *Public Perceptions and the Nuclear Industry in West Cumbria* (Lancaster: Lancaster University, 1993).
77. Easterling and Kunreuther, *The Dilemma of Siting a High-Level Nuclear Waste Repository*.
78. Ray Kemp, "Why Not in My Backyard? A Radical Interpretation of Public Opposition to the Deep Disposal of Radioactive-Waste in the United Kingdom," *Environment and Planning A* 22, no. 9 (1990): 1239–58; Andrew Blowers, David Lowry, and Barry Solomon, *The International Politics of Nuclear Waste* (London: MacMillan, 1991).
79. Steve Carver and Stan Openshaw, *Using GIS to Explore the Technical and Social Aspects of Site Selection for Radioactive Waste Disposal Facilities* (Leeds: School of Geography Working Paper 96/18, University of Leeds, 1996), http://eprints.whiterose.ac.uk/5043/1/96-18.pdf.
80. Cotton, *Nuclear Waste Politics*; Paul Slovic, James H. Flynn, and Mark Layman, "Perceived Risk, Trust and the Politics of Nuclear Waste," *Science* 254, no. 1603–7 (1991): 1603–7; Blowers and Pepper, "The Politics of Nuclear Power and Radioactive Waste Disposal"; Ray Kemp, *The Politics of Radioactive Waste Disposal* (Manchester: Manchester University Press, 1992).
81. Brian Hilton Flowers, *Sixth Report: Nuclear Power and the Environment: Presented to Parliament by Command of Her Majesty, September 1976*, vol. 6 (HMSO, 1976).
82. Flowers, *Sixth Report*.
83. Max Oelschlaeger, "The Myth of the Technological Fix," *Southwestern Journal of Philosophy* 10, no. 1 (1979): 43–53; Alvin M. Weinberg, "Technology and Democracy," *Minerva* 28, no. 1 (1990): 81–90.
84. Eugene A. Rosa and William R. Freudenburg, "The Historical Development of Public Reactions to Nuclear Power: Implications for Nuclear Waste Policy," in *Public Reactions to Nuclear Waste: Citizens' Views of Repository Siting*, ed. Riley E. Dunlap, Michael E. Kraft, and Eugene A. Rosa (London: Duke University Press, 1993): 32–63.
85. Gordon Beveridge, "The Work of a Radioactive Waste Management Watchdog: The Work of the Radioactive Waste Management Advisory Committee," *Interdisciplinary Science Reviews* 23, no. 3 (1998): 209–13.
86. Cotton, "Environmental Justice as Scalar Parity"; Matthew Cotton, "Ethical Assessment in Radioactive Waste Management: A Proposed Reflective Equilibrium-Based Deliberative Approach," *Journal of Risk Research* 12, no. 5 (2009): 603–18.

87. John R. Parkins and Ross E. Mitchell, "Public Participation as Public Debate: A Deliberative Turn in Natural Resource Management," *Society and Natural Resources* 18, no. 6 (2005): 529–40; Jason Chilvers, "Deliberating Competence: Theoretical and Practitioner Perspectives on Effective Participatory Appraisal Practice," *Science, Technology & Human Values* 33, no. 2 (2008): 421–51.
88. Mackerron and Berkhout, "Learning to Listen."
89. Department for Trade and Industry, *Meeting the Energy Challenge: A White Paper on Energy* (London: Department for Trade and Industry, 2007).
90. Walt Patterson, *Keeping the Lights On: Towards Sustainable Electricity* (London: Earthscan, 2007); House of Commons Environmental Audit Committee, *Keeping the Lights on: Nuclear, Renewables and Climate Change, Sixth Report of Session* 6 (London, House of Commons, 2005); Jeremy Robert Grossman, *Keeping the Lights on: Post-Apocalyptic Narrative, Social Critique, and the Cultural Politics of Emotion* (PhD diss., Colorado State University, 2011).
91. Anoeska Buijze, Tijn Kortmann, and Caroline Verwijs-van Fraassen, "Transparency and Nuclear Law: An Instrumental Perspective" in *Proceedings of the 21st Nuclear Inter Jura Congress "Nuclear Law in Progress"* (Buenos Aires: Legis Argentina, 2014): 157–82.
92. Karen Morrow, "On Winning the Battle but Losing the War . . . R (on the Application of Greenpeace Ltd) v Secretary of State for Trade and Industry [2007] EWHC 311, [2007] Env LR 29," *Environmental Law Review* 10, no. 1 (2008): 52–71.
93. John S. Dryzek, "Democratization as Deliberative Capacity Building," *Comparative Political Studies* 42, no. 11 (2009): 1379–1402.
94. Andy Stirling and Philip Johnstone, *A Global Picture of Industrial Interdependencies Between Civil and Military Nuclear Infrastructures* (Brighton: University of Sussex, 2018); Phil Johnstone and Andy Stirling, "Shining a Light on Britain's Nuclear State," *Guardian*, August 7, 2015, https://www.theguardian.com/science/political-science/2015/aug/07/shining-a-light-on-britains-nuclear-state.
95. Ray Kemp, "Planning, Legitimation and the Development of Nuclear Energy: A Critical Theoretic Analysis of the Windscale Inquiry," *International Journal of Urban and Regional Research* 4 (1980): 350–71.
96. William Walker, *Nuclear Entrapment: THORP and the Politics of Commitment* (London: Institute of Public Policy Research, 1999).
97. Timothy O'Riordan, "The Sizewell B Inquiry and a National Energy Strategy," *Geographical Journal* 150, no. 2 (1984): 171–82.
98. Espluga, Medina, and Konrad, "*Case Studies Reports*"; Konrad and Espluga, "*Comparative Cross-Country Analysis on Preliminary Identification of Key Factors Underlying Public Perception and Societal Engagement with Nuclear Developments in Different National Contexts.*"
99. Espluga, Medina, and Konrad, "*Case Studies Reports.*"

CHAPTER 8

Nuclear Energy in Europe: A Public Technology

Stathis Arapostathis, Robert Bud, and Helmuth Trischler

Introduction

Over the past seventy years, having and sharing views on nuclear power has been part of the performance of citizenship. The discussion of nuclear power has been as public as any other political issue less associated with technological expertise. The nature of this discussion has varied. Between powerful affirmation and worried protest there has been hopeful chat and skeptical dismissal. Like all public discourses, it has drawn on grand visions of history, fragmentary knowledge, anecdote, and scientific assertion. While privilege has been granted to experts in some places and at certain times, this has not been universal. Historians must reflect, too, upon the polarization of opinion that has been such an obvious feature of the nuclear power discourse in Europe. Without losing sight of the distinctive technological issues at stake, historians need to bring to bear analysis and understanding that engage with those sometimes incoherent public qualities, rather than seeking only the coherent and the expert.[1]

In this chapter, we therefore propose to draw on our own more general reflections on the concept of "public technology" to provide a structure for reflecting on the specifics of the history of nuclear power in Europe.[2] The concept of public science focuses on public engagement as essential to our understanding of science as a whole. This approach to the production and use of scientific knowledge provides a powerful methodological model for understanding technologies as discrete cultural entities, independent of science. By analogy to the methodological model of science, we suggest the term *public technology* as an analytical tool, useful for interpreting a "delimited, if large, class of individual technologies, in a particular historical period—the Long Twentieth Century (1870–2000). For all the complex ambiguity of categories, in their materiality and social existence these are distinctively 'technologies.'" The Long Twentieth Century was a period in which the experience of encountering new techniques and devices unparalleled in number and power was matched by public discourse in which such innovations held a special place across the (post) industrialized world. Specifically, the concept enables us to think of the history of nuclear power in the context of other public technologies, rather than as a unique phenomenon.[3]

The concept links to a formidable body of writing. In recent years, Sheila Jasanoff and her colleagues have deployed the term "sociotechnical imaginaries" to express societal interpretations of the future. They have argued that imaginaries are stabilized, dominant, and institutionalized societal visions about how social and political life should be organized through the promotion of technoscientific advances. They define "sociotechnical imaginaries" as the "collectively held, institutionally stabilized and publicly performed visions of desirable futures, animated by shared understandings of forms of social life and social order attainable through, and supportive of, advances in science and technology."[4] Through cross-national comparisons they have unraveled the ways different political cultures thought and framed technoscientific problems, risks, and technological innovations and how they have coproduced with collective visions on advances in science and technology. Scholars such as David Nye have, for instance, examined the technological sublime.[5] There is a rich body of literature on the public shaping and engagement with such technologies as the radio and the motor car, and indeed with more feared parallels such as the Zeppelin and poison gas. Each of these became part of public culture, beyond the control of the manufacturers and engineers concerned with their origination or the direct consumers.

It is within this context, we propose, that nuclear power can also be understood. Without overstressing the trope of a *nuclear exceptionalism*, we explore the specificities of nuclear power as a publicly shaped societal entity. We deal with the identification of nuclear energy as a symbol of a kind of technology so new as to merit widespread descriptions by contemporary actors as a renewed industrial revolution. Used widely by journalists, engineers, and politicians in the period from the 1950s to the 1970s, this term differed from the category of "Second Industrial Revolution" employed by economic historians and associated with the final third of the nineteenth century and the early twentieth century as a period of rapid industrial growth based on advancements in manufacturing and production technology as well as electrification.[6] This Second Industrial Revolution served to legitimize a visionary linkage of nuclear with social progress, while at the same time leaving space for society to address the problems posed by new technology.

We study the terms and conditions of the polarization of views between supporters and opponents among the general public in various societies and the mechanisms of coproduction of the nuclear industry with political cultures and social practices. Numerous visions and expectations culturally shaped and contributed to the political and social legitimization of nuclear energy in specific periods during the second half of the twentieth century. We argue that public and social movements were active in assessing and evaluating nuclear technologies as well as in shaping the sociopolitical setting for the valorization of relevant technologies. Transnational antinuclear social movements were far from being centers of opposition to technical knowledge in itself. They were activities centered on knowledge production and the assertion of the legitimacy of counter-expertise. Social opposition or unrest influenced the making of the nuclear power regime by compelling its actors to change their strategy, framings, and technical designs.[7] For some exemplary cases, we even trace the formative power of the public down to the level of concrete decisions in nuclear engineering.

We argue that the focus should expand beyond the antinuclear movements and incorporate implicit or explicit public support for and positive attitudes toward nuclear energy. Nuclear energy experienced massive support, positive attitudes, and social legitimacy in several European nations and post-Soviet countries. In this context, critical events like accidents, natural phenomena and disasters, political unrest, and energy crises affected the patterns of civic engagement and public attitudes.

This chapter covers the periods of the 1950s and 1960s, when generally the benefits of nuclear power were acclaimed; the 1970s even before the Three Mile Island accident; and the 1980s and 1990s, which were overshadowed by a series of incidents, above all Chernobyl. In the West, the public discourses emphasized the impact of nuclear power on the economic prosperity that in some cases was linked with the construction of national identity, and in others with the capacity for industrial growth and participation in international collaboration and competition. On the other side of the Iron Curtain, that impact was equally stressed and publicly linked with the establishment of the ideal socialist society. The dominant sociotechnical imaginary in the communist countries was established and enforced through command and control of public representations of nuclear by the state and the ruling party. This was so strong that it prevented the development of any opposition or ambivalence until the Chernobyl accident. At times, it was informed by national identities and a desire to question Soviet dominance and the dependence on Russian technologies. In the post-Soviet era, post-communist countries appropriated nuclear sociotechnical imaginaries not only as part of their economic liberalization and economic performance but also as part of their national identity formation. We examine here the cultural and sociopolitical shaping of common visions and sociotechnical imaginaries that were instituted through organized political strategies and transnational flows of information and expertise, as well as geopolitics and international industrial politics. The approach thus serves to provide a framework more coherent than the traditional representation of polarities such as those between early optimism and later pessimism across the Iron Curtain, and between science believers and science sceptics.

Nuclear Visions and Expectations as Part of a "New Industrial Revolution"

Nuclear technology was distinguished in postwar Europe by the visions and expectations shaped by politicians, policymakers, and engineers. Public discourse developed by journalists, scientists, and engineers prioritized the effect that nuclear energy technologies could have in covering the rapidly increasing demand. While investment in nuclear technology was strongly linked to the planet-threatening weapons and their use for geopolitical domination, Eisenhower's Atoms for Peace, on the one side, and the Soviet dominance and technocratic organization of societies and social order, on the other, legitimized nuclear energy beyond the obvious military uses. As an energy technology, it fit into a postwar public discourse on technological modernization and dreams of wealth in general, and on the shortage of energy and fear of destruction in particular—not only through its relationship to the environment, disasters, and weapons but also to the release from the bonds of energy shortage, tinkering, national pride, consumption, personal benefit, and the sublime. As early as the late 1940s, civil nuclear energy was becoming discursively constructed in the public realm as something that could be linked with societal progress. The post–atomic bomb years saw the emergence of a public scientific ethos among atomic scientists and the organization and promotion of exhibitions, demonstrations, lectures, and fairs that informed society of the "good or evil" of atomic energy and the Atomic Age.[8] The "Atom Train, conceived by the pacifist Joseph Rotblat, the only physicist to resign from the Manhattan Project on principle, toured Britain in 1947."[9] Audiences were large.[10]

At that time in Germany, the galleries of the Deutsches Museum in Munich were still closed due to war damage. By way of compensation, in 1949 Nobel Laureate Otto Hahn gave a speech in the museum's enormous congress hall, which despite its great size was overcrowded for the event. Hahn, a "discoverer of nuclear fission, lectured on the 'Utilization of the Energy of Atomic

Nuclei'—and complained about the US government's restrictive information policy concerning not only the military but also the civil use of the new power source."[11]

A few years later, the US government changed its course from restriction to support, from defense to offense. On December 8, 1953, President Eisenhower's famous speech to the UN General Assembly in New York City launched the Atoms for Peace campaign. This was something that could convey the image of an altruistic offer to assist the launching of a nuclear energy program while at the same time serving as a strategy for gaining control and hegemony over the nuclear research and development activities of America's European allies.[12] Each of these qualities helped to shape a public image for the technology. It was in this context that Disney made the animated film *Our Friend the Atom*, shown in cinemas in 1954 and later aired on the new Disney television channel in 1957.[13]

Eisenhower's speech was followed two years later by the First International Conference on the Peaceful Uses of Atomic Energy in Geneva, which took place August 8–20, 1955, and sparked a public craze for atomic energy. In 1958 the World's Fair in Brussels had as its icon the Atomium, "a giant model of a unit cell of an iron crystal that illustrated the fair's motto of 'progress of humanity through technological progress.'[14] Large-scale exhibitions featuring nuclear technology as a peacemaking and future-saving force toured all over Europe, from the Deutsches Museum in Munich to the Science Museum in London, seeking to promote an enthusiastic urban audience that would emphatically embrace atomic energy and enable a bright and prosperous future."[15] These exhibitions had a technopolitical function and purpose; they were designed to create public trust in an emerging technology.

The Atoms for Peace exhibition in Geneva was a well-orchestrated propaganda event deployed by the US to promote civilian uses of nuclear energy and related technology while concealing the risks and societal vulnerabilities that nuclear technology could introduce in the social fabric, whether through its association with nuclear weaponry or with radioactivity in its ordinary use. Focusing only on the peaceful use of nuclear energy, the exhibition featured two highlights: the twenty-minute animated cartoon *The Little Giant*, produced by General Electric, and a full-size replica of the graphite nuclear reactor at the Oak Ridge National Laboratory. In many ways, the emerging nuclear age emanated from narrations and visualizations in the popular media, ranging from postcards and magazines to films and cartoons.[16] Through these media, the sociotechnical imaginary was fueled by visions in which atomic energy would be ubiquitous and abundant and would drive new lifestyles through nuclear-propelled cars, ships, and airplanes and a revolution of the agri-food industry.[17]

Because of its association with the unprecedented destructive power of nuclear weapons, this energy technology could be seen as unique. Discussion of it, however, was widely linked with other technological changes and indeed with a new industrial revolution. In the US, the UK, Germany, and elsewhere, the promise of nuclear power was linked to automation in reflections on the new industrial revolution. In part, this followed from the coincidence in timing between the peak in discourse on automation and the summer 1955 Atoms for Peace conference. The approach of British newspapers was reviewed in 1955 by a Swedish journalist who looked principally at public attitudes to automation.[18] He reflected: "Behind the way the subject is presented there is a hope that Britain may, by a combined progress in nuclear energy and automation, lead a 'second industrial revolution' and appear as the leader, or at least secure her standing as a leading industrial power, in a new age. In this way, automation 'catches on' sometimes, perhaps, at the expense of a number of realities."[19] Atomic power was therefore seen within the framework of a larger set of new technologies.

In Germany, the former radar engineer Leo Brandt drafted a paper for the meeting of the Social Democrat Party (SPD) in spring 1956 titled the "Second Industrial Revolution."[20] With a Bohr atom on its cover and the bid for a new nuclear research center for Germany at its end, the brochure was clear in its interpretation of nuclear power as the poster child of a broader era of new technology.

The framing of atomic energy as a leading technology of the Second Industrial Revolution underlay the formation of Euratom in 1958, which sprang from the Messina Conference held at the critical moment of June 1955. Foreign ministers commissioned Paul-Henri Spaak to explore a new stage of economic integration with a resolution whose third paragraph read, "The development of atomic energy for peaceful purposes will soon open the perspective of a new industrial revolution incomparable with that of the past hundred years."[21] Whereas until then the industrial focus of integration had been on the mature industries of coal and steel, nuclear power would be the technological expression of an innovative society and community.

This Second Industrial Revolution discourse can be interpreted as part of the era's generalized optimism. But it came with a challenge. The failings of the earlier industrial revolution were well known. To Marxists it was a stage that had to be traversed to reach socialism, but one that had been fraught and miserable. To social democrats, who were the dominant users of the term, it expressed the need to do things better this time around, better for the workers, for citizens, and indeed for the environment. The engagement of trade unions, popular newspapers, and political parties showed how widespread the engagement with this challenge was.

The promise that atomic energy would secure social progress through cheap energy and technologies for the improvement of agriculture, food production, and public health emerged as an imaginary that legitimized further state intervention and the role of the state as a patron of scientific activities and new science research institutions. At the same time, it was an important push to the industry to bypass its initial reluctant stance. "Public pressure was a driving force for the rise of atomic power, which in turn in some cases also meant that the development of technical alternatives to nuclear energy were blocked or neglected."[22] In countries like Greece and Austria, research on atomic and nuclear physics emerged through institutions that were established exactly in the context of the post–Atoms for Peace era.

In Austria, public discourses promoted by physicists and politicians alike were dominated by an emphasis on the importance of nuclear physics research in promoting technological change and progress and, eventually, the improvement of people's daily lives. Scientific societies such as the Austrian Electrotechnical Society as well as the Ministry of Education supported the establishment and construction of a research reactor and the possibility for energy production from nuclear fission. Lectures and exhibitions were organized to link atomic physics research activities to societal ideals that departed from the National Socialist past, from which the local epistemic communities were keen to distance themselves.[23]

The same period, the 1950s, also witnessed developments in Greece. The establishment of the Greek Atomic Energy Commission (1954) and the nuclear research center *Demokritos* (initially conceived of in 1954) were important institutional innovations that aimed to build research facilities and the local scientific communities as part of the agenda of a modern state in postwar Europe. Throughout this period, the US sought to structure and influence scientific expertise and European epistemic communities according to its own interests, via reconstruction projects and directed aid programs.

The HoNESt project suggests: "In this context, and with the support of the US as well as the political patronage of Queen Frederica and the Greek Palace, the scientific community of nuclear and atomic physicists emerged as an important group of experts. It too promoted atomic and nuclear research for peaceful purposes and positioned itself as a critical group whose research would contribute to the modernization of the Greek state. State-building processes, the diffusion of Western values, and the formation of scientific institutions around the development of nuclear energy emerged synergistically."[24] The politics of artifacts during periods when the state was reconfigured as a political entity chiefly come to the fore during militaristic regimes like the Greek junta (1967–1974). The junta's public statements and plans prioritized the establishment of a nuclear power plant for energy purposes, as manifested by the 1972 plan to foster the large-scale use of nuclear energy with the establishment of ten nuclear power plants by 2000. Through an emerging techno-nationalism, the regime aimed to establish a public representation of Greece as an energy hub of the Balkans. The junta's public plans and imaginaries acquired momentum and influenced public discussion and public policies—albeit on a substantially smaller scale—during the early years of the restoration of democracy. Since the junta's years, the public had begun to imagine nuclear energy either as a means for the country's empowerment in relation to its geostrategic position or, later during the democratic period, as part of the nation's modernization in the context of its integration with Europe.

Nuclear in the East: Between Soviet and National Nuclearity

At the same time, on the other side of the Iron Curtain, and particularly in the USSR, a Russian normalcy was established by promoting the linkage of nuclear power with the upgrade of the standard of living and the population's well-being. A rationalistic perception of nuclear energy permeated public discourse. The establishment of nuclear power plants was linked with national pride and identity, and thus there were no connotations of risk and vulnerability. In Russia, the period from 1955 to 1986 was marked by the unquestioned integration of nuclear energy in development planning and economic policy. Russian society's stance on nuclear energy and the country's nuclear future was positive—something that would maintain a tremendous influence on future attitudes of the Russian population.[25]

The vision of a Russian normalcy was based on state policies that promoted technocratic problem-solving strategies, secretive regimes, and an emphasis on nuclear power to secure economic development. The establishment of the Obninsk nuclear power plant in 1954 was politically and socially legitimized by national pride. A technological advance of this kind would be of untold benefit to the nation and its people. Public discourse linked nuclear power engineering with rational control and organization of energy needs by the state. Russian nuclearity was represented as the ultimate Soviet success story, attesting to the nation's dominance in the international Cold War geopolitical competition.

Through secrecy and control of public information, any criticism was suppressed, and any ambivalences were pushed to the edges and thus marginalized, if not erased.

Secrecy contributed in an ironic way to the shaping of the public nuclear imaginaries. In 1957, the Soviet state took measures to seal any detailed information on the Kyshtym accident. The disaster occurred in Maya (Chelyabinsk-40, today Ozersk in the Chelyabinsk region), a plutonium production site for nuclear weapons and a nuclear fuel reprocessing plant. Thousands of workers and employees as well as the adjacent regions were contaminated. The public representation of the accident was based on secrecy that resulted in the diffusion of panic and continuous rumors that circulated predominantly locally and through face-to-face contact. The contamination chiefly affected the area around the production plant and the 10,000 people from the neighboring village, yet it was a concern for almost 250,000 people in the region. Publicity was minimal, and critical information about the accident and technical dimensions of the risk management remained exclusively within the industrial, military, and government nexus. For the affected territories, however, the Kyshtym accident was a public event that disturbed the local social and cultural order. Villages were evacuated, employees asked for early retirement from the plutonium production works, and contaminated people changed social practices and habits—refusing, for example, to pay public transport fares in lieu of the compensation to which they now considered themselves entitled.

This early accident defined the way the Soviet Union managed not only such critical events but also the distribution of relevant information in general. Until the mid-1980s, nondisclosure or very limited and controlled disclosure was deemed the appropriate strategy to shape a specific, state-controlled sociotechnical imaginary. Economy and geopolitics were the two major concerns, while safety and risks were absent from public discourse. On both sides of the Iron Curtain, secretive regimes coping with risks, uncertainties, and dangers were part of the state culture in the management of nuclear failures and accidents. In the West, however, environmental movements supported by activist scientists succeeded, through public pressure and engagement, in achieving degrees of openness in the management of relevant information.[26]

The emphasis on the economic impact of nuclear energy for power purposes persisted in the Soviet era. The case of Belarus provides a clear example of how, by the late 1960s, pronuclear arguments dominated public discourse, with rhetoric focused on the economic efficiency of nuclear energy and its role in the country's development.[27] The nuclear activities of the Belarusian Soviet Socialist Republic (BSSR) were publicly promoted as an integral part of the

Russian nuclear program. From the late 1960s to the early 1980s, they were discussed in various governmental and public forums, like the BSSR Council of Ministers, the Soviet Council of Ministers, the Ministry of Energy, and scientific institutes. The discussion became more frequent and intense, particularly in 1983, when the construction of the first nuclear thermal power plant started. The program was supported by the republic's scientific elites like the Joint Institute for Power and Nuclear Research, the BSSR Academy of Sciences, the Belarusian State University, the Institutes of Radiophysics, and the Belarusian Polytechnic Institute. Initial discussions and the support expressed by the scientific community were associated with the creation of experimental nuclear installations and the construction of a portable nuclear reactor (PAMIR) and a power plant with local nuclear technology (the BRIG 300). The scientific consent, the institutes' intervention, and support gave the plans social and political legitimacy, despite their experimental and small-scale character. From 1970 to 1985, the Joint Institute and the National Academy of Sciences of Belarus functioned as the prominent expert bodies promoting nuclear power generation as precondition for economic progress. Despite the mutual interest regarding Belarus's nuclear future and the public support given by both Russian and Belarusian scientists and politicians, there were conflicts over the identity of the technology. Belarusians represented BRIG 300 as a nuclear power plant that would be based on local technologies and would thus increase Belarus's technical capabilities and technological independence.[28]

From the 1960s to the mid-1980s, the framing of nuclear technology and energy were important for the active engagement of people and for securing their passive support. Concerns about regional and national pride intermingled and, in some cases, conflicted with an emphasis on the superiority of Soviet engineering ingenuity. Promoted by the central government and state institutions, nuclear energy was deemed as a coherent program based on Soviet engineering traditions. The Permanent Commission on the Peaceful Development of Atomic Energy, initiated in 1961 and led by the Russians, involved the USSR, Bulgaria, Hungary, Poland, GDR, Romania, and Czechoslovakia. It became an institution for the promotion of the Soviet nuclear regime in Eastern Europe, based on planned economic development and a technocratic ideology. Nuclear energy was linked to the Soviet approach that prioritized large-scale energy infrastructures and heavy industries. Moreover, it was socially and politically legitimized as a means of securing communist superiority over the capitalist world.[29] In the case of Lithuania, the construction of the Ignalina nuclear power plant in the early 1980s was hailed as a manifestation of the preeminence of Soviet, and more specifically Russian, engineering. Scientists, engineers, and bureaucrats represented the establishment; the construction and operation of the plant was welcomed as an indicator of the country's successful transition to communism. The Lithuanian daily press described the Ignalina power station as the outcome of the contribution of working-class people from different nationalities who were all under the coordination, supervision, and direction of Soviet power.[30]

Yet the first Bulgarian nuclear power plant became a symbol of both national pride and communist strength in science and technology. The power station in Kozloduy was publicly represented as an expression of Bulgarian technological achievement and capacities. Despite its Soviet origins, the power station was promoted as a technical infrastructure of national importance and linked to the modernization agenda of the 1960s and 1970s. Through the "First Atomic" rhetoric, Kozloduy nuclear power plant was celebrated as the first on the Balkans and the kickoff of a coherent program for the promotion of nuclear energy in the country. Glorifying the triumph of communist science and technology was intermingled with techno-nationalism. The latter represented the nuclear energy program as a key development for the country's empowerment. The 1877 liberation of the city of Kozloduy from the Ottoman Empire by the Romanian Army was linked in a linear way with the site selection for the nuclear power plant as part and parcel of national identity-building. Similarly, Romania's decision to reject Soviet nuclear technology and import the technology from other countries points to the limitations of Soviet hegemony in the Eastern bloc.[31]

The Public Shaping of Nuclear Energy in the West

In the West, even during the 1950s and 1960s, there were emerging anxieties and ambivalences about the risk and vulnerabilities involved in nuclear power plant construction. The planning enquiry into the UK Bradwell nuclear power station in 1956 revealed safety as an underlying anxiety.[32] The public's fear and reservations about the possibility of the establishment of massive nuclear weapons programs grew ever stronger and contributed to policies relevant to nuclear and atom research. In Germany, in 1956, at the prompting of Leo Brandt, the enthusiastic government of North Rhine-Westphalia sought to locate its newly founded nuclear research center in the forests close to Cologne but met fierce local resistance. "Overcrowded meetings at local municipalities happened in a 'highly explosive atmosphere' in which an alarmed public expressed severe concerns about the safety of nuclear technologies. The government had to give in and search for a new site (which eventually was found near the city of Jülich)."[33] Proponents of nuclear energy encountered similar local resistance all over Europe time and again, even in this early period of seemingly unlimited public trust.

Where the public debate and discussion was conducted among scientific experts, fear was not the dominant characteristic of the appropriation of nuclear energy technologies by the public. Yet arguments emerged that stressed the constraints, the economic risks, and the unknowns in relation to the risks that those technologies posed for society.

In Greece, in the 1960s, public discussion on nuclear power was initiated by the Public Power Company (PPC), a state-owned electricity provider. On the one hand, pronuclear energy experts argued that a nuclear power plant was feasible and important for the country's economic development. Nuclear power was represented as a power source cheaper than hydropower and even coal and lignite-based power generation. The prospects of the European Common Market spurred optimistic forecasts for the country's economic development, and thus experts publicly supported a nuclear power plant as a necessary infrastructure and an urgent project on which the PPC would take the lead. On the other hand, rather skeptical views and voices were expressed by the public. The major concerns and sources of ambivalence were the supposedly high costs and the dependency on both foreign technologies and resources. The critical voices from the public, in combination with the lack of a clear and directed state energy policy, help to explain why the envisioned nuclear power plant was never built.[34]

One can find similar arguments, views, and ambivalences in other countries, such as 1960s Austria, when an emphasis on economic rationalism spurred critical views on nuclear energy research and electricity generation infrastructures.

With the optimistic anticipation of construction and operating costs, a belief in the possibility of processing waste in fast breeder reactors,[35] and a determination to participate in the Cold War technopolitical struggle, the 1960s saw widespread growth in the construction and planning of new reactors across Western countries. In the absence of opinion polls, governments and companies had to consider the likely responses by virtual or imagined consumers to various alternative scenarios, such as electricity shortages and the introduction of nuclear power stations.[36]

By the end of the decade, the postwar boom was already waning. Alternative modernisms were emerging, with the rise of environmental movements and a more skeptical attitude toward technological progress. The late 1960s and early 1970s hence marked a turning point in nuclear energy as a public technology. In Western Europe, the public discourse on nuclear energy spurred the emergence of a new civil environmental movement. Both individual citizens and social groups became concerned about nuclear energy on account of their own particular interests. Their engagement in antinuclear protest changed the way they saw both the environment and democracy. In West Germany, the "greenest nation," as well as in other European countries, nuclear energy created a radical democratic potential of grassroots protests.[37] Both elitist groups and dedicated ordinary activists publicly denounced nuclear technologies, and in doing so contributed to the renewal and expansion of liberal democracy in West Germany.[38]

During the 1970s, antinuclear movements emerged not only on the national level all over Western Europe but also on the transnational level. We have previously suggested: "Cultural patterns of public protest crossed national borders; antinuclear actors cooperated internationally, exchanged experiences, discussed strategies, and jointly experimented with new practices to counter the nuclear establishment. They even created international associations and developed into a societal force of European integration. There was a significant transfer of ideas between ecological movements worldwide. These ideas were communicated not only through mediators such as activists but also through politicians, experts, the media, and social organizations. Practical cooperation, however, did take place predominantly among activists in neighboring countries. Obviously, collaboration was more difficult if protest groups lived far apart. Personal contacts and international travel opportunities were crucial, and geographical distance sometimes constituted an obstacle.

Although it might seem in retrospect that the public mood changed only after the 1979 Three Mile Island reactor accident in Pennsylvania, in several countries public anxieties had already proved to be sufficiently strong during the 1970s to halt the development of nuclear power." It is striking that even before the anxieties raised by Three Mile Island, there had been widespread debate across Europe. In Sweden, for example, public attitudes already began to turn strongly against nuclear power in the early 1970s.

Critical events—including oil crises, technological accidents, geopolitical tensions, and natural phenomena such as earthquake disasters—together impacted the formation of sociotechnical imaginaries, civic engagement, and participation. Critical events can trigger new governance practices, legitimize new regulatory regimes, or create pressure that provides competing actors with new legitimacy to further enforce their agency and role in intervening and shaping public participation and civic engagement. In the case of nuclear energy, the latter unfolded at local, national, and transnational levels.

In France, the crisis of the rapid rise in oil prices after 1973 spurred the growth of nuclear power. Whereas the British had gas and oil reserves, the

French drew on technical ingenuity and know-how: in 1974, the prime minister announced the construction of thirteen American-developed pressurized water reactors by 1980, and a longer-term vision of 100 reactors by 2000. Although the aspiration was not quite met, France became Europe's leading nuclear nation. Very little public consultation preceded the decision, which was chiefly a product of experts and technocrats.[39]

During this period in Britain, the critical event was an industrial accident. British engineers, with long-standing experience of disasters caused by brittle metal fracture, worried about the safety of the pressurized water reactors protected by steel pressure vessels. In 1974, when a new generation of reactors had to be chosen, it proved politically impossible to opt for such reactors in the aftermath of the explosion of the Flixborough chemical plant. Although the chemical plant, which made an ingredient of nylon, was technically unrelated to power stations, in the public mind the two became intermingled.[40] The government chose the steam generating heavy water reactor, which did not need a pressure vessel.[41]

Of course, the Three Mile Island accident was the most dramatic critical event of the 1970s. Sweden provides an instructive example on how this led to severe social anxieties that in turn triggered major changes in nuclear technologies. Public pressure was so intense that just four weeks after the accident in the United States, the Swedish government set up a commission to investigate the safety standards at its power plants.[42] The public wanted to know, first, if a serious analysis of the accident's causes should lead to a reassessment of the risks of nuclear power and, second, if the level of safety could be substantially improved by applying new technical means. The Nuclear Safety Commission (NSC), as the newly created body was called, consisted mainly of researchers with expertise not only in nuclear technology, physics, and chemistry but also in biology and psychology. The NSC worked rapidly and presented its final report only half a year later, in mid-November 1979. It worked in parallel with the so-called Kemeny Commission, set up by President Carter and named after John G. Kemeny, president of Dartmouth College. In the final report the NSC included a summary of the Kemeny Commission's findings.

One of the most creative outcomes of the NSC's discussions and investigations was the idea to build filter chambers at nuclear power plants. The purpose of these chambers is to reduce the consequences in case of certain kinds of accidents. If a reactor is damaged, radioactive gases and other gases, including hydrogen, are meant to be held within its containment. However, if pressure builds up or the hydrogen explodes, the containment may collapse (as happened in Fukushima). Filter chambers are intended to ensure that before the pressure becomes critical, a valve can be opened and the gases can be led through a filter chamber that will absorb most of the radioactivity in the gases.

The NSC commissioned two investigations on how such a chamber could be designed. In its final report, it recommended that the development of filter chambers should continue and functioning chambers be installed primarily in reactors close to large cities. That condition was particularly true for the boiling water nuclear power plant in Barsebäck in the far south of Sweden, only twenty kilometers from the Danish capital, Copenhagen. It is unsurprising that both the Danish government and Danish society pressed for the closure of the plant during its entire operating lifetime. The Swedish Nuclear Power Inspectorate continued the investigations on filter chambers, and in 1981 the government decided that from 1985 the Barsebäck nuclear power plant could continue operation only on the condition that a filter chamber would be installed.

The chamber that was built at Barsebäck between 1983 and 1985 consisted of a concrete cylinder thirty-five meters high and with a diameter of twenty-three meters containing 10,000 cubic meters of small stones (about thirty millimeters each). In case of a release through the filter, the gases would be condensed on the stones, to which particles and iodine would then attach. It was calculated that 99.9 percent of the particles and iodine would be captured by the filter chamber. The newly installed safety device has luckily never been tested in real circumstances, and will not be tested in the future either, as the two reactors at Barsebäck were shut down in 1999 and 2005. The filter chamber itself was a public technology. It had evolved from the search for new safety devices in response to the public anxiety in Sweden (and Denmark) as a consequence of the Three Mile Island accident. In reaction to enormous public pressure, the German government followed the Swedish pattern and demanded that all nuclear power plants be equipped with the filtered venting valves known as Wallmann valves, named after Walter Wallmann, the hastily appointed new federal Minister for Environment, Nature Conservation and Nuclear Safety.[43]

The responses to public anxieties were not just technical. As Michael Schüring has shown in a meticulous study on the public relation campaigns of German nuclear energy providers, industry responded by stressing the reliability, safety, and economic necessity of the technology. Its public campaigns shed light on the rhetorical strategies and cultural encodings of an engineering culture under attack, revealing a disposition of technocratic assertiveness and presumption.[44] In many other European countries, when the nuclear industry geared up for the full-scale commercial use of nuclear power plants, it faced increasing public resistance and political obstacles.

With such marketing efforts to foster a positive image of nuclear energy, industry responded—in close alliance with state administrations—to an ever-intensifying public discourse. Europe witnessed the rise of a public critical to the promises that atomic energy would guarantee technological progress. Antinuclear movements emerged on the national level all over Western Europe. In some cases, these movements had strong local and regional characteristics, such as in Greece in the 1970s.[45] In other cases, the movements emerged on the transnational level. Protest cultures crossed national borders; antinuclear actors cooperated transnationally, exchanged experiences, discussed strategies, and jointly experimented with new practices to counter the nuclear establishment.[46] They even created international associations and developed into a societal force of European integration.[47]

Public Responses to Chernobyl

Drawing on in-depth studies on the nuclear discourse in Germany, historian of technology and the environment Joachim Radkau suggested already in the late 1970s, and hence even before Chernobyl, that a rich debate on nuclear energy was catalyzing West Germany's transition into a new kind of civic society. Later, after Chernobyl, he reassessed the nuclear controversy as a "new enlightenment."[48] This controversy engaged all parts of society and fostered societal change on all levels, including the transformation of Protestant churches into arenas open for critical social debates.[49]

The Chernobyl disaster marked the public discussions over the safety of reactors as well as the civic engagement in safety issues and relevant technological design details. Safety of the graphite-moderated RBMK reactors (Reaktor *Bolshoy Moshchnosti Kanalnyy*) became a transnational technopolitical concern to which expert panels, scientific authorities, industrial experts, politicians, and antinuclear movements contributed.[50] Europe saw a variety of different appropriations of the incident. Yet a common characteristic was the revival of antinuclear movements and attitudes in the West. By the time of the accident, environmentalism had been declining in Europe, but the "Chernobyl cloud" reignited opposition that resulted in the cancellation or postponement of various national nuclear programs. The Dutch postponed and canceled any plans for nuclear power construction for at least fifteen years.[51] In Sweden, the disaster led to a temporary revival of the antinuclear rhetoric that favored the immediate phaseout of reactors. The antinuclear camp reunited and demanded the immediate cancellation of the nuclear program, while mass media coverage further boosted public antinuclear attitudes. This led to the amendment of the country's energy policy toward a gradual phaseout. Rather soon, however, public interest declined, permitting stakeholders in nuclear power to acquire a stronger voice in the public debates. The argument that the Swedish reactors were fundamentally different and safer than their Russian counterparts gained public acceptance. Eventually the Swedish energy policy changed once more in the mid-1990s, and the phaseout of the nuclear reactors was postponed for an undefined period.[52]

In France, the Chernobyl disaster failed by some measure to have a long-lasting impact on the country's energy policy. It resulted, however, in a decline of public trust in state institutions. The lack of appropriate, trustworthy information during the incident led to public criticism of secretive and opaque information management and boosted the development of counter-expertise. Citizen-led expert organizations, notably ACRO (*Association pour le Contrôle de la Radioactivité dans l'Ouest*) after Chernobyl and CRIIRAD (*Commission de Recherche et d'Information Indépendantes sur la Radioactivité*), were created to monitor and report the results of the analysis for contamination levels in more than 100 communes. Their research activities questioned official contamination levels, the methodologies to compile toxicological analysis, and the public policies in relation to the risks from radioactive pollution.[53]

In the post-Soviet democracies, repercussions varied and were combined with the implications of the collapse of the USSR, which triggered further concerns and reactions. In Belarus, the Chernobyl disaster empowered local authorities and hitherto excluded social groups to become involved in the evaluation of the technological integrity of Russian reactors. The impact on Belarus bypassed the impact on nuclear culture and energy policy. It influenced its internal politics, as the accident contributed to a significant political transformation beginning with independence, political and social mobilization in the early 1990s, the further development of the political system, and the emergence of new political parties and electoral processes. Any considerations and plans for new nuclear power plants were postponed. Only in 1992 did the country's energy plans make provisions for new nuclear power stations. Since then, however, nuclear energy has acquired new political support and has been integrated in the technopolitical planning by the political elites. In almost twenty years, the public's stance shifted from ambivalence to strong support. Opinion surveys of Belarusians have followed this shift during the last two decades. While in 1997 only 17.6 percent of the Belarusians were supportive, the numbers increased to 28.3 percent in 2005 and to 54.8 percent in 2008.[54]

In Bulgaria, the emphasis on transparency and societal involvement in the management of critical events was linked to the democratization process and the emergence of the post-Soviet and more liberal political order.[55] In Lithuania, the antinuclear movement was engaged both in the transfer of Soviet infrastructures to Lithuanian control and in the emergence, legitimization, and accountability of new political and technological actors.

Conclusion

Thinking about nuclear power as a public technology helps us understand why such a variety of discourses intersects in its analysis. As we have seen in the analysis of several national cases, discourses about modernity, industrialization, and progress shaped the public representations of nuclear energy on both sides of the Iron Curtain. While the West approached nuclear power as part and parcel, if not the cause, of a new industrial revolution, the East sought to achieve an ideal communist society through heavy industrialization and energy abundance based on nuclear power. Ideological trust in nuclear power, relevant technologies, and institutions, as defined in chapter 6 of this volume, characterized both sides of the Iron Curtain, obviously for different reasons. State organizations played key roles in promoting public discourses that linked nuclear energy with the aspired social order. In this setting, nuclear power was tied to the capitalist imaginary of industrial revolutions, initiated to promote social and economic progress. In Eastern European countries, nuclear power generation was identified with the establishment of either the Soviet communist sociopolitical order or varieties thereof with some predominantly national characteristics.

In other cases, as Brian Wynne has shown, dominant public discourses were purely about the technical evaluation of engineering processes, the credibility of experts, and their technocratic problem-solving strategies.[56] Our own analysis has shown that in several countries public discourses and representations were deeply concerned with the qualities of institutions, embracing both their promise and perceived trustworthiness. Such institutions included not just organizations of the nuclear sector but also political parties and indeed national and transnational social movements.

Antinuclear campaigns helped Greenpeace, for instance, to stabilize its reputation as a trusted institution.[57] We have seen here how critical events shaped public attitudes toward the trustworthiness of such institutions and thus the meaning of the nuclear energy "brand." In the case of a public technology as complex as nuclear power, this reassessment of the trustworthiness of institutions at times of crisis was critical. As shown by the Three Mile Island accident, reassessment was only incidentally related to the technical outcome of the event. In the case of Chernobyl, the disaster influenced—albeit temporarily—citizens' evaluation of state institutions in managing risk and information in respect of critical events and the vulnerability of societies.

Time and again, talk about nuclear power has been framed by larger discourses about the risky nature of modernity. These have involved judgment about the management of difficult challenges, which have gone beyond science and engineering to involve national independence and development, experience of technologies as systems involving people as well as machines, national landscapes, and personal aspirations. Even when people seemed "optimistic" in the 1950s, the metaphor of a Second Industrial Revolution carried connotations of the societal risks inherent in technological change as well as its attractions. Citizens have sought to reach judgments by drawing not just on their understanding of the particularities of nuclear power or even of energy generation more broadly but also on their visions of trustworthiness of institutions and visions of the future and meaning of technology and its place in society. The Chernobyl cloud shook public certainties about institutions and experts, revitalized environmental movements in the West, and triggered environmentalism in the East. Despite its international influence, its symbolic legacy, and its impact on the European public memory, the resilience of the imaginary that linked energy with development and industrialization helped to revitalize nuclear power in the early years of the twenty-first century in a variety of nation-states.

Notes

1. On the heterogeneous character of the public response to nuclear energy, see chapter 3.
2. See Helmuth Trischler and Robert Bud, "Public Technology: Nuclear Energy in Europe," *History and Technology* 34, no. 4 (2018): 187–212. Hereafter "PT."
3. PT, 189. For the complex meaning and history of the word *technology*, see Eric Schatzberg, *Technology: Critical History of a Concept* (Chicago: University of Chicago Press, 2018).
4. Sheila Jasanoff, "Future Imperfect: Science, Technology and the Imaginations of Modernity," in *Dreamscapes of Modernity: Sociotechnical Imaginaries and the Fabrication of Power*, ed. Sheila Jasanoff and Sang-Hyun Kim (Chicago: University of Chicago Press, 2015), 1–33, here 4; see also, related to the case of nuclear energy, Sheila Jasanoff and Sang-Hyun Kim, "Containing the Atom: Sociotechnical Imaginaries and Nuclear Power in the United States and South Korea," *Minerva* 47 (2009): 119–46. For similar STS approaches to technical imaginaries, see e.g. Gert Verschraegen et al., eds., *Imagined Futures in Science, Technology and Society* (London and New York: Routledge, 2017), and Armin Grunwald, "Energy Futures: Diversity and the Need for Assessment," *Futures* 43 (2011): 820–30.
5. David C. Nye, *American Technological Sublime* (Cambridge, MA: MIT Press, 1994).
6. See Joel Mokyr, "The Second Industrial Revolution, 1870–1914," in *Storia*

dell'economia Mondiale, ed. Valerio Castronovo (Rome: Laterza Publishing, 1999), 219–45; and Ernst Homburg, "De 'Tweede Industriële Revolutie.' Een Problematisch Historisch Concept," *Theoretische Geschiedenis* 13, no. 3 (1986): 367–85. We are grateful to Ernst Homburg (Maastricht) for bringing his article to our attention.

7. Samuel Walker, *Containing the Atom: Nuclear Regulation in a Changing Environment, 1963–1971* (Berkeley: University of California Press, 1992); David Okrent, *Nuclear Reactor Safety: On the History of the Regulatory Process* (Madison: University of Wisconsin Press, 1981); Mirela Gavrilas et al., *Safety Features of Operating Light Water Reactors of Western Design* (Boca Raton, FL: CRC Press, 1995).
8. For the concept of nuclearity, see Gabrielle Hecht, "Nuclear Ontologies," *Constellations* 13 (2006): 320–31.
9. PT, 196. See Reiner Braun et al., eds., Joseph Rotblat, *Visionary for Peace* (Weinheim: Wiley-VCH, 2007); Andrew Brown, *Keeper of Nuclear Conscience: The Life and Work of Joseph Rotblat* (Oxford: Oxford University Press, 2012); Christoph Laucht, "Atoms for the People: The Atomic Scientists' Association, the British State and Nuclear Education in the Atom Train Exhibition, 1947–1948," *British Journal for the History of Science* 45 (2012): 591–608.
10. Even on Christmas Eve 1947, several hundred people attended the exhibition in Dundee. See *Dundee Courier*, December 25, 1947; the same paper had reported the speech by Norman Feather in "Atom Train in City Next Week," December 18, 1947, and the speech by the mayor at the opening in "Steam Vies with Atom Train," December 23, 1947; the popularity of various sections was reported in "Last Day of Atom Train Exhibition," December 24, 1947.
11. PT, 196–97. Otto Hahn, "Die Nutzbarmachung der Energie der Atomkerne," *Abhandlungen und Berichte des Deutschen Museums* 18, no. 2 (1950): 3–37.
12. John Krige, *American Hegemony and the Postwar Reconstruction of Science in Europe* (Cambridge, MA: MIT Press, 2006); John Krige, "Atoms for Peace, Scientific Internationalism, and Scientific Intelligence," *Osiris* 21 (2006): 161–81; John Krige, "Building the Arsenal of Knowledge," *Centaurus* 52 (2010): 280–96.
13. Walker Mechling and Jay Mechling, "The Atom according to Disney," *Quarterly Journal of Speech* 81 (1995): 436–53.
14. See Brigitte Schroeder-Gudehus and David Cloutier, "Popularizing Science and Technology during the Cold War: Brussels 1958," in *Fair Representations. World's Fairs and the Modern World*, ed. Robert W. Rydell and Nancy Gwinn (Amsterdam: VU University Press, 1994), 157–80.
15. PT, 197. See Karen Königsberger, *"Vernetztes System"? Die Geschichte des Deutschen Museums 1945–1980 dargestellt an den Abteilungen Chemie und Kernphysik* (München: Utz, 2009), 247–51.
16. Dirk van Lente, ed., *The Nuclear Age in Popular Media: A Transnational History, 1945–1965* (Basingstoke: Palgrave Macmillan, 2012); see also John O'Brian, *Atomic Postcards: Radioactive Messages from the Cold War* (Bristol: Intellect, 2011); John O'Brien, ed., *Camera Atomica* (London: Black Dog Publishing, 2015); Spencer R. Weart, *Nuclear Fear: A History of Images* (Cambridge, MA: Harvard University Press, 1988); Sophie Forgan, "Atoms in Wonderland," *History and Technology* 19 (2003): 177–96; Catherine Jolivette, ed., *British Art in the Nuclear Age* (Burlington: Ashgate, 2014).

17. Margot A. Henriksen, *Dr. Strangelove's America: Society and Culture in the Atomic Age* (Berkeley: University of California Press, 1997).
18. Daniel Viklund, "The Automatic Factory—What Does It Mean?," *Institution of Production Engineers Journal* 34, no. 12 (1955): 817.
19. Viklund, "The Automatic Factory—What Does It Mean?," 817.
20. Leo Brandt, *Die zweite industrielle Revolution* [Anlässlich des SPD-Parteitages vom 10. bis 14. Juli 1956 in München vorgestelltes Referat. Anhang: Atomplan der SPD, erarbeitet vom Ausschuss für Fragen der Atomenergie beim Parteivorstand der SPD] (Bonn: Vorstand der SPD, 1956). See Bernhard Mittermaier and Bernd-A. Rusinek, eds., *Leo Brandt (1908–1971): Ingenieur—Wissenschaftsförderer—Visionär. Wissenschaftliche Konferenz zum 100. Geburtstag des Nordrhein-Westfälischen Forschungspolitikers und Gründers des Forschungszentrums Jülich* (Jülich: Forschungszentrum, 2009).
21. "Le développement de l'énergie atomique à des fins pacifiques ouvrira à brève échéance la perspective d'une nouvelle révolution industrielle sans commune mesure avec celle des cent dernières années." See https://www.cvce.eu/obj/resolution_adoptee_par_les_ministres_des_affaires_etrangeres_des_etats_membres_de_la_ceca_messine_1er_au_3_juin_1955-fr-d1086bae-0c13-4a00-8608-73c75ce54fad.html. See Ilina Cenevska, *The European Atomic Energy Community in the European Union Context: The "Outsider" Within* (Leiden: Brill, 2016), 289–90.
22. PT, 198. See Joachim Radkau, *Aufstieg und Krise der deutschen Atomwirtschaft 1945–1975: Verdrängte Alternativen in der Kerntechnik und der Ursprung der nuklearen Kontroverse* (Reinbek bei Hamburg: Rowohlt, 1983), 87.
23. Christian Forstner, "The Failure of Nuclear Energy in Austria: Austria's Nuclear Energy Programmes in Historical Perspective," in *Pathways into and out of Nuclear Power in Western Europe: Austria, Denmark, Federal Republic of Germany, Italy, and Sweden*, ed. Astrid M. Kirchhof (Munich: Deutsches Museum, 2019), 36–73; Christian Forstner, "Nuclear Energy Programs in Austria," *Jahrbuch für Europäische Wissenschaftskultur* 7 (2012): 413–32.
24. Stathis Arapostathis et al., "Greece Short Country Report (version 2018)," in *History of Nuclear Energy and Society (HoNESt) Consortium Deliverable no. 3.6* (2019), 7–8, https://hdl.handle.net/2454/38269; Maria Rentetzi, "Gender, Science and Politics: Queen Frederika and Nuclear Research in Post-war Greece," *Centaurus* 51 (2009): 63–87; Krige, "Atoms for Peace."
25. See also, for the following, Natalia Melnikova et al., "Russia Short Country Report (version 2018)," in *History of Nuclear Energy and Society (HoNESt) Consortium Deliverable no. 3.6* (2019), https://hdl.handle.net/2454/38269.
26. See Bernd Helmbold, "German Democratic Republic (GDR) Short Country Report (version 2018)," in *History of Nuclear Energy and Society (HoNESt) Consortium Deliverable no. 3.6* (2019), https://hdl.handle.net/2454/38269; Stathis Arapostathis et al., " 'Tobacco for Atoms': Nuclear Politics, Ambivalences and Resistances about a Reactor that Was Never Built," *History of Technology* 27 (2017): 205–27; Forstner, "Failure of Nuclear Energy"; Jan-Henrik Meyer, " 'Atomkraft—Nej tak': How Denmark Did Not Introduce Commercial Nuclear Power Plants," in Kirchhof, *Pathways into and out of Nuclear Power in Western Europe*, 74–123.
27. Andrei Stsiapanau, "Belarus Short Country Report (version 2018)," in *History of Nuclear Energy and Society (HoNESt) Consortium Deliverable no. 3.6* (2019), https://

hdl.handle.net/2454/38269; see also Sonja D. Schmid, "Nuclear Colonization? Soviet Technopolitics in the Second World," *Entangled Geographies: Empire and Technopolitics in the Global Cold War*, ed. Gabrielle Hecht (Cambridge, MA: MIT Press, 2011), 139.

28. Stsiapanau, "Belarus Short Country Report," 8–9.
29. Paul Josephson, "Atomic-Powered Communism: Nuclear Culture in the Postwar USSR," *Slavic Review* 55 (1996): 297–324; Paul Josephson, *Red Atom: Russia's Nuclear Power Program from Stalin to Today* (Pittsburgh: University of Pittsburgh Press, 2005); see also Schmidt, "Nuclear Colonization."
30. Andrei Stsiapanau, "Lithuania Short Country Report (version 2018)," in *History of Nuclear Energy and Society (HoNESt) Consortium Deliverable no. 3.6* (2019), https://hdl.handle.net/2454/38269.
31. Ivan Tchalakov and Ivaylo Hristov, "Bulgaria Short Country Report (version 2019)," in *History of Nuclear Energy and Society (HoNESt) Consortium Deliverable no. 3.6* (2019), https://hdl.handle.net/2454/38269; Ivaylo T. Hristov, *The Communist Nuclear Era: Bulgarian Atomic Community During the Cold War, 1944–1986* (Eindhoven: Eindhoven University Press, 2014), 70–75, 80–89, 100–110. See also Schmidt, "Nuclear Colonization."
32. Elizabeth Rough, "Policy Learning through Public Inquiries? The Case of UK Nuclear Energy Policy 1955–61," *Environment and Planning C: Government and Policy* 29 (2011): 24–45. See also K. A. Parkhill et al., "From the Familiar to the Extraordinary: Local Residents' Perceptions of Risk When Living with Nuclear Power in the UK," *Transactions of the Institute of British Geographers* 35, no. 1 (2010): 39–58.
33. PT, 198. The meeting was described in: *Kölnische Rundschau*, August 7, 1956, in Bernd-A. Rusinek, *Das Forschungszentrum: Eine Geschichte der KFA Jülich von ihrer Gründung bis 1980* (Frankfurt a.M. and New York: Campus, 1996), 226.
34. Arapostathis et al., "Tobacco for Atoms."
35. See e.g. Stuart Butler and Robert Bud, "United Kingdom Short Country Report (version 2018)," in *History of Nuclear Energy and Society (HoNESt) Consortium Deliverable no. 3.6* (2019), https://hdl.handle.net/2454/38269.
36. PT, 199; Alessandro Maranta et al., "The Reality of Experts and the Imagined Lay Person," in *Acta Sociologica* 46, no. 2 (2003): 150–65; Kate Burningham et al., "Industrial Constructions of Publics and Public Knowledge: A Qualitative Investigation of Practice in the UK Chemicals Industry," *Public Understanding of Science* 16, no. 1 (2007): 23–43; see also Tomas Moe Skjølsvold, "Publics in the Pipeline. On Bioenergy and its Imagined Publics in Norway and Sweden," *Past and Present Energy Societies: How Energy Connects Politics, Technologies and Cultures*, ed. Nina Möllers and Karin Zachmann (Bielefeld: transcript, 2012), 277–304; Julie Barnett et al., "Imagined Publics and Engagement around Renewable Energy Technologies in the UK," *Public Understanding of Science* 21, no. 1 (2012): 36-50; Ian Welsh and Brian Wynne, "Science, Scientism and Imaginaries of Publics in the UK: Passive Objects, Incipient Threats," *Science as Culture* 22 (2013): 540–66; Kjetil Rommetveit and Brian Wynne: "Technoscience, Imagined Publics and Public Imaginations," *Public Understanding of Science* 26, no. 2 (2017): 133–47.
37. Frank Uekötter, *The Greenest Nation? A New History of German Environmentalism* (Cambridge, MA: MIT Press, 2014).
38. Stephen Milder, *Greening Democracy: The Anti-Nuclear Movement and Political*

Environmentalism in West Germany and Beyond, 1968–1983 (Cambridge: Cambridge University Press, 2017); Stephen Milder, "Protest and Participation," *Modern Germany in Transatlantic Perspective*, ed. Michael Meng and Adam Seipp (New York: Berghahn Books, 2017), 217–41; Stephen Milder, "From Wyhl to Wall Street: Occupation and the Many Meanings of 'Single Issue' Protest," *Moving the Social* no. 56 (March 2017): 5–26; Andrew S. Tompkins, *Better Active than Radioactive: Anti-Nuclear Protest in 1970s France and West Germany* (Oxford: Oxford University Press, 2016). For the following quote, see PT, 200.

39. See e.g. Gabrielle Hecht, *The Radiance of France: Nuclear Power and National Identity after World War II* (Cambridge, MA: MIT Press, 2009); Michael Bess, *The Light Green Society. Ecology and Technological Modernity in France, 1960–2000* (Chicago: University of Chicago Press, 2003); Markku Lehtonen, "France Short Country Report (version 2019)," in *History of Nuclear Energy and Society (HoNESt) Consortium Deliverable no. 3.6* (2019), https://hdl.handle.net/2454/38269.
40. Butler and Bud, "United Kingdom Short Country Report," 23.
41. Butler and Bud, "United Kingdom Short Country Report," 23–25.
42. For the following, see *Säker Kärnkraft? Betänkande av reaktorsäkerhetsutredningen*, SOU, 1979:86 (Stockholm: LiberFörlag/Allmänna förlaget, 1979); *Regeringens proposition 1984/85: 120 om riktlinjer för energipolitiken* (Stockholm: Sveriges Riksdag, 1985); *Rapport från den dansk-svenska kommittén 1983–84 om Barsebäcksverket*, DsI: 1985 (Stockholm: LiberFörlag/ Allmänna förlaget 1985), 1; Henrik Borg and Helen Sannerstedt, *Barsebäcks kärnkraftverk. Dokumentation* (Kritstianstad: Regionmuseet Kristianstad/Landsantikvarien i Skåne, 2006). The authors are grateful to Arne Kaijser for his input on the Barsebäck case.
43. Anna V. Wendland, "Reaktorsicherheit als Zukunftskommunikation. Nuklearpolitik, Atomdebatten und kerntechnische Entwicklungen in Westdeutschland und Osteuropa 1970–2015," in *"Security Turns Its Eye Exclusively to the Future": Zum Verhältnis von Sicherheit und Zukunft in der Geschichte*, ed. Christoph Kampmann, Angela Marciniak, and Wencke Meteling (Baden-Baden: Nomos, 2018), 305–52.
44. Michael Schüring, "Advertising the Nuclear Venture: The Rhetorical and Visual Public Relation Strategies of the German Nuclear Industry in the 1970s and 1980s," *History and Technology* 29 (2013): 369–98.
45. See also, for the following, Arapostathis et al., "Tobacco for Atoms."
46. Stephen Milder, "Thinking Globally, Acting (Trans-)Locally: Petra Kelly and the Transnational Roots of West German Green Politics," *Central European History* 43 (2010): 301–26; Astrid Mignon Kirchhof and Jan-Henrik Meyer, eds., *"Global Protest against Nuclear Power: Transfer and Transnational Exchange in the 1970s and 1980s," Historical Social Research* 39, no. 1 (2014), special issue. For specific examples of transnational support and cooperation, see Astrid Mignon Kirchhof, "Spanning the Globe: Australian Protest against Uranium Mining and their West-German Supporters," *Historical Social Research* 39 (2014): 254–73.
47. Jan-Henrik Meyer, "Challenging the Atomic Community: The European Environmental Bureau and the Europeanization of Anti-nuclear Protest," in *Societal Actors and European Integration. Polity-building and Policy-making 1958–1992*, ed. Wolfram Kaiser and Jan-Henrik Meyer (Basingstoke: Palgrave Macmillan, 2013), 197–220.

48. Joachim Radkau, "Kernenergie-Entwicklung in der Bundesrepublik: ein Lernprozeß? Die ungeplante Durchsetzung des Leichtwasserreaktors und die Krise der gesellschaftlichen Kontrolle über die Atomwirtschaft," *Geschichte und Gesellschaft* 4 (1978): 195–222; Joachim Radkau, "Die Kernkraft-Kontroverse im Spiegel der Literatur. Phasen und Dimensionen einer neuen Aufklärung," in *Das Ende des Atomzeitalters? Eine sachlich-kritische Dokumentation*, ed. Armin Hermann and Rolf Schumacher (München: Moos & Partner, 1987), 307–34; see also Joachim Radkau, "Hiroshima und Asilomar: Die Inszenierung des Diskurses über die Gentechnik vor dem Hintergrund der Kernenergie-Kontroverse," *Geschichte und Gesellschaft* 14 (1988): 329–63.
49. Michael Schüring, "West German Protestants and the Campaign against Nuclear Technology," *Central European History* 45 (2012): 744–62.
50. Thomas R. Wellock, "The Children of Chernobyl: Engineers and the Campaign for Safety in Soviet-designed Reactors in Central and Eastern Europe," *History and Technology* 29, no. 1 (2013): 3–32.
51. Eric Berkers, "The Netherlands Short Country Report (version 2018)," in *History of Nuclear Energy and Society (HoNESt) Consortium Deliverable no. 3.6* (2019), https://hdl.handle.net/2454/38269, 4; Frank Geels and Bram Verhees, "Cultural Legitimacy and Framing Struggles in Innovation Journeys: A Cultural-Performative Perspective and a Case Study of Dutch Nuclear Energy (1945–1986)," *Technological Forecasting and Social Change* 78 (2011): 910–30.
52. Arne Kaijser, "The Referendum that Preserved Nuclear Power and Five Other Critical Events in the History of Nuclear Power in Sweden," in Kirchhof, *Pathways into and out of Nuclear Power in Western Europe*, 238–93.
53. Lehtonen, "France Short Country Report."
54. Stsiapanau, "Belarus Short Country Report."
55. Tchalakov and Hristov, "Bulgaria Short Country Report," 18–21; Hristov, *Communist Nuclear Era*, 70–75, 80–89, and 100–10.
56. Brian Wynne, "Risk and Environment as Legitimatory Discourses of Technology: Reflexivity Inside-out," *Current Sociology* 50 (2002): 459–77; Ian Welsh and Brian Wynne, "Science, Scientism and Imaginaries of Publics in the UK: Passive Objects, Incipient Threats," *Science as Culture* 22 (2013): 540–66.
57. Frank Zelko, *Make It a Green Peace! The Rise of Countercultural Environmentalism* (New York: Oxford University Press, 2013).

CHAPTER 9

Nuclear Installations at European Borders: Transboundary Collaboration and Conflict

Arne Kaijser and Jan-Henrik Meyer

Introduction

In Europe a remarkable number of nuclear power plants, nuclear waste facilities, reprocessing plants, and uranium mines are—or were planned to be—located close to national borders. Some of these have become deeply contested. In fact, some of the largest and most dramatic antinuclear protests have been directed against such facilities, often involving participants from both sides of the border. Four examples illustrate this.

On August 7, 1976, about 10,000 Swedish and Danish demonstrators marched against the Swedish nuclear power plant in Barsebäck, located only twenty kilometers away from Copenhagen, the Danish capital. It was the largest antinuclear demonstration at the time in Scandinavia, and it was followed by similar, even larger, demonstrations in the following years.[1]

On July 31, 1977, 30,000 protesters from France, West Germany, and Switzerland tried to invade the construction site of the fast breeder reactor officially called Superphénix Creys-Malville, close to the Swiss border. One demonstrator was killed, and twenty were seriously injured in clashes with the police.[2]

In the summer of 1983, West German antinuclear activists established camps close to the village of Gorleben, a few kilometers from the East German border, to protest plans to build a reprocessing plant and a final repository. Some of the activists even entered into the forbidden border area, the so-called no-man's land, which legally belonged to East Germany but was to the west of East Germany's border fences. There they sat, playing guitar, singing, and swaying a banner. The protest event was prominently covered by West German television.[3]

On April 2, 1987, the vice president of the provincial council of Salamanca was held captive for thirty hours in the town hall of the small Spanish town of Aldeadávila de la Ribera, four kilometers from the Portuguese border, in protest against plans to build a pilot plant for nuclear waste disposal there. The vice president was finally freed by the police, but large demonstrations with both Spanish and Portuguese participants were organized in the following weeks.[4]

The construction of nuclear facilities has not, however, only been connected with conflict. In many cases it also involved extensive cross-border cooperation. Such cooperation was characteristic not least of the early years, when nuclear developments were still mainly associated with visions of modernity, prosperity, and progress (see chapters 1 and 8). Some of these facilities were planned and built from the very beginning in close collaboration between power companies, research institutions, and regulating agencies on both sides of the border. So far, however, most of the research concerning nuclear facilities close to borders has focused almost entirely on the protests against them.[5] Our aim is to broaden the scope of analysis and to study both the cooperative and the conflict-ridden relations among the various actors involved in or engaged with nuclear facilities close to borders: power companies; research institutions; regulating agencies; antinuclear organizations; local, regional, and national governments; and international organizations. A nuclear facility at the border may promise shared economic benefits, but it also raises issues of transboundary risks and of environmental externalities for those on the other side of the border, turning nuclear facilities into a "transboundary issue."[6]

By looking at a number of different facilities from the late 1960s until today, we set out to analyze a number of issues: Which factors influenced siting decisions, and why were so many facilities placed close to national borders? How did the perception of "closeness" change over time? And, in particular, what kind of cross-border relationships emerged when these facilities were built? How did they differ in character, depending on context? How have these relationships changed over time, notably in the wake of major nuclear accidents or in response to overarching political changes?

We have deliberately chosen the concept *transboundary relationships* to characterize the specific conditions linking partners and adversaries across national boundaries. The rise of transnational history has contributed to a growing confusion about concepts regarding relations beyond the nation-state, for which "transnational" has become a convenient but increasingly fuzzy shorthand.[7] In this chapter, we will distinguish between international, transnational, and transboundary (or cross-border) relations.

Conventionally, *international relations* refer to relations between state actors and international organizations, and tend to be dominated by intergovernmental relations. By contrast, *transnational relations* involve nonstate actors.[8] However, this chapter focuses on the examination of nuclear installations as transboundary issues relating to a shared border between contiguous countries. In the case of environmental problems, transboundary issues usually result from cross-border contamination—or risk of pollution—due to geographical proximity.[9] Transboundary issues are thus distinct from broader transnational issues, many of which environmental historians have studied, such as fallout from nuclear weapons testing or long-range air pollution.[10] Transnational issues may result from impacts at a distance, and do not necessarily require a common border. Transboundary issues can thus be considered a subset of transnational issues in which the presence of the border is a constitutive part of the problem definition. Transboundary issues trigger a variety of transboundary relations, conflictual or cooperative, between different actors and at different geographical and political scales. Such relations are at the heart of our analysis.

Empirically, the chapter mainly draws on a comparative discussion of four case studies.[11] The first focuses on the Iberian Peninsula, where different kinds of nuclear facilities were located (or were planned to be located) close to the Spanish-Portuguese border.[12] The second case deals with a Swedish nuclear power plant at Barsebäck on the coast of the Öresund only twenty kilometers from Copenhagen.[13] The third case is about the fast breeder reactor Superphénix, co-owned by France, Italy, and Germany, near the small French village of Malville, on the Rhône river upstream from Lyons about 100 kilometers from the Swiss and Italian borders.[14] The fourth case is the planned final nuclear waste repository in Gorleben, West Germany, very close to the East German border, while a similar repository was in Morsleben, some 120 kilometers farther south on the other side of the border.[15] These cases have been covered in detail in the four journal articles referenced here and will provide the primary evidence for this analysis, unless other references to further literature are given.

First, we examine what influences the siting of nuclear installations and why so many nuclear facilities in Europe are located close to borders. Second, we discuss the perception of closeness and how it has changed over time. The third and main part of the chapter addresses how the siting close to borders spurred conflict and cooperation between different kinds of actors. The conclusions will highlight the relevance of borders for the relations between nuclear energy and society.

The Siting of Nuclear Installations: Why Are So Many Nuclear Facilities in Europe Located Close to Borders?

As international awareness of issues of nuclear safety was growing in the wake of the 1979 Three Mile Island accident, the renowned science journal *Nature* drew attention to the phenomenon of "border problems" caused by the ongoing expansion of nuclear power. In its December 1980 issue, *Nature* reported that a substantial number of nuclear installations in Europe were located near national borders. "Of the 120 or so nuclear power stations at present in operation, under construction or planned in the European Community, 33 are less than 25 miles from national borders and another 15 are less than 6 miles from borders."[16]

Why are so many nuclear power plants and other nuclear installations in Europe located close to borders? A first reason has to do with the *political geography* of Europe: it is a small continent with many countries. This means that a rather large part of the entire territory is close to a national border as compared with, for example, North America. Thus, even if nuclear facilities had been distributed at random across Europe, a fair number would have been located close to borders. However, such facilities were not distributed randomly. On the contrary, a number of factors were taken into consideration when decisions on localization were made.[17] These factors differ somewhat among different kinds of facilities.

For nuclear power plants, most of the *economic and technical considerations* had to do with existing infrastructure. They needed to be located at a place with good transport facilities—ideally a harbor—as very large and heavy components needed to be installed during construction and spent fuel needed to be transported in heavy containments. Abundant access to cooling water was also a necessity. Moreover, they had to be strategically located in relation to the power grid to strengthen security of supply and minimize transmission costs. This factor often favored locations not too far from larger cities. However, this had to be balanced against *safety considerations*. For safety reasons, the population density close to the plant had to be low.

In particular, two of the above factors favored locations close to borders. Abundance of cooling water and access to harbors made sites by the sea or along major rivers desirable. Coastal locations are by definition close (twelve miles at most) to national borders but not necessarily close to a neighboring country. Major rivers sometimes form national borders and—notably in continental Europe—often flow through several countries, creating environmental justice issues of upstream locations potentially adversely impacting downstream water users (on environmental justice, see chapter 7).

Similar criteria applied to other nuclear facilities, like reprocessing plants, fuel factories, and waste facilities, which also required good transport facilities and low population densities. Placing such installations close to nuclear power plants was also seen as an advantage, and decisions to co-locate different kinds of nuclear installations were not uncommon. For final repositories, however, the geological conditions were a very important factor. Desirable geological formations included solid bedrocks, salt mines, and special kinds of clay. The location of uranium mines was of course also crucially dependent on geological conditions, namely the presence of sufficiently substantial uranium deposits.

For all kinds of nuclear facilities there were also *political considerations* at play when localization decisions were made. For example, at times politicians proposed placing nuclear installations in economically struggling rural or postindustrial regions. Investment in nuclear facilities was intended to combat local unemployment and foster regional development. In a number of cases, such economically deprived regions were close to borders, such as the former coal mining districts in Lorraine in France, where Cattenom is located.[18] Similarly, the northeast of Bavaria, bordering the communist Czechoslovakia, was selected for the planned Wackersdorf reprocessing plant, which was intended to provide new jobs after coal mining stopped in 1982.[19] It was also considered suitable because the region had been an "energy landscape" for a long time.[20] Moreover, German planners assumed that predominantly Catholic and conservative populations in rural districts were less likely to protest. Critics of nuclear power raised similar suspicions about the political rationale behind the Bavarian government's decision for Wackersdorf. Such suspicions stirred up substantial regional resentment, which concerned issues of environmental justice and respect for the Upper Palatinate region within the government in Munich.[21] Conveniently far away from the bustling Bavarian capital and the Alpine tourist landscapes, the sleepy, overwhelmingly Catholic, conservative, and economically deprived region bordering the Iron Curtain seemed least likely to resist.[22]

Political reasons also mattered for the suggested siting of a reprocessing plant and disposal facility in Gorleben near the inner German border. Next to the geological conditions (the existence of salt domes), low population density, and economic marginality, the border was also a political concern. This played out in two ways: as a technocratic planning consideration of government, and once again—something unanticipated by the planners—as a case for local resentment, which helped antinuclear mobilization. From a planning perspective, it was convenient that in case of protest or accident, the area could be easily sealed off, surrounded as it was on three sides by the Iron Curtain. This, however, would leave the locals trapped—and was thus a specter they feared and resented. An additional advantage from the planners' point of view was that the predominantly westerly winds ensured that most of the emissions of the reprocessing plant's chimneys would rain down on

East German territory.[23] According to such a rationale, the border siting would reduce the number of citizens of one's own country potentially affected by the plant, while those on the other side of the border were irrelevant as voters and had little opportunity to take action, living in a socialist country with no free speech or right of assembly. Nonetheless, the West German federal government initially disapproved of the siting because it feared diplomatic problems at the inner German border, which was already fraught with conflict.[24]

Finally, there were some nuclear facilities for which closeness to a national border was an important factor in itself, namely for reactors that were built in cooperation between several countries. For example, the breeder reactor Superphénix was built by a French, West German, and Italian consortium and located in France near both Germany and Italy. At the practical level, this was intended to help contractors from the partner countries to access the site. Electricity generated there was to be fed into the participating countries' systems. At the symbolic level, the project was meant to demonstrate the achievement of peaceful European integration through technological projects. By coincidence of European geography, the site happened to be close to Switzerland as well. The Swiss, however, did not participate in and thus potentially benefit from the project. Many Swiss critics of nuclear power resented the plutonium-producing plant in their vicinity. Likewise, the breeder reactor in Kalkar, built by Germany, the Netherlands, and Belgium, was located in Germany close to the Dutch and Belgian borders for similar symbolic and practical reasons.[25]

The Changing Perception of What It Means to Be "Close"

What does it mean to be "close" to a nuclear installation? The perception of closeness has changed considerably over time. In the 1960s and 1970s, when most European nuclear power plants were built, few countries had explicit siting criteria for nuclear plants. However, most regulating agencies in Western Europe sought to follow the criteria developed by the United States Nuclear Regulatory Commission (NRC), which were first formulated in 1962.[26] According to these criteria, there should be an exclusion area around a nuclear plant, with no inhabitants at all, followed by a low population zone. The first zone should have a radius of between 200 meters and two kilometers, with an average of 650 meters. The second zone should have a radius of about five kilometers. In addition, the distance to the nearest urban area with more than 25,000 inhabitants should not be less than about eight kilometers. These criteria were based on an assessment of the possibilities for evacuating the population in case of an accident. In 1975, the US NRC further developed these criteria in a guide titled *General Site Suitability Criteria for Nuclear Power Stations*. While this guide has since been updated and revised a number of times, the distance criteria for exclusion areas and low population zones have remained the same.[27]

In hindsight, nuclear power plants like the Swedish one in Barsebäck, twenty kilometers from Copenhagen, or the French plant in Cattenom, twenty-four kilometers from Luxembourg, may seem dangerously close to big cities, which are, moreover, their neighbor's capitals. But when these plants were planned in the late 1960s and early 1970s, they were in accordance with the siting criteria used at the time. Nevertheless, the Swedish nuclear authorities informed their Danish counterparts about the plans for a nuclear power plant in Barsebäck as soon as they were informed about them in 1968. This illustrates that the agency thought that the location might be questioned, or at least that there might be potential concerns across the border.

Within a few years, when antinuclear movements emerged in both countries, Barsebäck's siting was indeed very strongly questioned. The Three Mile Island accident in 1979, with its experience of mass evacuation of the area close to the plant, reinforced the awareness of transboundary risks to the Danish population posed by the Swedish plant. By contrast, however, a Danish pronuclear organization dismissed such safety concerns outright as irrational, scientifically unfounded fears, not only in the wake of Three Mile Island but also after Chernobyl, with a publication affirmatively titled *Barsebäck—A Safe Neighbor*.[28] Yet to most Europeans, Chernobyl demonstrated that a nuclear meltdown could have important effects at very long distances. This led the Danish antinuclear Organization for Nuclear Information, or more literally, "for Enlightenment about Atomic Power" (Organisationen til oplysning om Atomkraft, OOA), to take its protest across borders not only of Sweden but also of those of West and East Germany. In its "Radiating Neighbors" campaign, the OOA targeted all reactors within 150 kilometers of Danish borders.[29]

In a similar way, perceptions of closeness also changed on the Iberian Peninsula. In 1980 Spain and Portugal signed an agreement on the safety of border nuclear installations, agreeing to fully inform each other about any intentions to build nuclear facilities less than thirty kilometers from the border. However, six years later—after Chernobyl—the Portuguese decided to request an extension of the thirty-kilometer limit and negotiate with the Spanish about a nuclear power plant in Almaraz, 100 kilometers from the border but with the Tagus River as its cooling source—a river that meets the sea in Lisbon.

These examples illustrate that common perceptions of closeness of nuclear installations indeed changed over time, not least in response to major accidents.

However, the official, technical guidelines for siting have hardly changed. In light of the continued insistence on the part of nuclear energy advocates that the technology is safe, and thus poses no local and transboundary risks, this may seem less surprising. Moreover, if siting criteria for new reactors were to be reformulated and become stricter, this would probably lead to demands to close existing installations that do not conform to the new norms.

Collaboration and Conflict

Whenever—and wherever—a nuclear facility was planned, many different actors became involved: power companies, manufacturing companies, nuclear regulating authorities, local communities, regional planning authorities, national parliaments, and governments. In many Western European countries, antinuclear organizations were also highly active from the mid-1970s, and in Eastern Europe from the late 1980s. Relationships among the various actors ranged from cooperation to conflict. The role of regulating agencies was to monitor the power companies and the manufacturers they contracted to build a given facility. Authorities at all levels had to accommodate the facility into their plans and grant permissions of different kinds. Antinuclear organizations tried to halt the plans by organizing protests to inform the general public of the safety issues and exert pressure on policymakers.

When a nuclear facility was planned close to a border, the pattern of relationships became even more complex as actors across the border became involved too. Those on the other side, however, had more limited possibilities to influence the process. In some cases, they tried to use their peers across the border as mediators or allies. In other cases, they mainly interacted with actors in their own country, for example, by urging their national or subnational government to intervene. We will now discuss the various kinds of transboundary relations, analyzing how they emerged and were utilized to influence the process of building and operating facilities on the other side of the border.

Power Companies

It was usually power companies that took the initiative to build nuclear power plants. In the 1960s and 1970s, when most of Europe's nuclear facilities were planned and built, most large utilities in Europe were state-owned, and their building programs responded to national energy planning priorities. Many of these power companies maintained close relations to power companies in neighboring countries, at least if they were members of a power pool. A number of power pools were established in Europe in the postwar decades, particularly in the western, southern, and northern countries, but also in Eastern Europe under the auspices of the Soviet-led Council for Mutual Economic Assistance (Comecon) (see also chapter 4). Power pools had the aim of enabling power exchange across national borders, in order to both increase the reliability and decrease the cost of power supply. They coordinated the building of cross-border power lines and strived for standardization. In this process, the managers of the power companies developed considerable mutual trust.[30] This was facilitated by the fact that they shared professional attitudes, values, and worldviews from similar technical educations, forming what political scientists call "epistemic communities."[31]

On the Iberian Peninsula, Spanish and Portuguese power companies had been cooperating for a long time and, since 1962, belonged to the same power pool. This cooperation paved the way for joint projects. In the late 1960s, two serious attempts were undertaken to build nuclear reactors in Spain—to be jointly owned by Spanish and Portuguese utilities. Both were to be located at the border, one on the bordering Guadiana River and the other around 100 kilometers from the border at Almaraz on the Tagus River, which flows down to Lisbon. However, both attempts failed, primarily due to insufficient financial resources and political support on the Portuguese side.

The Barsebäck nuclear power plant was also the culmination of a long tradition of cooperation between power companies—in this case, on each side of Öresund. The Swedish Sydkraft and Danish Kraftimport had exchanged power via submarine cables built in 1915. In the 1960s, both companies developed plans to construct nuclear power plants. When Sydkraft decided to erect a nuclear power plant at Barsebäck, Kraftimport's reaction was positive. The two companies signed an agreement in 1970, in which Kraftimport committed to buying 25 percent of the power from Barsebäck. The companies have continued to be close partners in the following decades, and power generated at the Barsebäck plant remained an important part of their power exchange, even after the Danish government started demanding its closure.

The fast breeder reactor (FBR) Superphénix involved three large utilities from three countries that also had a long tradition of cooperation: the French Electricité de France (EDF), the West German Rheinisch-Westfälisches Elektrizitätswerk (RWE), and the Italian Ente Nazionale per l'Energia Elettrica (ENEL). Their common ambition was to build an industrial-scale prototype to demonstrate the future potential of FBRs. The project became intensely contested for almost thirty years, but throughout all this time the French, Italian, and German co-owning power companies continued cooperating to solve the technical, political, and economic challenges the reactor faced. In this sense the project is an example of an unusually trustful transnational collaboration. By contrast, the companies in charge of planning and building the Gorleben and Morsleben facilities in West and East Germany had no contact whatsoever with each other before the Berlin Wall came down in 1989.

Nuclear Authorities

Nuclear authorities played an important role during the planning, building, and operation of nuclear facilities. In Western Europe these nuclear authorities were modeled after the US, while in Eastern Europe they followed the Soviet model. Officials in these authorities formed "epistemic communities," often developing relations of mutual trust with colleagues in other countries.[32] This is clearly shown in the Barsebäck case. When the Barsebäck plant was planned and built, there were not yet any international agreements or rules concerning nuclear power plants close to borders, but the Swedish nuclear authorities were nonetheless careful to exchange information and consult with their Danish counterparts from an early stage. During the entire planning and building phase, the officials at the Swedish and Danish regulating authorities stayed in close contact. The Danish officials had a high level of trust in their Swedish colleagues, and made no objections to the location of the plant. The Swedish authorities also trusted their Danish colleagues, with whom they shared all of the classified documents on the plant. When Danish nuclear experts criticized important aspects of the reactor construction, this critique was taken seriously and led to changes in the design of the Swedish reactor.[33]

During the dictatorial regimes in Spain and Portugal, nuclear authorities had a promotional role, and in 1971 the two countries signed a formal cooperation agreement for the development of civil uses of nuclear power. After the fall of the dictatorships in the two countries, the nuclear authorities were transformed into regulatory bodies. In 1980 they signed a new agreement on the safety of border nuclear installations. A binational technical commission was created. It organized technical visits to nuclear installations close to the border. Each of the installations was on the Spanish side of the border, not least because the Portuguese nuclear program never took off. In a number of instances, the Spanish nuclear authorities showed a willingness to accommodate the safety concerns of their Portuguese counterparts.

A Franco-Swiss agreement was concluded on the Superphénix FBR in 1979 regarding the exchange of information on events with possible radiological consequences. When the plant was ready for startup in 1985, meetings were held between the relevant French and Swiss nuclear authorities. The Swiss delegation concluded that the breeder reactor "meets the stringent safety requirements set for modern light water reactors." In late 1988 a Franco-Swiss Commission for Nuclear Safety and Radiation Protection was established, extending the 1979 agreement. The task of this commission was to discuss not only the Superphénix plant but also other French and Swiss plants near the common border. Thus, the French and Swiss nuclear authorities also cooperated closely and showed mutual trust.

Gorleben and Morsleben, however, provide a very different case, in which the two nuclear authorities on either side of the Cold War border did not exchange any information on the waste repositories they were building.

Antinuclear Organizations

As illustrated by the examples at the outset of this chapter, antinuclear organizations often joined forces across borders in their struggle against nuclear power facilities close to national boundaries. Social movement organizations of this kind emerged in Western Europe at the beginning of the 1970s, when many nuclear power plants were already under construction. From an early stage, antinuclear movements cooperated and learned from each other, and gained some of their strength from both wider transnational and direct cross-border cooperation.[34]

On the Iberian Peninsula in particular, antinuclear movements emerged only after the fall of the dictatorships in 1974 and 1975. When groups started protesting against nuclear facilities, they benefited greatly from the expertise and experience of environmental activists from countries like France and Germany. Portuguese and Spanish antinuclear activists began cooperating in the late 1970s in joint struggles against nuclear facilities close to their common border. At times they were also able to enlist local authorities from both sides of the border in their opposition to nuclear installations. Cross-border cooperation has continued ever since, seemingly without significant tensions or conflicts. In November 2015, Spanish and Portuguese activist groups even established the transnational Iberian Antinuclear Movement (MIA), targeting the closure of all Iberian nuclear facilities.

From 1976 onward, the Danish and Swedish antinuclear organizations organized "Nordic" demonstrations against the Barsebäck plant almost every year. However, at times this cooperation met with considerable challenges due to diverging political priorities. The Danish antinuclear organization, the OOA, wanted its Swedish counterpart, the People's Campaign against Atomic Power (FMA), to focus on closing down Barsebäck. Strategically, the FMA prioritized the struggle against building and commissioning additional nuclear plants in Sweden, rather than attempting to close an existing facility—a much more difficult case to argue within Swedish politics.[35] The politically fragmented Swedish antinuclear movement grew very quickly in the late 1970s and, in the wake of Three Mile Island, was able to find parliamentary support for a referendum on the future of nuclear power in the country. During the referendum campaign, OOA supported their partners across the Öresund, for instance, by sending numerous letters to editors of Swedish newspapers.[36] On losing this referendum, however, many Swedish activists left the movement, which meant that the Danish OOA no longer had a significant counterpart in Sweden.

Differences in political and protest cultures between the French and German antinuclear movements played an important role in the struggle against the Superphénix FBR. Some of the German activists were much more militant than their French counterparts. When they brought along the same protective gear they typically used at demonstrations in Germany (such as motorbike helmets), the French police viewed this as provocative, and intervened using heavy gear and tear gas. In the resulting confrontation, one activist was killed and three were severely injured. After this the French antinuclear movement gave up large-scale demonstrations as a strategy. Instead, scientists at CERN in Geneva—seventy kilometers from the Superphénix site—took over the role as main opponents based on their "expert critique" of the plant's design.

After the end of the Cold War, in the 1990s, similar problems of transboundary and transnational cooperation emerged in the cross-border, Austrian-Czech mobilization against the Czech plant at Temelin. Unwittingly, Austrian antinuclear groups eroded the credibility of their Czech partners. To begin with, Austrian NGOs contributed massively to the financing of their Czech counterparts, compromising the latter's independence. Even more damagingly, the Austrians exported the whole range of protest practices and messages that had worked in Austria. This package was not adapted to the Czech context, however, and did not resonate with Czech audiences. Czech utilities could thus easily dismiss the protests as uncalled-for interventions from abroad.[37]

Governments

In most European countries, governments played an important role as major funders and supporters of nuclear research programs. These programs resulted in the building of nuclear power plants and other facilities from the 1960s onward. When nuclear installations were built by neighboring countries, governments normally accepted this without any objections. A general principle for the conduct of states in relation to environmentally sensitive activities was formulated at the United Nations Conference on the Human Environment in Stockholm in 1972: "States have, in accordance with the Charter of the United

Nations and the principles of international law, the sovereign right to exploit their own resources pursuant to their own environmental policies, and the responsibility to ensure that activities within their jurisdiction or control do not cause damage to the environment of other States or of areas beyond the limits of national jurisdiction."[38] This principle placed a certain responsibility on the state that was the source of transboundary risks. Nevertheless, reasserting national sovereignty, the principle did not grant any specific rights to neighboring countries if they were harmed. Before Chernobyl, it had been difficult to demonstrate the potential transboundary harm posed by nuclear facilities. Even though the risk seemed rather small, the potential damage was vast.

However, if a nuclear facility in a neighboring country gave rise to great anxiety and opposition among its own population, this posed a dilemma for the government. The Danish government's treatment of the Barsebäck plant is an illustrative case. Denmark and Sweden have enjoyed very close and cordial relations with each other and the other Nordic countries since the mid-nineteenth century. When the Barsebäck nuclear plant was planned and built, the Danish government made no objections to it. In the mid-1970s the Danish antinuclear movement began criticizing the plant—which in turn influenced many Danes, not least many of the inhabitants in the Copenhagen region. After the Three Mile Island accident, OOA activists collected more than 300,000 signatures urging the Danish government to ask the Swedish government to close Barsebäck. The Danish government refrained from putting such an emphatic demand to the Swedes but did contact the Swedish government concerning the matter. This led two successive joint Swedish-Danish expert commissions to make overall assessments of the safety of the Barsebäck plant, which concluded that the probability of an accident with a major release of radioactivity was extremely small. However, they also stated that the plant's proximity to Copenhagen was problematic.

In 1985 the Danish parliament decided to abandon nuclear power definitively as a part of Denmark's future energy supply. One year later, a majority in the Danish parliament demanded that the government urge the Swedish government to close the Barsebäck plant, Chernobyl having demonstrated the magnitude of the damage that a major nuclear accident could cause. The Danish antinuclear movement also directly targeted the Swedish government in an effort of NGO diplomacy. The OOA took the 160,000 signatures they had collected in 1986 in their "Radiating Neighbors" campaign to the Swedish, West German, and East German embassies in Copenhagen, demanding the banning of nuclear power within 150 kilometers of the Danish border. The Danish government became increasingly critical of Barsebäck vis-à-vis the

Swedish government, and this issue became a bone of contention between the two countries. However, Denmark protested only in words; the controversy did not influence cooperation in other spheres. Thus, in 1991, the two governments agreed to build a bridge across Öresund, between Copenhagen and Malmö, to strengthen transboundary regional cooperation. The same year, Sweden submitted an application for membership in the European Community, and Denmark—already a member since 1973—was extremely supportive. Nonetheless, disagreement concerning Barsebäck remained a source of tension between the two countries, and when the Swedish government in 1998 decided to phase out the two reactors, it did so largely to please the Danish.

A similar development occurred on the Iberian Peninsula. The governments of Portugal and Spain enjoyed cordial relations during the era of dictatorships and have continued to do so in the subsequent era of democracy. Moreover, both countries became members of the EU in 1986. The two countries had maintained nuclear research programs since the late 1950s. However, while Spain built ten commercial nuclear reactors, Portugal never saw its nuclear ambitions materialize. In 1987 Portugal decided to abandon its nuclear plans due to strong domestic opposition. From then on, the Portuguese government became more outspoken in its criticism of Spanish nuclear facilities close to its borders, and when regular bilateral diplomacy did not produce any tangible results, on several occasions the Portuguese government requested the European Commission intervene to stop such Spanish plans. In the case of a nuclear waste pilot plant, the European Community mediation did in fact contribute to the abandoning of the plant in 1987. Yet in two later projects, one concerning waste storage and one the reopening of uranium mines, the European Union mediation did not prevent Spain from deciding to build a storage facility and possibly also to reopen uranium mines. In contrast to Denmark and Sweden, Portugal and Spain have thus not been able to resolve their nuclear controversies without EU intervention.

In federal systems, or countries with strong regions, regional governments have also played an important role concerning nuclear facilities close to borders. In the case of the West German waste facilities, there was a struggle in the late 1970s between the state and federal level. Ernst Albrecht, the Christian Democratic Prime Minister of Lower Saxony, did not consult the federal government, led by the Social Democratic Chancellor Helmut Schmidt, before he decided where to locate a planned nuclear waste facility. Albrecht wanted to site it at Gorleben, very close to the inner German border, while Schmidt feared that this might disturb relations with East Germany. However, in the end Schmidt agreed to Gorleben, as the Lower Saxony salt domes were seen as an ideal disposal site, and other locations in Lower Saxony had proved politically untenable. In the case of the Superphénix breeder reactor, there was a similar struggle within Switzerland between the Canton of Geneva and the federal government in Bern. The Canton of Geneva was outspoken in its critique of Superphénix, even amending its constitution and filing lawsuits in its effort to stop the plant. By contrast, the federal government refused to protest against Superphénix, exercising great caution regarding its friendly relations with France—as well as regarding the domestic impact any such protest would have on nuclear projects within Switzerland.

Interestingly, there seems to have been little cooperation between neighboring socialist countries in Central and Eastern Europe in respect to nuclear facilities close to borders. Even within the Soviet Union, no cross-border information was exchanged between neighboring Soviet republics; for instance, when the Chernobyl plant was located in Ukraine very close to the border with Belarus, Belarus was not consulted.[39] Most of the Central and Eastern European countries had a close bilateral cooperation with the Soviet Union and multilateral cooperation through Comecon but little bilateral cooperation among each other (see chapter 4). For instance, when Bulgaria built a Soviet-designed nuclear power plant at Kozloduy on the banks of the Danube River, which was commissioned in late 1974, there was no agreement with Romanian authorities regarding cross-border issues such as emergency plans, even though the plant was very close to the border. In March 1977 a major earthquake near the plant caused severe damage in the region but seemingly spared the nuclear power plant. Some months later, Romania asked Comecon to "specify the procedure of timely information in case of a nuclear accident of a nuclear power plant with the aim to enable efficient measures on the territory close to our borders."[40] The representative of the Bulgarian Nuclear Committee responded that bilateral agreements in this matter were improper and that this was a legal matter involving foreign policy, and hence should be solved at Comecon level. He further declared that Bulgarian nuclear authorities were not willing to inform Romanian authorities about accidents in the Kozloduy nuclear power plant. Bulgaria was only prepared to send information through Comecon, and refused to reveal anything unofficially or bilaterally. This demonstrates a remarkably low level of government cooperation, and at the same time the importance of international organizations.[41]

International Organizations

In the early 1970s, there were no international rules concerning nuclear installations close to the border. The Nordic Council, an intergovernmental organization for the Nordic countries, was the first international organization to formulate rules of conduct for its members. In November 1976, it adopted a set of guidelines for contacts concerning nuclear facilities close to borders between Denmark, Finland, Norway, and Sweden. This agreement was largely a result of the experiences of the Danish-Swedish relations concerning Barsebäck.[42]

In the European Communities, the predecessor of today's European Union, rules on cross-border impacts were limited to nuclear waste. Article 37 of the 1957 Euratom Treaty only required the member states to inform the European Commission about "any plan for the disposal of any kind of radioactive waste . . . likely to involve radioactive contamination of the water, soil or airspace of another member State."[43] The Commission, in turn, had an expert body draw up an opinion on these plans. Member states duly submitted information on their nuclear facilities to the Commission, for the article also covered liquid and gaseous wastes released into rivers and the air.[44] This article was invoked only from the 1970s onward, as the nuclear waste issue and the siting of plants became increasingly controversial.

Different actors within the European Communities responded in different ways to the growing cross-border antinuclear protest on the Upper Rhine in the mid-1970s, where a large number of Swiss, French, and West German nuclear power plants were to be located to use the Rhine's ample cooling water. In his 1975 report on the European Union, which was to provide a motivating vision of the future of European integration, the Belgian Premier Leo Tindemans acknowledged the large-scale presence of cross-border environmental impacts—including those arising from nuclear power—and advocated European solutions. An independent, strict, and open European regulatory authority, modeled after the US Nuclear Regulatory Commission, with sweeping responsibilities for the "siting, construction and operation of the power stations, of the power stations, the fuel cycles and the disposal of radioactive and thermic waste" would help build citizen trust, which national authorities did not enjoy. This, he argued, would defuse protest, something that he (reproducing a contemporary pronuclear pattern of argument) ascribed to "psychological reactions." This highly ambitious idea of a common European nuclear regulatory agency was never implemented, however.[45]

An initiative by the European Parliament (EP)—still unelected and largely powerless at the time—proved more effective. In 1975, the EP committee on energy, research, and technology wrote a report on the siting of nuclear power stations, "taking account of their acceptability for the population."[46] This was followed by a resolution calling for a community siting policy to inform the public (§16) and for "collaboration with the local and regional authorities concerned" (§17). The overwhelming majority of members of the EP speaking on that day clearly saw no viable alternative to the expansion of nuclear power, and the resolution thus highlighted that the public "be given a clear understanding of the alternatives, which entail an impoverishment of the quality of life" (§16).[47] By contrast, the European Environmental Bureau (EEB), the newly established representation of the emerging transnational network of environmentalists in Brussels, started to address their concerns about the problems of nuclear power to the European Commission.[48] In 1976 the Commission hence proposed legislation on the "introduction of a Community consultation procedure in respect of power stations likely to affect the territory of another member state." While the EP supported this proposal,[49] several governments, notably France, blocked it, even when the European Commission attempted to reintroduce it after Three Mile Island.[50] Nevertheless, the neighbors across the border continued to oppose French nuclear plants in their vicinity, drawing on existing European law. The West German Saarland's case at a French court in Strasbourg was referred to the European Court of Justice (ECJ), the highest Court of the European Communities. The ECJ ruled in 1988 that the French authorities had not informed the European Commission on time, as required by article 37, before giving the go-ahead for the Cattenom nuclear power plant.[51] Only in the aftermath of Chernobyl, legislation on cross-border effects and information was enacted in the so-called Chernobyl directive of 1989.[52] The European Commission became involved as a mediator in controversies between Portugal and Spain concerning nuclear facilities close to the border, drawing on article 37, with varying degrees of success.

In autumn 1986, following the Chernobyl accident, the International Atomic Energy Agency (IAEA) also adopted two conventions: one on early notification of a nuclear accident and one on assistance in case of a nuclear accident.[53] In 1994, the IAEA also adopted a convention on nuclear safety. None of these specifically addressed nuclear facilities close to borders, but they did contribute to a closer cooperation between governments on nuclear safety, which in turn made it easier for governments to address issues on nuclear facilities close to their borders with the responsible government.

Conclusions

This chapter aimed to highlight the relevance of borders for the relations between nuclear energy and society by analyzing nuclear facilities close to national borders. Many more actors became involved in such cases than under purely national circumstances, as actors on the other side of the border were also potentially affected. This increased the complexity of cooperative and conflictual relations between the relevant actors. By looking mainly at four cases, we analyzed a number of issues.

The first concerned which factors actually influenced siting decisions and why so many facilities were placed close to borders. Here we identified technical factors favoring locations by the sea and along rivers, which often entailed proximity to a border. But at times specific political considerations also contributed to such locations. Locating a nuclear facility close to a border could seem politically attractive, as potential opponents on the other side of the border were irrelevant as voters in a democratic political system and had few possibilities to oppose it, particularly when the border was with nondemocratic countries.

We further analyzed whether and to what extent the perception of closeness changed over time. Unsurprisingly, we found that the general perception of closeness has indeed changed, not least in response to major accidents like Three Mile Island and Chernobyl. However, the technical guidelines for siting new facilities have hardly changed at all. One of the reasons might be that the introduction of stricter siting criteria would have legal and political implications not only for future but also for existing installations, and lead to demands for closures.

Our primary interest has been in transboundary relationships between the main types of actors—power companies, nuclear authorities, antinuclear movements, and national and regional governments—as well as the involvement in (and attempts at) rule-making by international organizations. Except for the West and East German case, we found that power companies in neighboring countries have mostly cooperated closely regarding nuclear facilities near borders, based on a high degree of trust. In some cases, such plants were co-owned—or intended to be co-owned. Electricity generated by these nuclear plants continued to be sold to the other side of the border even where governments had disagreements about these facilities. Furthermore, nuclear authorities have mostly cooperated closely with their colleagues in neighboring countries, based on mutual trust and existing transnational epistemic communities. In this light, the nonexistent relations between the Bulgarian and Romanian nuclear authorities with regard to the plant in Kozloduy seem exceptional. It would require further research to establish whether this was characteristic of the situation in the Soviet bloc more generally, with the dominant and central role of the Soviet Union in nuclear affairs (see chapter 4).[54]

Antinuclear organizations have also cooperated across borders and have often gained more influence through such cooperation. However, at times there have also been frictions between antinuclear movements due to differences in protest culture or political priorities. In such instances, they have strived to influence their own government to intervene diplomatically or protested directly to the government in the neighboring country or countries, as in the Danish OOA's "Radiating Neighbors" campaign. Governments were mostly careful not to criticize their colleagues in neighboring countries for building nuclear facilities close to their borders—at least not if they maintained nuclear programs themselves and had friendly relations with their neighboring country. It was only after they had abandoned their own nuclear plans that they started to openly question and act diplomatically against nuclear facilities close to their borders, as the Denmark and Portugal examples illustrate. Existing trustful neighborly relations, however, ensured that such pressure was exerted politely and diplomatically. International organizations were aware of and discussed issues relating to nuclear installations at the border. Hampered in particular by national veto powers, they were relatively late to make rules and mediate on siting, safety, and other issues involving transboundary impacts. Only after the Chernobyl accident did the EC and IAEA adopt rules and regulations of this kind.

The study of cooperation and conflicts regarding nuclear facilities close to borders is indeed a fruitful lens for studying transnational and transboundary relations in the nuclear realm, and thus contributes very specific insights to the social, political, and environmental history of nuclear power, and the history of technology more broadly. Three of our cases concern countries with basically friendly relations, which gives this analysis a certain bias. Future studies of nuclear facilities close to borders in Eastern Europe, and also beyond Europe, will provide additional insights for more comprehensive comparisons and conclusions. It would also be interesting to compare the nuclear situation to that concerning other kinds of polluting facilities, contributing to broader debates in environmental history. Such studies would fruitfully engage with Kate Brown's recent assertion that nuclear technology just adds another set of potentially harmful pollutants characteristic of the modern world we have lived in since the twentieth century.[55]

Notes

1. Arne Kaijser and Jan-Henrik Meyer, " 'The World's Worst Located Nuclear Power Plant': Danish and Swedish Perspectives on the Swedish Nuclear Power Plant Barsebäck," *Journal for the History of Environment and Society* 3 (2018): 71–105, https://doi.org/10.1484/J.JHES.5.116795.
2. Claire Le Renard, "The Superphénix Fast Breeder Nuclear Reactor—Cross-Border Cooperation and Controversies," *Journal for the History of Environment and Society* 3

(2018): 107–144, https://doi.org/10.1484/J.JHES.5.116796; Andrew Tompkins, “Transnationality as a Liability? The Anti-Nuclear Movement at Malville,” *Revue Belge de Philologie et d’Histoire/Belgisch Tijdschrift voor Filologie en Geschiedenis* 89, no. 3–4 (2011): 1365–79; Andrew Tompkins, *“Better Active than Radioactive!” Anti-Nuclear Protest in 1970s France and West Germany* (Oxford: Oxford University Press, 2016), 166–73.

3. Astrid Mignon Kirchhof, “East-West German Transborder Entanglements through the Nuclear Waste Sites in Gorleben and Morsleben,” *Journal for the History of Environment and Society* 3 (2018): 145–78, https://doi.org/10.1484/J.JHES.5.116797. See also Astrid M. Eckert, *West Germany and the Iron Curtain: Environment, Economy and Culture in the Borderlands* (Oxford: Oxford University Press, 2019), 201–44.
4. Mar Rubio-Varas, António Carvalho, and Joseba de la Torre, “Siting (and Mining) at the Border: Spain-Portugal Nuclear Transboundary Issues,” *Journal for the History of Environment and Society* 3 (2018): 33–69, https://doi.org/10.1484/J.JHES.5.116794.
5. Sandra Tauer, *Störfall für die gute Nachbarschaft? Deutsche und Franzosen auf der Suche nach einer gemeinsamen Energiepolitik (1973–1980)* (Göttingen: V&R Unipress, 2012); Birgit Müller, “Anti-Nuclear Activism at the Czech-Austrian Border,” in *Border Encounters: Asymmetry and Proximity at Europe’s Frontiers*, ed. Jutta Lauth Bacas and William Kavanagh (New York: Berghahn, 2013), 68–89; Cécile Oberlé, “Civil Society and Nuclear Plants in Cross-Border Regions: The Mobilisation against Fessenheim and Cattenom Nuclear Power Stations,” *Progress in Industrial Ecology—An International Journal* 10, no. 2/3 (2016): 194–208; Natalie Pohl, *Atomprotest am Oberrhein: Die Auseinandersetzung um den Bau von Atomkraftwerken in Baden und im Elsass (1970–1985)* (Stuttgart: Steiner, 2019); Tompkins, *“Better Active than Radioactive!”*; Stephen Milder, *Greening Democracy: The Anti-Nuclear Movement and Political Environmentalism in West Germany and Beyond, 1968–1983* (Cambridge: Cambridge University Press, 2017).
6. Arne Kaijser and Jan-Henrik Meyer, “Nuclear Installations at the Border: Transnational connections and international implications: An Introduction,” *Journal for the History of Environment and Society* 3 (2018): 10, https://doi.org/10.1484/J.JHES.5.116793.
7. Joseph E. Taylor III, “Boundary Terminology,” *Environmental History* 13, no. 3 (2008): 454–81, https://doi.org/10.1093/envhis/13.3.454.
8. Joseph S. Nye Jr. and Robert O. Keohane, “Transnational Relations and World Politics: An Introduction,” *International Organization* 25, no. 3 (1971): 329–49.
9. Taylor, “Boundary Terminology,” 462.
10. Christoph Laucht, “Scientists, the Public, the State and the Debate over the Environmental and Human Health Effects of Nuclear Testing in Britain, 1950–1958,” *Historical Journal* 59, no. 1 (2016): 221–51, https://doi.org/10.1017/S0018246X15000096; Rachel Emma Rothschild, *Poisonous Skies: Acid Rain and the Globalisation of Pollution* (Chicago: University of Chicago Press, 2019); Rachel Rothschild, “Environmental Awareness in the Atomic Age: Radioecologists and Nuclear Technology,” *Historical Studies in the Natural Sciences* 43, no. 4 (2013): 492–530, https://doi.org/10.1525/hsns.2013.43.4.492.
11. For a more detailed account of these case studies, conducted within the History of Nuclear Energy and Society (HoNESt) project, see Arne Kaijser and Jan-Henrik

Meyer, "Siting Nuclear Installations at the Border: Special Issue," *Journal for the History of Environment and Society* 3 (2018): 1–178, https://www.brepolsonline.net/toc/jhes/2018/3/+.

12. Rubio-Varas, Carvalho, and Torre, "Siting (and Mining) at the Border."
13. Kaijser and Meyer, " 'The World's Worst Located Nuclear Power Plant.' "
14. Renard, "The Superphénix Fast Breeder Nuclear Reactor."
15. Kirchhof, "East-West German Transborder Entanglements."
16. Jasper Becker, "Nuclear Power: Border Problems," *Nature* 288, no. 4 (1980): 424–25.
17. In 1975, the European Commission distinguished between economic, social, safety, technical, and environmental siting criteria. Hanna Walz, "Report Drawn Up on Behalf of the Committee on Energy, Research and Technology of the European Parliament on the Conditions for a Community Policy on the Siting of Nuclear Power Stations Taking Account of Their Acceptability for the Population, doc 392/75, PE 40.985/fin, 26 November 1975," Historical Archives of the European Parliament, Luxembourg, 22.
18. Oberlé, "Civil Society and Nuclear Plants in Cross-Border Regions."
19. Janine Gaumer, " 'Was hat die Wiederaufbereitungsanlage mit Frieden zu tun?' AtomkraftgegnerInnen, FriedensaktivistInnen und der gemeinsame 'Widerstand' gegen nukleare Bedrohungsszenarien in den 1980er Jahren," in *Gespannte Verhältnisse. Frieden und Protest in Europa während der 1970er und 1980er Jahre*, ed. Claudia Kemper (Essen: Klartext, 2017), 223–47; NN, "Als in Wackersdorf noch Kohle abgebaut wurde," *Mittelbayrische Zeitung* (Regensburg), February 15, 2017, https://www.mittelbayerische.de/region/schwandorf/gemeinden/wackersdorf/als-in-wackersdorf-noch-kohle-abgebaut-wurde-21492-art1487389.html; Janine Gaumer, *Wackersdorf: Atomkraft und Demokratie in der Bundesrepublik 1980–1989* (Munich: Oekom, 2018), 48–56.
20. M. Pasqualetti and S. Stremke, "Energy Landscapes in a Crowded World: A First Typology of Origins and Expressions," *Energy Research & Social Science* 36 (2018): 94–105, https://doi.org/10.1016/j.erss.2017.09.030; Dolly Jørgensen and Finn Arne Jørgensen, "Aesthetics of Energy Landscapes: Special Issue," *Environment, Place, Space* 10, no. 1 (2018): 1–153.
21. On environmental justice, see chapter 7. On regionalism see Jan-Henrik Meyer, "Nature: From Protecting Regional Landscapes to Regionalist Self-Assertion in the Age of the Global Environment," in *Regionalism and Modern Europe: Regional Identity Construction and Regional Movements from 1890 until the Present*, ed. Xosé M. Núñez Seixas and Eric Storm (London: Bloomsbury, 2019), 65–82.
22. Gaumer, *Wackersdorf*, 48–70.
23. Dieter Rucht, *Von Wyhl nach Gorleben: Bürger gegen Atomprogramm und nukleare Entsorgung* (Munich: C.H. Beck, 1980), 110, 250 fn. 35.
24. Rucht, *Von Wyhl nach Gorleben,* 109; Eckert, *West Germany and the Iron Curtain,* 203.
25. Astrid Mignon Kirchhof and Helmuth Trischler, "The History behind West Germany's Nuclear Phase-Out," in *Pathways into and out of Nuclear Power in Western Europe, Austria, Denmark, Federal Republic of Germany, Italy, and Sweden*, ed. Astrid Mignon Kirchhof (Munich: Deutsches Museum, 2019), 140–43; Eric Berkers, "The Netherlands Short Country Report (version 2018)," in *History of Nuclear Energy and Society (HoNESt) Consortium Deliverable no. 3.6* (2019), 47–52, https://hdl.handle.net/2454/38269.

26. Barsebäckkommittén, "Rapport från den dansk-svenska kommittén 1983–84 om Barsebäckverket," *Departmentskrift Industridepartementet (DsI)*, no. 1 (1985): 231f.
27. US Nuclear Regulatory Commission, "Regulatory Guide 4.7. General Site Suitability Criteria for Nuclear Power Stations. Revision 3, March 2014, Draft was issued as DG-4021 on December 30, 2011," https://www.nrc.gov/docs/ML1218/ML12188A053.pdf.
28. Uffe Korsbech, *Notat med nogle oplysninger om det, man kalder det værst tænkelige uheld på Barsebäck* (Aarhus: Reel Energi Oplysning, 1982); Uffe Korsbech, *En vurdering af den danske beredskabsplan for Barsebäck-uheld: i lyset af den nyere viden*, ed. P. L. Ølgaard, REO informationsblad no. 3 (Århus: REO, Reel Energi Oplysning, 1983); Uffe Korsbech, *Barsebäck-værket: en sikker nabo: en redegørelse om de sikkerhedsmæssige forhold ved Barsebäckværket på baggrund af Tjernobyl-ulykken*, ed. P. L. Ølgaard (Mariager: REO, 1986).
29. Siegfried Christiansen, "Letter to GDR ambassador Norbert Jaeschke, 23 August 1986, Materiale vedr. internationalt arbejde: Sovjet/Rusland Henvendelser 1983—Henvendelser 1986," *Rigsarkivet, Copenhagen* OOA 10451, no. 134 (1986).
30. Vincent Lagendijk, *Electrifying Europe: The Power of Europe in the Construction of Electricity Networks* (Aksant: Warsaw, 2009).
31. Peter M. Haas, "Introduction: Epistemic Communities and International Policy Coordination," *International Organization* 46, no. 1 (1992): 1–35; Arne Kaijser, "Trans-border Integration of Electricity and Gas in the Nordic Countries, 1915–1992," *Polhem: Tidskrift för teknikhistoria* 15 (1997): 4–43.
32. Arne Kaijser et al., *The Past and the Present of Nuclear Safety Regulation in Europe: Transcript of the Discussions of the Witness Seminar in Barcelona, 16 October 2018* (Mimeo, Barcelona: Pompeu Fabra University, 2019).
33. See also Jan-Henrik Meyer, "To Trust or Not to Trust? Structures, Practices and Discourses of Transboundary Trust around the Swedish Nuclear Power Plant Barsebäck near Copenhagen," *Journal of Risk Research* 24 (2021).
34. Astrid Mignon Kirchhof and Jan-Henrik Meyer, "Global Protest Against Nuclear Power. Transfer and Transnational Exchange in the 1970s and 1980s," *Historical Social Research* 39, no. 1 (2014): 165–90, https://doi.org/10.12759/hsr.39.2014.1.165-190.
35. Siegfried Christiansen, *Interview with Siegfried Christiansen, founding member of OOA, conducted by Jan-Henrik Meyer, 28 September 2017*, Copenhagen; Jørgen Steen Nielsen, *Interview with Jørgen Steen Nielsen, founding member of OOA, conducted by Jan-Henrik Meyer, 11 May 2016*, Copenhagen, 2016; Björn Eriksson et al., *Det förlorade försprånget* (Stockholm: Miljöförbundet, 1982).
36. E.g., Susan Mortensen, Ole Christiansen, and Helle Carlsson, "Letter to the editor of Helsingborgs Dagblad, Ishoj, 13 March 1980," Rigsarkivet, Copenhagen, Fond OOA 10451, no. 25 (1980). The Austrian antinuclear movement supported the Swedish referendum campaign, too, by sending postcards to Sweden (see chapter 3).
37. Birgit Müller, "Anti-Nuclear Activism at the Czech-Austrian Border," in *Border Encounters: Asymmetry and Proximity at Europe's Frontiers*, ed. Jutta Lauth Bacas and William Kavanagh (New York: Berghahn, 2013), 68–89.
38. United Nations, *Report of the United Nations Conference on the Human Environment, Stockholm, 5–16 June 1972, A/CONF.48/14/Rev.1, 21.09.1972* (New York: United Nations, 1972).

39. Serhii Plokhy, *Chernobyl: History of a Tragedy* (London: Penguin, 2019), 32.
40. Ivaylo T. Hristov, *The Communist Nuclear Era: Bulgarian Atomic Community During the Cold War, 1944–1986* (Amsterdam: Amsterdam University Press, 2014), 131.
41. Hristov, *The Communist Nuclear Era*, 129–34.
42. Barsebäckkommittén, "Rapport från den dansk-svenska kommittén 1983–84 om Barsebäckverket."
43. Euratom, "Treaty Establishing the European Atomic Energy Community (Euratom), signed 17 April 1957," *Eur-Lex* (document 11957A/TXT), http://data.europa.eu/eli/treaty/euratom/sign.
44. European Commission, Directorate-General for Social Affairs, "Application of Article 37 of the Euratom Treaty. Survey of Activities. Experience Gained 1959–1972," Luxembourg, 1972, *Archive of European Integration*, http://aei.pitt.edu/49455/1/B0018.pdf.
45. Leo Tindemans, "European Union. Report by Mr. Leo Tindemans, Prime Minister of Belgium, to the European Council (commonly called the Tindemans Report), 29 December 1975," *Bulletin of the European Communities* Suppl. no. 1 (1976): 27, *Archive of European Integration* http://aei.pitt.edu/942/1/political_tindemans_report.pdf.
46. Walz, "Report Drawn Up on Behalf of the Committee on Energy, Research and Technology of the European Parliament on the Conditions for a Community Policy on the Siting of Nuclear Power Stations Taking Account of Their Acceptability for the Population, doc. 392/75, 26 November 1975."
47. European Parliament, "Resolution on the Conditions for a Community Policy on the Siting of Nuclear Power Stations taking account of their Acceptability for the Population," *Official Journal of the European Communities (OJEC)* 19, no. C28 (February 9, 1976): 12–14. European Parliament, "Debates on 'Community Policy on the Siting of Nuclear Power Stations' (13 January 1976)," *Official Journal of the European Communities, Annex: Debates of the European Parliament* 1975/76 Session, No. 198: 36–66.
48. Jan-Henrik Meyer, " 'Where Do We Go from Wyhl?' Transnational Anti-Nuclear Protest targeting European and International Organisations in the 1970s," *Historical Social Research* 39, no. 1 (2014): 212–35; Jan-Henrik Meyer, "Challenging the Atomic Community: The European Environmental Bureau and the Europeanization of Anti-Nuclear Protest," in *Societal Actors in European Integration: Polity-Building and Policy-Making 1958–1992*, ed. Wolfram Kaiser and Jan-Henrik Meyer (Basingstoke: Palgrave, 2013), 197–220.
49. Hanna Walz, "Report Drawn Up on Behalf of the Committee on Energy and Research of the European Parliament on the Draft Council Resolution Concerning the Consultation at Community Level on the Siting of Power Stations, and on the Proposal from the Commission of the European Communities to the Council (Doc 506/76) for a Regulation Concerning the Introduction of a Community Consultation Procedure in Respect of Power Stations Likely to Affect the Territory of Another Member State, doc. 145/76, PE 47.939fin, 14 June 1977," *Historical Archives of the European Parliament, Luxembourg.*
50. H. U. Beelitz, "Vermerk für Herrn Narjes. Betr. Kernkraftwerke in Grenznähe. 30.11.1981," *Historical Archives of the European Commission* BAC 35/1980 Cabinet Narjes (1975–1982), no. 2 Cross-border consultation on power stations (1981): 637–38; European Commission, "Proposal for a Council Regulation Concerning the

Introduction of a Community Consultation Procedure in Respect of Power Stations Likely to Affect the Territory of Another Member State (submitted to the Council on 17 May 1979)," *Official Journal of the European Communities (OJEC)* C149 (June 15, 1979): 2–4.

51. European Court of Justice, "Judgment of the Court of 22 September 1988. Land de Sarre and others v Ministre de l'Industrie. Reference for a preliminary ruling: Tribunal administratif de Strasbourg, France. Nuclear power stations: Opinion of the Commission under Article 37 of the EAEC Treaty. Case 187/87, ECLI:EU:C:1988:439," *European Court Reports* 1988 (05013), https://eur-lex.europa.eu/legal-content/EN/TXT/?uri=CELEX%3A61987CJ0187.
52. Euratom, "Council Directive of 27 November 1989 on informing the general public about health protection measures to be applied and steps to be taken in the even of a radiological emergency," *Official Journal of the European Communities (OJEC)* 32, no. L357 (December 7, 1989): 31–34.
53. On the IAEA's response to Chernobyl, see the relevant chapter in Elisabeth Röhrlich, *Dual Mandate: A History of the IAEA* (Baltimore: Johns Hopkins University Press, 2021).
54. Sonja D. Schmid, "Nuclear Colonization? Soviet Technopolitics in the Second World," in *Entangled Geographies: Empire and Technopolitics in the Global Cold War*, ed. Gabrielle Hecht (Cambridge, MA: MIT Press, 2011), 125–54.
55. Kate Brown, *Manual for Survival: A Chernobyl Guide to the Future* (New York: W.W. Norton, 2019).

CONCLUSIONS

Future Challenges for Nuclear Energy and Society in a Historical Perspective

Arne Kaijser, Markku Lehtonen, Jan-Henrik Meyer, Mar Rubio-Varas

Introduction

The chapters in this book are one outcome of the international, interdisciplinary research project HoNESt, History of Nuclear Energy and Society. The project activities included dissemination and engagement events, in which we discussed and engaged with a wide range of stakeholders from the nuclear sector, politics, business, research, and society.[1] This chapter draws on these discussions and on key findings from the preceding chapters to address three central questions concerning the past, present, and future of nuclear. First, looking back on the European (and United States) experience, we draw some key lessons regarding nuclear-societal relations and engagement practices. Second, we stress the fact that not only did public engagement with nuclear technology evolve in the context of postwar Europe but this very engagement itself contributed to shaping Europe. This influence also reached beyond the nuclear sector and societal engagement practices, contributing to a diversity of attitudes, discourses, and institutional structures throughout the continent. The final section of the book is more exploratory. Against the background of the historical experiences outlined in the earlier chapters, it addresses some of the current and future challenges that the nuclear sector faces in its relations with society.

What We Learned

This volume enquired into the complex interaction between the nuclear sector—encompassing industry, government, and research—and the societies in which it was embedded. Initially, in the 1950s, enthusiasm about the prospect of nuclear power prevailed both among political elites across the political spectrum and within society at a time when nuclear energy was still more of a technological vision than a reality. In Western countries, this gradually gave way to increasing criticism once nuclear power plant projects were built in large numbers, from the late 1960s onward. Critics included individual citizens, (counter-)experts, social movement activists, and other representatives of civil society. In the HoNESt project, we adopted a transdisciplinary and transnational social science approach to analyze this historical development—the evolution of perceptions, engagement practices, and their consequences in the nuclear sector. We explored how engagement between the nuclear sector and society in Europe varied across countries, how its forms and nature changed over time, which actors and networks contributed to such changes, and how these dynamics shaped societies and nuclear sector research, industry, and politics. The chapters provide examples of different types of engagement. These included both actions undertaken by industry and authorities—from the exclusion of the public to one-way information provision to consultation and efforts to encourage active participation—and citizen-led engagement, that is, the various ways in which civil society actors organized themselves to protest, criticize, and appraise nuclear projects and policies.[2]

Present-day nuclear energy debates and decisions do not start from a clean slate. Past experience is crucial for understanding how the public and policymakers today perceive issues relating to nuclear energy and for designing responses to future challenges. Nuclear energy was born during an era characterized by a Promethean confidence in the ability of science and technology to overcome any problem or obstacle.[3] However, the evolution of the nuclear sector turned out to be a continuous process of learning by doing, adapting to new contexts, and searching for fixes to new challenges generated by technical problems, unexpected accidents, societal pressures, political upheavals, and socioeconomic trends. Over time, such adaptations led actors within the nuclear sector to realize that the various publics did not view and address nuclear energy purely—or even primarily—as a technological matter. Instead, the interaction between nuclear and society turned out to be a multifaceted complex made up of interactions between technological, economic, social, and political elements, and ultimately relating to deeply held values, worldviews, and identities.[4] In particular, the mushrooming of diverse social movements and critical expert analysis in the 1970s demonstrated how profoundly nuclear energy was entangled with broader societal debates, notably those relating to the nature and quality of democracy.[5] Ever since then, development in this sector has been subject to continuous tacking back and forth between those seeking to separate the technical from the societal and others stressing the unavoidable interconnections between these two spheres. Tensions remain, despite the growing awareness of the sociotechnical nature of the matters in question. Furthermore, as this book's contributors illustrate, the recognition of the inseparability of the technical and the societal has varied widely across countries and between the subcommunities within the nuclear sector. For instance, actors in the fields of nuclear waste management and radiation safety have learned from past failures of excessively technical approaches, and today routinely seek to include citizens in deliberations concerning technical choices. Similar awareness and its translation into practice has been far less evident among the promoters of nuclear new-build programs.

The journey of nuclear technology from secluded research laboratories to large-scale commercial deployment, and ultimately to the public sphere, involved a variety of actors at local, national, and international levels. The political institutions responsible for authorizing nuclear installations varied considerably across European countries and over time—ranging from democratic to authoritarian or even dictatorial political systems, and from centralized to federal and decentralized polities. The commercial organization of the electricity sector similarly ranged from public provision to privately owned utilities, established at national, regional, or local levels, operating as monopolies or not, in regulated or liberalized markets. The societal reactions to nuclear energy also displayed varying origins, structures, and strategies, depending in part on location and on time. These differences contributed to the diversity of societal engagements with nuclear energy. In general, democratic and decentralized settings were more conducive to early engagement—via communication by government and industry, bottom-up protest from civil society, critique from independent experts, top-down participatory processes initiated by project promoters, or the channels of representative democracy. Centralized and authoritarian approaches, in turn, resisted and delayed engagement, paying little attention to societal concerns and offering few if any opportunities for the expression of those concerns. In reality, both tendencies often coexisted. For a long time, democratic governments lacked transparency and openness in nuclear decision-making and sometimes employed considerable force to quell opposition.

Just as the actor constellations and institutional settings are highly context- and time-specific, so do the preconditions for public engagement vary widely across different localities and situations. To use the old adage, context matters. Regardless of, and indeed thanks to, this context-dependence, we can draw a number of lessons from past practices. First, open dialogue and deliberation must start early and demonstrate a sincere desire to take seriously the concerns of affected stakeholders and communities. This is essential both for democratic legitimacy and for successful project planning. Second, regardless of how well designed and well intended, such engagement guarantees neither the smooth implementation of projects nor easy and harmonious decision-making processes.[6]

Third, the key to understanding conflicts on nuclear energy lies in the sphere of trust in institutions, on the one hand, and sociocultural contextual factors, on the other.[7] What matters is not only trust in those very institutions responsible for nuclear safety but also trust in institutions at large. Focus on trust and mistrust also helps to illustrate how the pathways and forms of engagement, as well as its outcomes and broader impacts, crucially depend on the historically defined context. For example, similar kinds of dialogue and participatory practices have met with considerable success in the Nordic high-trust societies—notably Finland and Sweden—but have engendered widespread suspicion and frustration in countries with a more mistrust-based political tradition and history. Likewise, the forms and outcomes of engagement depend on ownership structures and on the historically defined trust relations among private industry, government, and society. For example, private nuclear operators may appear as fully legitimate organizers of engagement events and practices in one country, while in another context or at a different point in time only public authorities would be acceptable and considered trustworthy. Yet trust is not a panacea, and excessive, unwarranted trust may undermine both nuclear safety and the democratic quality of decision-making. Crucially, experience also shows that earning trust requires embracing *mis*trust. Open and sincere dialogue with those who oppose and disagree by no means guarantees agreement, yet ignoring such dialogue risks making such conflicts and mistrust endemic, leading to paralysis, and in the worst case generating irresolvable and potentially violent conflict.

Fourth, the chapters of this book help explain why nuclear technology has a particular tendency to generate social controversies while other technologies seem to be adopted widely without further ado. Explanations have to do with differences between technologies in terms of their propensity to alter the social fabric and ways of life. Nuclear technologies have been perceived as highly transformative of local societies and landscapes, and as such they have faced strong societal opposition. To understand cross-country differences, one needs to go beyond risk, safety, and communication. Reasons for controversy have just as much—or more—to do with how citizens believe the technology will affect their ways of life; their communities, traditions, and cultures; their physical living environment; their political or economic opportunities; their children; or even their survival as a nation. Ultimately, the specific character of nuclear energy stems from how it resonates with people's fundamental values and worldviews, including considerations around ethics, responsibility, moral judgments, identity, accountability and legitimacy, and principles of justice, equity, and fairness.

Last, the chapters in this book demonstrate the unavoidably international and transnational character of nuclear energy. This is most evident in the operation of intergovernmental organizations (IGOs) set up in the late 1950s to promote the promises of the new technology. Both in the West and in the East, these organizations employed communication strategies designed to persuade the opinion-building elites and the public at large of the virtues of the peaceful uses of the atom. The promotional role has mellowed over time but nevertheless continues to hamper efforts at developing public engagement. Yet, for forerunner countries, these organizations have provided a forum to help diffuse engagement practices internationally.[8] Some such countries, which had faced challenges of public acceptance and thus developed advanced engagement practices, played a central role in these processes of diffusion. Transboundary conflicts generated by nuclear facilities close to national borders also required approaches built on trust and collaboration that would produce solutions integrating the concerns of the other side. By contrast, where transborder relations were already tense, nuclear projects near the border could further aggravate tensions.[9]

How Has Public Engagement with Nuclear Technology Changed Europe?

We have argued that the public has influenced the shaping of nuclear technology as a "public technology."[10] In many European countries in the 1970s and 1980s, the development of nuclear power and the conflicts that it raised fundamentally affected how central political issues were addressed and resolved. In this sense, the interaction between nuclear energy and society—the conflicts, controversies, and various engagement practices—contributed to the shaping of Europe.[11] Such shaping took place at different geographical levels.

At the international level, collaboration via organizations such as Euratom, ENEA/NEA, and Comecon, or via various joint endeavors (Eurodif, CERN, JINR) turned nuclear issues into a means for advancing European cooperation and integration and constructing European identities on both sides of the Iron Curtain. After the collapse of the USSR, the integration of Eastern Europe into the Western nuclear community—not least into the Western safety regulation regime—also served the goal of political integration.[12]

Multilateral international cooperation in joint research and development projects, such as for fast breeders in Western Europe, and fission reactors—such as in a Nordic context between Denmark and Sweden in the 1960s—and subsequently between sellers and buyers of nuclear technology, fostered the integration of markets, scientific, technological, and business elites too. Such cooperation and commercial relations also strengthened Western Europe's transatlantic integration with the United States, in particular because almost all West European countries building nuclear reactors adopted the American light-water technology in the 1970s. The Soviet Union played a similar role in an even more unequal exchange with its Central and Eastern European partners.[13] In the Cold War context, relations between providers and recipients of such technologies were deeply political. Finland managed the delicate balance between East and West by buying nuclear technology from both Sweden and the Soviet Union.[14]

At the national level, decisions to either commit to or reject nuclear power have not only shaped energy and industry infrastructures and markets but had profound and long-lasting consequences on the organization of societies more generally. In the 1950s, the commitment to nuclear technology among political elites across Europe was an issue of national pride, and nuclear technology became closely attached to projects of national modernization.[15] In many countries, being a nuclear or nonnuclear nation has become part of national self-representation and identity. France stands out as one of the nations most deeply committed to nuclear power for both military and civilian ends. Historian Gabrielle Hecht argues that nuclear power played a vital role in shaping and even creating modern France: pursuing excellence in both military and civil nuclear technology came to incarnate a new version of the "greatness" of France, whose traditional foundations had been shattered in twentieth-century wars and decolonization. However, more than just becoming a constituent part of French identity, nuclear technology profoundly shaped France's material and immaterial infrastructures, through its impact on electricity networks, industrial structures, higher education and training, trade union activities, and the like. A mid-1970s publicity campaign captured the ethos of modernization through the nuclear sector in an expression that soon became a widely adopted slogan: "France doesn't have oil, but it has ideas."[16]

In a similar manner, although to a lesser degree, nuclear shaped societies in the Soviet Union, the United Kingdom, and Sweden. All three developed nuclear technology for both civilian and military purposes. West Germany was a major developer of nuclear power, too, but without overt military aims. Others—like Bulgaria, Finland, and Spain—also invested heavily in nuclear power but imported the technology from abroad. In all these countries, nuclear technology was a major source of national pride, with Finland as one of the most striking examples: ever since its introduction in the 1970s, the "Finlandization" of Soviet reactor technology has frequently been taken as incontestable proof of the superiority and near infallibility of the "Finnish engineer."[17]

Even in countries that never adopted commercial nuclear power, the promoters of nuclear research convinced national governments to invest heavily in national nuclear research programs in the 1950s and 1960s. Their arguments were based on "technological nationalism," the desire not to be left behind in international competition in this important new area of innovation. Even small countries like Denmark aimed for their own national nuclear reactor lines, with great hopes of exporting such high technology to international markets.[18] In the early 1970s, most countries had plans for future deployment of nuclear power, which varied according to their ambition. Governments continued to support exports of technology, but these became increasingly controversial in the light of growing international awareness of proliferation risks.[19] However, in the following decade, a number of countries left the "normal" path toward a nuclear future and instead advocated an energy transition to renewables. Especially in Austria, but to a lesser extent also in Denmark and Portugal, being nonnuclear increasingly became part of the national identity. The countries even began seeking to limit nuclear power in their neighboring states, and within Europe more generally.[20] In 2011, Germany joined this group of countries.

Country-specific differences shaped relations at the supranational level. Within today's European Union (EU), conflicting views between member states committed to and those rejecting nuclear power are highly divisive, not only straining French-German relations but also influencing EU politics on key issues such as climate change. The controversies around the preparation of the European Parliament statement for the 2019 UN Climate Change Conference (COP 25) in Madrid provides an illustration. The draft text stated: "nuclear power is neither safe, nor environmentally or economically sustainable."[21] However, a group of forty-six Members of Parliament (MEPs) managed to pass an amendment that completely reversed the meaning: "nuclear energy can play a role in meeting climate objectives because it does not emit greenhouse gases, and can also ensure a significant share of electricity production in Europe." The proposal for the amendment was primarily led by fifteen French MEPs, and seconded by those of seven other nuclear countries—Czech Republic (8), Sweden (7), the Netherlands (5), Belgium (4), UK (3), Finland (2), and Slovenia (1)—and one MP from Greece, which had withdrawn its plans for constructing nuclear power plants because of earthquake risks. In the final vote, the pronuclear formulation won, with 332 votes in favor and 298 against (45 abstentions).[22] The result reflects the historical positions of the MEPs' home country on nuclear technology rather than the positions of their respective parties. Thus, while the shared commitment to nuclear technology was intended to foster integration in the 1950s and 1960s,[23] in recent years, it has aggravated internal conflicts within the European Union.

At the local level, antinuclear citizen movements were instrumental to the emergence of the new social movements—notably the environmental movement—which radically transformed societal engagement in Europe. Indeed, many of today's international environmental NGOs, such as Greenpeace[24] and Friends of the Earth, have their roots in this early antinuclear activism. The movements also contributed to change in party systems, with the founding of new "green" political parties.[25]Antinuclear groups not only highlighted the potential health and environmental impacts of nuclear but also criticized the decision-making processes for secrecy and lack of democratic legitimacy, called for more public participation and dialogue on energy and nuclear waste policy, and thus challenged the technocratic consensus.[26] The 1970s critique also placed on the agenda the question of the very compatibility of the highly centralized and risky nuclear energy technology with a democratic society.[27]

The antinuclear movements faced a diverse range of counterparts, depending on national conditions. In some countries, they confronted powerful nuclear-industrial complexes that had invested heavily, both economically and politically, in nuclear technology. In others, they opposed politicians across the political spectrum and researchers who advocated ambitious visions of a nuclear future. In nondemocratic countries, antinuclear movements could not develop until dictatorships ended—in the mid-1970s in Southern Europe, and in the late 1980s and early 1990s in Central and Eastern Europe. When such movements emerged in the latter countries, they often adopted elements of the language, strategies, and structures of their Western counterparts, not all of which would resonate with Eastern European publics. The late-Soviet antinuclear activism originally linked its action and demands to a critique of state-socialist structures and Soviet/Russian dominance, to demands for transparency (*glasnost*) and democracy, and—in many cases—to the reassertion of nationalism. As chapter 3 demonstrates, antinuclear activism since the 1970s also paved the way for transnational NGO collaboration in Europe—overcoming differences between national contexts and traditions of civic action.[28]

The conflict over nuclear power changed European democracies in highly distinct ways, depending on national specificities. In Austria, Sweden, and Italy, for instance, the conflict led to novel experiments with direct democracy via referendums, while the Danish antinuclear movement defended representative democracy, demanding that Parliament rather than government should decide about nuclear power. In many countries, elements of deliberative democracy were introduced, for instance via extensive consultations through the establishment of government commissions, multistakeholder expert committees, citizen juries, and local information and liaison bodies.[29] In a number of countries, newly founded green parties for the first time gave a parliamentary voice to antinuclear and other alternative views on society.[30]

Major nuclear accidents had a considerable impact on public engagement and safety regulation that reached beyond the nuclear sector. The accidents at Three Mile Island in 1979 and Chernobyl in 1986 hastened clearer separation between responsibilities for the promotion and regulation of nuclear technology—which had been called for and at least formally implemented in a number of European countries and the US in the 1970s. These accidents also accelerated reforms and new thinking, such as awareness about the importance of regulatory independence in respect of risk technologies more generally.[31] Chernobyl was particularly instrumental in the emergence of antinuclear, environmental, and prodemocracy movements in the Soviet bloc countries.[32] The degree to which the accident helped precipitate the fall of the Soviet Union can be debated, yet Chernobyl incontestably played a role in what was to become the most significant political and societal transformation in Europe in the second half of the twentieth century. The more recent Fukushima catastrophe (2011) also had considerable repercussions in European societies, triggering debate on the tools of public engagement. Discussions of planned or actual nuclear phaseout have in turn accelerated wider policies and debates concerning energy and sustainability transitions, with wide-ranging impacts on society at large.[33]

Future Challenges for Nuclear Energy and Society in Europe and Beyond

Regardless of whether countries decide to launch new nuclear reactor projects, maintain the current capacity, or phase out nuclear altogether, nuclear issues will remain on the political agenda. Most notably, significant investments in human and financial resources will be needed for many decades to come. Nuclear countries need to tackle questions such as whether to extend the lifetime of existing reactors, how to decommission old reactors, what to do with radioactive waste and spent nuclear fuel, how to set priorities for R&D on new reactor technologies, and whether and how to speed up the market deployment of the possible chosen priority technologies. In Europe, the magnitude of the sociotechnical challenges is formidable: in December 2019, 126 nuclear power reactors were in operation in fourteen EU countries, ninety had been shut down, and three decommissioned; eighty-two research reactors in nineteen EU member states were in operation, long-term shutdown, or under decommissioning.[34] Meanwhile, in July 2020, only six nuclear reactors were under construction in the EU, whereas most of the new projects were situated in Asia.[35] This closing section reflects on the future challenges for nuclear and society in Europe and beyond.

Extending Reactor Lifetimes or Decommissioning Aging Reactors?

About two-thirds of all operating nuclear power plants in the world are more than thirty years old. Hence, they are either already beyond or approaching their initially planned lifetime—usually thirty years for the oldest reactors and forty to sixty years for the newer ones. Keeping aging reactors operating as long as possible entails significant economic benefits for a number of stakeholders. As the investment costs have already been written off for a long time, the plants provide stable revenue to their owners and cheap electricity to large customers—in particular for energy-intensive industries. Furthermore, many host municipalities have over time become "nuclear communities" highly dependent on the plants for providing jobs and prosperity.[36] Decommissioning is costly, presents technical and social challenges, and entails a host of unknowns.[37] Unless heavy investment in new energy supply projects is undertaken, alternatives tend to narrow down, especially in countries highly reliant on nuclear for their electricity supply, and hence unable to rapidly shift to the now cheaper renewable sources. Belgium and France, sourcing respectively about 40 and 70 percent of their electricity from nuclear, are among the most prominent examples. Both are committed to shutting down reactors—with Belgium intending to phase out nuclear by 2025, and France intending to bring the share of nuclear electricity down to 50 percent by 2035. Both also face pressure from neighboring Germany, where different levels of government along the border have called for the early shutdown of aging plants they consider unsafe.[38] Incidentally, the first on the French list of planned closures was the country's oldest operating plant in Fessenheim, only 1.5 kilometers from the German border, where antinuclear protest on the Upper Rhine started in 1971, before crossing the border to Germany.[39] Economic interests, commitments to cut down CO_2 emissions, and pressures from local communities fearing job losses repeatedly postponed the closure—scheduled originally for 2016 but finally completed only in 2020.[40]

In Sweden, too, the phasing out of old reactors has recently been politically contested. In 1999 and 2005, the two reactors at Barsebäck, only twenty kilometers from Copenhagen, were closed down after political decisions by the Swedish government, mainly to please Denmark.[41] Two old reactors at the privately owned Oskarshamn nuclear power plant, in turn, were shut down in 2015 and 2017 for economic reasons. But when the state-owned Vattenfall decided to close down two old reactors it operates at the Ringhals plant, four opposition parties questioned the decision in Parliament, and demanded that Vattenfall reverse its decision and keep the plants in operation. In the parliamentary vote in January 2020, the opposition was only one vote short of forcing Vattenfall to reverse its decision.[42]

The national safety authorities hold crucial power over decisions concerning the renewal of operating licenses for nuclear power plants. Following outcomes from regular inspections, these authorities can demand further repairs and safety investments or ultimately call for the closure of a plant that no longer fulfills the safety requirements. Decisions on plant closures and life extensions are likely to involve hardening battles among regulators, reactor owners, and governments. In France, clashes have become increasingly commonplace between the largely state-owned operator Electricité de France (EDF)—already struggling with mounting economic difficulties and hence reluctant to undertake ever costlier upgrade measures—and the national safety authority that is keen to demonstrate its independence. On the other hand, given the political and economic stakes at play, neither governments nor safety authorities are keen to take on the responsibility for shutdown decisions. Responding to German concerns, the safety authority declared already in 2016 that the two Fessenheim reactors continued to be safe to operate.[43] In doing so, it passed the hot potato to the politicians. The integrity of the national safety regulators will be repeatedly put to the test, as legal battles over the plant closures and license extensions intensify. Such questions are particularly critical in countries with a weak tradition of transparency, and where powerful vested interests push for continued plant operation.

Safety of the long-term operation of nuclear plants requires what could be called civic vigilance.[44] Critical scrutiny via demonstrations, counter-expertise, and opposition movements often emerges when nuclear facilities are planned or under construction, but once plants enter in operation, citizens often tend to accept them as a taken-for-granted part of their living environment and as a welcome source of jobs and prosperity. Many local opponents may by then have fled what increasingly turns into a nuclear community. Maintaining active civic vigilance of aging reactors therefore requires tailor-made measures. The local information and surveillance committees that are mandatory on each nuclear site in France constitute an example of attempts to institutionalize civic vigilance.

For the industry, external critique appears as a potential threat to the continued operation of the installations. However, a large-scale accident in an old reactor would be a major blow to the entire nuclear industry, and could ultimately threaten its very existence. The industry should therefore have a vital interest in not only maintaining high levels of competence and nurturing a strong safety culture within its own ranks but also in fostering regulator competence and integrity, as well as citizen awareness, vigilance, and knowledge.

Nuclear Waste

Because of its unique characteristics as an intergenerational problem, the nuclear waste issue has turned out to be a key driver for advanced public engagement approaches and practices in the nuclear sector and beyond.[45] All EU countries face the conundrum of what to do with the spent fuel from nuclear reactors and radioactive waste from medical, industrial, and research facilities and decommissioned power plants. Paradoxically, the faster the advancement of nuclear phaseout and decommissioning programs, the more acute the waste problem. The European Commission recently estimated that, by 2030, the planned reactor shutdowns and decommissioning will double the amount of very-low-level waste and increase that of the other types of waste by 20 to 50 percent.[46]

Already in the 1970s, the antinuclear movement pointed out the lack of a safe solution to the management of high-level nuclear waste as the Achilles' heel of the nuclear industry.[47] Since the late 1970s, the international nuclear community has advocated disposal in deep geological repositories as the preferred solution, instead of the previously practiced dumping at sea.[48] However, progress has been slow. Repository siting has given rise to some of the most intractable and long-lasting conflicts, sometimes even with cross-border implications.[49] Finland is the only country in the world with a repository under construction, scheduled to start operation in the mid-2020s. Pending outcome of the complex licensing procedure, the construction of the Swedish repository project is due to start soon and to be operational in the early 2030s.[50] France has advanced plans but has yet to overcome numerous hurdles for its repository to start operation in 2035. All other countries are lagging far behind.

In the 1980s and 1990s, vigorous local opposition to waste repository site investigations triggered a "participatory turn" within the radioactive waste management community.[51] Among the Finnish and Swedish private nuclear waste management companies, this implied a shift in strategy: the search for the geologically "best" site gave way to identifying communities with a sufficiently good geological conditions and, above all, a certain willingness to host a repository. The companies hence targeted existing nuclear communities, where citizens were already familiar with nuclear installations, enjoyed the economic benefits from hosting these facilities, and held little concern for the possible risks. The configuration is very different in France, as the process is in the hands of a public agency, the planned repository site is located in a remote nonnuclear region that is in desperate need for jobs and investments, and the project faces significant opposition. The West German Gorleben repository site, selected in the 1970s partly for similar reasons—then a remote border location next to the Iron Curtain and suffering from high unemployment—attracted formidable opposition for decades until 2020, when the government excluded Gorleben from the list of possible sites.[52]

The extremely long timescales involved make the waste problem particularly intractable. Ensuring that a repository can isolate the waste from living organisms for up to 100,000 years constitutes an unprecedented engineering challenge. However, even just planning and getting a repository built is a formidable endeavor and displays essential features of a complex "megaproject,"[53] entailing long timescales, the involvement of numerous stakeholders at multiple levels of governance, high political and economic stakes, significant environmental and social impacts on the involved communities, often irreducible uncertainties, and high propensity for conflicting values and interests. Even in the most advanced countries, Finland and Sweden, where R&D on nuclear waste disposal began in the late 1970s, the repositories have not started operation, forty years later. These projects therefore need to adapt to a host of political, economic, technical, and social changes that will be forthcoming during the century-long period of planning, implementation, and operation.

The "participatory turn" pioneered in nuclear waste management has been fundamental for the rest of the nuclear sector but also for the advancement of participatory policymaking more generally.[54] In the nuclear waste sector, the Swedes have been particularly active in spreading their expertise in citizen dialogue, including in countries of Eastern Europe.[55] Exchange of experience and knowledge also takes place via the OECD Nuclear Energy Agency, through interdisciplinary research projects within the EU, and other international forums.[56] Other international bodies, such as the IAEA, Euratom, and the World Association of Nuclear Operators (WANO), could also help to spread the lessons from the waste sector to arenas where decisions on nuclear as well as other energy policy and infrastructure projects are made.

If anything, experience has shown the importance of demonstrating that the participatory turn implies more than mere lip service. Truly democratic deliberation requires an evening out of asymmetries of power—in a situation where the industry has a vested interest in demonstrating that it has found a "solution"—and ensuring that any debate addresses issues that matter to the participants. It also requires showing how the views of the various publics were taken into account in decision-making. The possibility to participate but without the possibility to influence outcomes is a recipe for failure. Clearly, the introduction of elements of public deliberation and direct democracy tests the limits of representative democratic systems.

Last, the waste problem raises the question of national vs. international responsibilities. The principles of "national responsibility" and "prime responsibility of the license holder" are enshrined in current European legislation[57] concerning nuclear waste. In practice, this has translated into an attempt to develop a national waste repository in each country using nuclear power. Finding a willing host community is difficult in any country, but constructing a national repository is economically and politically challenging in countries that have only operated research reactors (e.g., Austria, Denmark) and in small, densely populated countries. Calls for international solutions—possibly international repositories—have hence become increasingly forthright.[58] EU legislation, indeed, does not exclude such a possibility. Hurdles exist, however, not least the reluctance of citizens to accept waste from reactors outside their own country; the choice by some countries[59] to reprocess their own and imported spent fuel rather than dispose of it directly; and the likely shifts in national boundaries during the long lifetime of repositories.

New Reactors: From Generation III "Problem Reactors" to the Promise of Generation IV and Small Modular Reactors

The nuclear industry began designing the so-called third-generation reactor technology in the late 1980s and early 1990s to spearhead the hoped-for nuclear renaissance. Rather than representing a revolutionary new design, these reactors promised incremental improvements in fuel technology, efficiency, lower construction and operating costs, and, crucially, greatly improved safety. The Franco-German 1600 MW European Pressurized Reactor (EPR), developed by Areva and Siemens, was to embody the renaissance in Europe. The reality turned out to be very different. All three European EPR projects—in Olkiluoto (Finland), Flamanville (France), and Hinkley Point (UK)—have faced substantial cost and schedule overruns and technical problems, with only the Olkiluoto plant approaching completion—over twelve years behind the original schedule. The Olkiluoto debacle was a major contributor to the collapse of the technology provider Areva, which was bailed out by EDF and split up in early 2018. The American Gen III design, APR 1000, has faced similar problems. Critics have dubbed the EPR the "European Problem Reactor," a white elephant symbolizing the impending demise of the Western nuclear industry. It is uncertain whether large Gen III nuclear reactor projects will have a future in the West. EDF still envisages replacing aging French reactors with EPRs and hopes to export EPRs outside of Europe (notably to India), but these plans remain an exception in the European landscape.

Instead, as throughout the history of nuclear energy, the promise of new, unproven technologies remains as strong as ever. Authorities, experts in many countries, and wealthy individuals, notably Bill Gates, have put their faith in the future deployment in large numbers of small modular reactors (SMRs) with a capacity of up to 300 MW.[60] Considerably smaller than present-day reactors, SMRs would help bring down costs via economies of scale, as standardized assembly line manufacturing would provide the constituent modules for the reactors. The small reactors would be relatively quick and simple to build, and could be deployed gradually, according to electricity demand. They would be particularly adapted to off-grid remote areas but also for combined heat and electricity production in urban areas. Licensing a given design instead of each individual plant separately could speed up deployment. As with Gen III designs, improved safety is a central marketing point: passive cooling systems would make SMRs independent of the availability of electric power. Unlike intermittent wind and solar, SMRs could provide a steady electricity output, which could, however, be easily adjusted according to fluctuating demand, thanks to the large number of reactor units. According to industry estimates, the first commercially operating SMR is still at least five years away. However, the economic case for SMRs remains highly contested. Moreover, SMRs would need to overcome a number of hurdles related to licensing, regulation, and, most notably, public opinion. If anything, the history of nuclear energy has shown that, for the nuclear industry in the Western countries, the existing nuclear power plant sites are precious assets, safely embedded in "nuclear communities." By contrast, any new reactor project on a greenfield site, however small, is likely to face fierce public opposition.

Beyond SMRs, innovation in other advanced reactor technologies retains its attraction in the nuclear sector. Since the 1950s, the community has viewed fast breeder reactors—in theory, able to generate more fuel than they consume and hence allowing a closed fuel cycle—as the logical culminating point of any nuclear program.[61] If the past is any indication, revolutionary Generation IV designs, most of which are based on fast breeder technology, let alone nuclear fusion, are unlikely to help with the urgent climate change challenges anytime soon. Fast breeder programs enthusiastically launched in the 1950s faced insurmountable technical, safety, economic, and political obstacles, and were all but terminated by the 1980s.[62] In Europe, the French were the most determined to pursue their breeder program until the late 1990s, despite the violent clashes between police and protesters over the Superphénix industrial prototype reactor, and despite the technical problems that plagued the plant until its early closure in 1998.[63] The French fast breeder dream was revived with the Astrid reactor project, which was nevertheless again suspended in summer 2019. While optimists still believe commercial Gen IV reactors to be deployable in the 2040s, the promise of commercially viable fusion reactors—as a truly inexhaustible source of energy—is unlikely to be fulfilled until well into the second half of the century, at best.

What seems to offer the greatest hope for the deployment of new nuclear in Europe is the growing public concern over climate change. While in the past, nuclear advocates portrayed their chosen technology as a key solution to the world's energy problems, they now argue that the battle against climate change cannot be won without nuclear energy. The climate argument is not new—indeed, it has underpinned much of the rhetoric around the nuclear renaissance since the late 1990s, and with increasing intensity as the impacts of climate change gain visibility, and with the emergence of social movements such as School Strike for Climate's Fridays for Future campaign.[64]

Nevertheless, what is new is the unexpectedly rapid boom of renewable energy. Thus, in the discourse of its advocates, nuclear has gone from being the only technologically available solution to becoming an indispensable complement—or, at the minimum, a "bridging technology"—without which the needed massive investments in a low-carbon future cannot be effective and cost-efficient. According to this argument, renewables and nuclear are low-carbon brothers in arms in the fight against climate change. The claim is contentious, and the relative virtues and mutual (in)compatibilities of nuclear and renewables within the electrical grid as well as in combating climate change are hotly debated. Today's pronuclear environmentalism,[65] or "ecomodernism,"[66] faces an uphill battle against the renewables constituency, whose historical roots are in the antinuclear movement of the 1970s and 1980s.[67] Yet, with Fukushima now a decade behind us, the public resistance to nuclear is mellowing, just as it did over the course of the 1990s, as the memory of the Chernobyl catastrophe gradually receded. The shift in public opinion is clearly detectable only in certain countries (e.g., Sweden, Finland, the UK), while many remain staunchly antinuclear. Even where the public has turned more favorable, antinuclear opinions still make up either a majority or a very significant minority.[68]

Last but not least, all of the challenges outlined above—safe reactor lifetime extensions, shutdowns, waste management, and possible new-build projects—face a crucial problem: how to attract talented students and the sorely needed skilled employees to an industry whose best years seem to lie in the past? This is not only an industry problem but one for society as a whole, because a skilled and dedicated workforce is vital for the organizations in charge of safe plant operation, decommissioning, and waste management, as well as for the regulatory authorities monitoring the safety of these activities.

Beyond Europe

While in recent decades Western nuclear constructors have not even come close to being able to deliver large reactor projects on time and within budget, their Asian counterparts certainly have. In July 2020, India had seven reactor projects under way, and the French hope to sell India half a dozen EPRs. South Korea had four reactors under construction, and their industry is seeking further markets abroad. Russia had three reactors under construction, and continuously looks for export opportunities. Crucial for Russia's ambitions in Eastern Europe and potential newcomer countries in Asia and the Gulf region[69] is the Fennovoima project in Finland, in which the Russian state corporation, Rosatom, is both the main shareholder and the technology supplier. Passing the test of the reputedly extremely strict Finnish licensing process would also help to open further European markets for Rosatom.

However, the key for the global nuclear industry is China. Although downscaled after Fukushima, its nuclear program remains the most ambitious in the world, with fifteen reactors under construction in 2020.[70] Half (thirty-one) of the total sixty-two reactors construction projects started, and not abandoned, worldwide since 2009 were in China. Its national nuclear industry is eyeing export markets, with two of its HPR-1000 reactors under construction in Pakistan and one to be built in Argentina. Through its participation in the UK Hinkley Point C and the possible Sizewell C projects, and with its hopes of constructing an HPR-1000 in Bradwell, China is establishing a foothold in European markets. In particular, the Chinese seem to have mastered the EPR technology that has proven so problematic in Europe: EDF's two EPR projects in Taishan were completed in 2018 and 2019, respectively—late and over budget, but nothing compared to the rather nightmarish Olkiluoto and Flamanville projects.

The nuclear sector likes to point to Chinese and also to Korean projects as success stories. These successes are said to demonstrate that the problems of the nuclear projects in the West have less to do with the specific technologies than with political and administrative institutions, markets, and safety regulation, as well as political and public support. Western nuclear advocates explain the problems by arguing that we in the West no longer know how to construct any large projects whatsoever because of the loss of skills and supply chains, and due to an excessive bureaucracy that supposedly characterizes today's Western project management culture.[71] They further argue that nuclear has been punished by "market-distorting" subsidies to renewables, and by modern society's obsessive yet futile demand for zero risk, which underpins the allegedly unreasonably rigid and complex regulation of safety.[72] These claims may be contested, yet they reveal a lot about the dominant sentiment within the Western nuclear sector, torn as it is between a pessimistic account of the flaws of "modern society" and cautious optimism about climate change concerns as a potential savior of the nuclear industry.[73] Lobbying for support, the nuclear industry advocates quick action, using as an argument the fear of international competition that already spurred the early investments in nuclear technology in the 1950s and 1960s: in the words of a senior nuclear-sector insider, unless European governments make crucial new-build decisions within the next five years, the market will be completely taken over by the Chinese, Korean, and Russian technology suppliers.[74]

The fact that most of the current nuclear construction takes place outside of the West, and often in countries with weak democratic traditions, an absence of critical civil society, and sometimes fragile politico-administrative institutions, raises its own challenges and unknowns. Lessons from European and US history suggest that rule of law, relative transparency, and willingness to allow public engagement are crucial components of safe and socially acceptable nuclear programs. If the interaction between the nuclear sector and society has at times been problematic in the West, and proved crucial in the fall of the Soviet Union, what is to be expected from nuclear development under more authoritarian regimes and in countries with weak political institutions?

Relevance of Humanities and Social Science

Tackling these challenges will require not only significant investment in science and engineering but also—and from the very outset—the incorporation of input from the humanities and social sciences. This book is but one example of how historical and social science research can inform decisions and assist nuclear practitioners in their work. In particular, the contributors have illustrated the various ways in which the seemingly technical nuclear sector matters are intertwined with the historically evolving social world. Just like any technological undertaking, a successful nuclear-sector project or policy needs "contextualization"—it must enroll the various stakeholders and publics, convince diverse constituencies of the relevance and interest of the technology, and ensure that the undertaking remains aligned with its ever-changing context.[75] Crucially, the need for such alignment concerns not only an individual nuclear project but the entire *raison d'être* of the industry. The recurrent conflicts around nuclear projects not only have concerned the details of those projects but have been rooted in distinct interests, knowledge, visions, political preferences, and ethical convictions. Arguably, it is the lack of contextualization that has time and again constituted a major stumbling block for nuclear technologies. Nuclear sector endeavors are likely to stumble time and again in the face of insurmountable conflicts and resistance unless public engagement is given a chance to reach beyond project planning and to instead address the entire range of societal issues at stake—including the very existence of the technology.

Notes

1. Jan-Henrik Meyer, et al., "Final Dissemination and Engagement Report (version 2019)," in *History of Nuclear Energy and Society (HoNESt) Consortium Deliverable no. 6.3* (2019), https://perma.cc/NQN6-F8JH; Matthew Cotton, "Engagement Futures for Nuclear Energy in Europe: Policy Brief II of the History of Nuclear Energy and Society Project (HoNESt)" (2019), https://perma.cc/44RU-LPQN.
2. Wilfried Konrad and Josep Espluga, "Comparative Cross Country Analysis on Preliminary Identification of Key Factors Underlying Public Perception and Societal Engagement with Nuclear Developments in Different National Contexts," in *History of Nuclear Energy and Society (HoNESt) Consortium Deliverable no. 4.2* (2018), 12, https://perma.cc/LJ4Q-HT3F. See also chapter 5.
3. Max Oelschlaeger, "The Myth of the Technological Fix," *Southwestern Journal of Philosophy* 10, no. 1 (1979): 43–53.
4. See chapters 3 and 8.
5. See chapter 3.
6. Cotton, "Engagement Futures for Nuclear Energy in Europe"; Ioan Charnley-Parry and John Whitton, "Principles for Effective Engagement," in *History of Nuclear Energy and Society (HoNESt) Consortium Deliverable no. 5.1* (2018), https://perma.cc/FXQ2-Y5M2.
7. See chapter 6.
8. See chapter 4.
9. See chapter 9; Jan-Henrik Meyer, "To Trust or Not to Trust? Structures, Practices and Discourses of Transboundary Trust around the Swedish Nuclear Power Plant Barsebäck near Copenhagen," *Journal of Risk Research* 24 (2021).
10. See chapter 8. See also Helmuth Trischler and Robert Bud, "Public Technology: Nuclear Energy in Europe," *History and Technology* 34, no. 3–4 (2018): 187–212, https://doi.org/10.1080/07341512.2018.1570674.
11. On the role of technology and infrastructures in shaping Europe more generally, see e.g. Per Högselius, Arne Kaijser, and Erik van der Vleuten, *Europe's*

Infrastructure Transition: Economy, War, Nature (Basingstoke: Palgrave, 2016); Wolfram Kaiser and Johan Schot, *Writing the Rules for Europe: Experts, Cartels, International Organizations* (Basingstoke: Palgrave, 2014).

12. See also Astrid Mignon Kirchhof and Jan-Henrik Meyer, "Revealing Risks: European Moments in Nuclear Politics and the Anti-Nuclear Movement," in *Protecting the Environment: Handbook on Contemporary European History*, ed. Patrick Kupper and Anna Katharina Wöbse (Berlin: De Gruyter Oldenbourg 2021), 352–53.
13. Sonja D. Schmid, "Nuclear Colonization? Soviet Technopolitics in the Second World," in *Entangled Geographies: Empire and Technopolitics in the Global Cold War*, ed. Gabrielle Hecht (Cambridge, MA: MIT Press, 2011), 125–54. See also chapter 1.
14. Karl Erik Michelsen and Aisulu Harjula, "Finland Short Country Report (version 2018)," in *History of Nuclear Energy and Society (HoNESt) Consortium Deliverable no 3.6* (2019), 38–44, https://hdl.handle.net/2454/38269.
15. Also beyond Europe, notably in Korea: Sheila Jasanoff and Sang-Hyun Kim, "Containing the Atom: Sociotechnical Imaginaries and Nuclear Power in the United States and South Korea," *Minerva* 47, no. 2 (2009): 119–46.
16. Gabrielle Hecht, *The Radiance of France: Nuclear Power and National Identity after World War II* (Cambridge, MA: MIT Press, 2009), 319.
17. E.g., Harri Lammi, "Tarinat kovasta ytimestä," in *Ydinvoima, valta ja vastarinta*, ed. Matti Kojo (Helsinki: Like, 2004), 11–50.
18. Henry Nielsen and Henrik Knudsen, "The Troublesome Life of Peaceful Atoms in Denmark," *History and Technology* 26, no. 2 (2010): 91–118, https://doi.org/10.1080/07341511003750022.
19. Dennis Romberg, *Atomgeschäfte: die Nuklearexportpolitik der Bundesrepublik Deutschland 1970–1979* (Paderborn: Ferdinand Schöningh, 2020).
20. See chapter 9.
21. European Parliament resolution on the 2019 UN Climate Change Conference in Madrid, Spain (COP 25), B9-0174/2019—Am 38, available at: https://www.europarl.europa.eu/doceo/document/B-9-2019-0174_EN.html.
22. Roll call vote on Amendment 38, pp. 126–27, available at: https://www.europarl.europa.eu/doceo/document/PV-9-2019-11-28-RCV_FR.pdf.
23. Kirchhof and Meyer, "Revealing Risks." In practice, national commitment to the common research projects within Euratom varied and actually led to the demise of the common nuclear reactor project by the end of the 1960s. Paul Bähr, "Was wird aus Euratom? Die Europäische Atomgemeinschaft in der Krisenzone," *Europa-Archiv* 25, no. 3 (1970): 81–90; see also chapter 4.
24. Frank Zelko, *Make It a Green Peace! The Rise of Countercultural Environmentalism* (Oxford: Oxford University Press, 2013).
25. Stephen Milder, "Between Grassroots Activism and Transnational Aspirations: Anti-Nuclear Protest from the Rhine Valley to the Bundestag, 1974–1983," *Historical Social Research* 39, no. 1 (2014): 191–211.
26. See also Dolores L. Augustine, *Taking on Technocracy: Nuclear Power in Germany, 1945 to the Present* (New York: Berghahn, 2018).
27. Robert Jungk, *The Nuclear State* (Parchment, MI: Riverrun Press, 1984 [1977]).
28. On the challenges, see Astrid Mignon Kirchhof and Jan-Henrik Meyer, "Global Protest Against Nuclear Power: Transfer and Transnational Exchange in the 1970s and 1980s. Focus Issue," *Historical Social Research* 39, no. 1 (2014): 163–273, https://doi.org/10.12759/hsr.39.2014.1.165-190.

29. See chapter 3.
30. Silke Mende, *"Nicht links, nichts rechts, sondern vorn." Eine Geschichte der Gründungsgrünen* (Oldenbourg: München, 2011); Stephen Milder, "Between Grassroots Protest and Green Politics: The Democratic Potential of the 1970s Antinuclear Activism," *German Politics & Society* 33, no. 4 (2015): 25–39.
31. Arne Kaijser et al., *The Past and the Present of Nuclear Safety Regulation in Europe: Transcript of the Discussions of the Witness Seminar in Barcelona, 16 October 2018* (Mimeo, Barcelona: Pompeu Fabra University, 2019).
32. E.g. see chapter 1; also Tetiana Perga, "The Fallout of Chernobyl: The Emergence of an Environmental Movement in the Ukrainian Soviet Socialist Republic," in *Nature and the Iron Curtain: Environmental Policy and Social Movements in Communist and Capitalist Countries, 1945–1990*, ed. Astrid Mignon Kirchhof and John R. Mc Neill (Pittsburgh: University of Pittsburgh Press, 2019), 55–72.
33. Ute Hasenöhrl and Jan-Henrik Meyer, "The Energy Challenge in Historical Perspective," *Technology and Culture* 61, no. 1 (2020): 295–306.
34. European Commission, *Report from the Commission to the Council and the European Parliament on progress of implementation of Council Directive 2011/70/EURATOM and an inventory of radioactive waste and spent fuel present in the Community's territory and the future prospects. Second report. 17 December 2019, COM(2019) 632 final* (Brussels: European Commission, 2019), https://ec.europa.eu/transparency/regdoc/rep/1/2019/EN/COM-2019-632-F1-EN-MAIN-PART-1.PDF.
35. Mycle Schneider et al., *The World Nuclear Industry Status Report 2020* (Paris and London: Mycle Schneider Consulting, 2020), 47.
36. Astrid M. Eckert, *West Germany and the Iron Curtain: Environment, Economy & Culture in the Borderlands* (Oxford: Oxford University Press, 2019), 223; Maria Rosaria Di Nucci and Achim Brunnengräber, "In Whose Backyard? The Wicked Problem of Siting Nuclear Waste Repositories," *European Policy Analysis* 3, no. 2 (2017): 295–327. See also Kate Brown, *Plutopia: Nuclear Families, Atomic Cities, and the Great Soviet and American Plutonium Disasters* (Oxford: Oxford University Press, 2013).
37. Diletta C. Invernizzi, Giorgio Locatelli, and Naomi J. Brookes, "Managing Social Challenges in the Nuclear Decommissioning Industry: A Responsible Approach Towards Better Performance," *International Journal of Project Management* 35, no. 7 (2016): 1350–64.
38. E.g., Christian Part, "Kernkraftwerk Tihange: Die Angst vor dem GAU," *Die Zeit—Zeit Online*, May 2, 2018, https://www.zeit.de/wirtschaft/2018-04/kernkraftwerk-tihange-anwohner-widerstand-angst/komplettansicht.
39. See chapter 3.
40. Christian Schubert, "Atomkraftwerk: Adieu Fessenheim," *Frankfurter Allgemeine Zeitung* February 21, 2020, https://www.faz.net/-gqe-9wqrm.
41. Arne Kaijser and Jan-Henrik Meyer, " 'The World's Worst Located Nuclear Power Plant': Danish and Swedish Perspectives on the Swedish Nuclear Power Plant Barsebäck," *Journal for the History of Environment and Society* 3 (2018): 71–105, https://doi.org/10.1484/J.JHES.5.116795; Anna Storm, *Post-Industrial Landscape Scars* (Basingstoke: Palgrave, 2014), 47–73.
42. Johan Ronge, "Sverige: Riksdagen röstade nej till fortsatt drift av Ringhals," *Expressen*, January 22, 2020, https://www.expressen.se/nyheter/klimat/riksdagen-rostar-om-karnkraftens-framtid/.

43. "L'ASN répond à l'Allemagne sur la sûreté de la centrale de Fessenheim," *Les Echos*, March 4, 2016, https://www.lesechos.fr/2016/03/lasn-repond-a-lallemagne-sur-la-surete-de-la-centrale-de-fessenheim-203581. The French safety authority confirmed the statement in 2018: https://www.asn.fr/L-ASN/L-ASN-en-region/Grand-Est/Installations-nucleaires/Centrale-nucleaire-de-Fessenheim.
44. Markku Lehtonen et al., "Healthy Mistrust or Complacent Confidence? Civic Vigilance in the Reporting by Leading Newspapers on Nuclear Waste Disposal in Finland and France," *Risk, Hazards & Crisis in Public Policy* 12, no. 2 (2021): 130–57, https://doi.org/10.1002/rhc3.12210.
45. Achim Brunnengräber and Maria Rosaria Di Nucci, eds., *Conflicts, Participation and Acceptability in Nuclear Waste Governance: An International Comparison Volume III* (Wiesbaden: Springer VS, 2019); Achim Brunnengräber et al., eds., *Challenges of Nuclear Waste Governance: An International Comparison Volume II* (Wiesbaden: Springer VS, 2018); Achim Brunnengräber et al., eds., *Nuclear Waste Governance: An International Comparison* (Wiesbaden: Springer VS, 2015).
46. European Commission, *Report from the Commission to the Council and the European Parliament on progress of implementation of Council Directive 2011/70/EURATOM.*
47. Anselm Tiggemann, *Die "Achillesferse" der Kernenergie in der Bundesrepublik Deutschland: zur Kernenergiekontroverse und Geschichte der nuklearen Entsorgung von den Anfängen bis Gorleben, 1955 bis 1985* (Lauf an der Pegnitz: Europaforum Verlag, 2004); Jasanoff and Kim, "Containing the Atom," 129–30.
48. Jacob Darwin Hamblin, *Poison in the Well. Radioactive Waste in the Oceans at the Dawn of the Nuclear Age* (New Brunswick, NJ: Rutgers University Press, 2008).
49. Mar Rubio-Varas, António Carvalho, and Joseba de la Torre, "Siting (and Mining) at the Border: Spain-Portugal Nuclear Transboundary Issues," *Journal for the History of Environment and Society* 3 (2018): 33–69, https://doi.org/10.1484/J.JHES.5.116794; Astrid Mignon Kirchhof, "East-West German Transborder Entanglements through the Nuclear Waste Sites in Gorleben and Morsleben," *Journal for the History of Environment and Society* 3 (2018): 145–78, https://doi.org/10.1484/J.JHES.5.116797; Eckert, *West Germany and the Iron Curtain*, 201–43. The Euratom Treaty of 1957 already anticipated cross-border impacts, requiring member states to notify the commission of planned repositories near intra-European borders; see chapter 9.
50. Mark Elam and Göran Sundqvist, "Meddling in Swedish Success in Nuclear Waste Management," *Environmental Politics* 20, no. 2 (2011): 246–63.
51. Mark Elam and Göran Sundqvist, "Public Involvement Designed to Circumvent Public Concern? The 'Participatory Turn' in European Nuclear Activities," *Risk, Hazards & Crisis in Public Policy* 1, no. 4 (2010): 203–29, https://doi.org/10.2202/1944-4079.1046.
52. Kirchhof, "East-West German Transborder Entanglements"; Eckert, *West Germany and the Iron Curtain*, 201–43; Gorleben-Archiv, ed., *"Mein lieber Herr Albrecht . . ." Wie der Gorleben-Konflikt eine Region veränderte. 34 Gespräche mit Zeitzeugen* (Lüchow: Gorleben-Archiv 2019). A similar reasoning was behind the choice of Wackersdorf for the West German reprocessing plant: Janine Gaumer, *Wackersdorf: Atomkraft und Demokratie in der Bundesrepublik 1980–1989* (Munich: Oekom, 2018), 62.
53. E.g., Bent Flyvbjerg, Nils Bruzelius and Werner Rothengatter, *Megaprojects and Risk: An Anatomy of Ambition* (Cambridge: Cambridge University Press, 2003);

Benjamin K. Sovacool and Christopher J. Cooper, *The Governance of Energy Megaprojects* (Cheltenham: Edward Elgar, 2013); Markku Lehtonen, Pierre-Benoit Joly, and Luis Aparicio, eds., *Socioeconomic Evaluation of Megaprojects: Dealing with Uncertainties* (London: Routledge, 2017).

54. Jason Chilvers, "Environmental Risk, Uncertainty, and Participation: Mapping an Emergent Epistemic Community," *Environment and Planning A* 40, no. 12 (2008): 2990–3008. Hannah Strauss, "For the Good of Society: Public Participation in the Siting of Nuclear and Hydro Power Projects in Finland" (PhD thesis, Oulu, Faculty of Education, 2010), http://jultika.oulu.fi/Record/isbn978-951-42-9507-2.
55. Lehtonen, Joly, and Aparicio, *Socioeconomic Evaluation of Megaprojects.*
56. E.g., the Implementing Geological Disposal of Radioactive Waste Technology Platform (https://igdtp.eu/), and the OECD NEA Forum on Stakeholder Confidence (https://www.oecd-nea.org/rwm/fsc/); see chapter 4.
57. Euratom, "Council Directive 2011/70/Euratom of 19 July 2011 Establishing a Community Framework for the Responsible and Safe Management of Spent Fuel and Radioactive Waste," *Official Journal of the European Union*, L199 (February 8, 2011): 48–56.
58. Jasanoff and Kim, "Containing the Atom," 142.
59. France remains the main European country today operating reprocessing facilities, while the UK recently suspended its extensive reprocessing program.
60. For overviews, see, e.g., B. Mignacca and Giorgio Locatelli, "Economics and Finance of Small Modular Reactors: A Systematic Review and Research Agenda," *Renewable and Sustainable Energy Reviews* 118 (2020), https://doi.org/10.1016/j.rser.2019.109519; *Small Modular Reactors: Nuclear Energy Market Potential for Near-term Deployment*, NEA No. 7213 (Paris: OECD Nuclear Energy Agency, 2016), https://www.oecd-nea.org/ndd/pubs/2016/7213-smrs.pdf; Steve Thomas, Paul Dorfman, Sean Morris, and M.V. Ramana, *Prospects for Small Modular Reactors in the UK & Worldwide* (Nuclear Consulting Group and Nuclear-Free Local Authorities, July 2019), https://www.nuclearconsult.com/wp/wp-content/uploads/2019/07/Prospects-for-SMRs-report-2.pdf.
61. Claire Le Renard, "The Superphénix Fast Breeder Nuclear Reactor: Cross-Border Cooperation and Controversies," *Journal for the History of Environment and Society* 3 (2018): 107–144, https://doi.org/10.1484/J.JHES.5.116796; Maja Fjæstad, "Fast Breeder Reactors in Sweden—Vision and Reality," *Technology and Culture* 56, no. 1 (2015): 86–114.
62. Proliferation concerns led President Carter to halt the US fast breeder program already in the late 1970s, see Michael Camp, " 'Wandering in the Desert': The Clinch River Breeder Reactor Debate in the U.S. Congress, 1972–1983," *Technology and Culture* 59, no. 1 (2018): 31–33.
63. Renard, "The Superphénix Fast Breeder Nuclear Reactor"; Markku Lehtonen, "France Short Country Report (version 2019)," in *History of Nuclear Energy and Society (HoNESt) Consortium Deliverable no. 3.6.* (2019), https://hdl.handle.net/2454/38269.
64. William J. Nuttall, *Nuclear Renaissance: Technologies and Policies for the Future of Nuclear Power* (New York: Taylor & Francis Group, 2015).
65. Ren van Munster and Casper Sylvest, "Pro-Nuclear Environmentalism: Should We Learn to Stop Worrying and Love Nuclear Energy?" *Technology and Culture* 56, no. 4 (2015): 789–811.

66. E.g., Ted Nordhaus and Michael Shellenberger, *Break Through: Why We Can't Leave Saving the Planet to the Environmentalists* (Boston: Mariner Books, 2009).
67. Niels I. Meyer, "Danish Pioneering of Modern Wind Power," in *Wind Power for the World: The Rise of Modern Wind Energy*, ed. Preben Maegaard, Anna Krenz, and Wolfgang Palz (Boca Raton, FL: CRC Press, 2013), 163–78.
68. For opinion surveys in Finland, France, and Sweden, see *Suomalaisten energia-asenteet 2018* (Helsinki: Energiateollisuus ry, November 2018), https://energia.fi/files/3278/Energia-asenne_2018_MATERIAALIPANKKIKUVAT.pdf; *Les Français et l'énergie nucléaire* (Paris: Ifop, April 2016), https://www.ifop.com/publication/les-francais-et-lenergie-nucleaire/; Sören Holmberg, *Swedish Opinion on Nuclear Power 1986–2019* (Gothenburg: SOM-institutet, June 2020), https://www.gu.se/valforskningsprogrammet/pagaende-projekt/mek-den-svenska-miljo-energi-och-klimatopinionen. Obtaining Europe-wide comparative data is very difficult, given that *Eurobarometer* discontinued its surveys on nuclear power after Fukushima. The latest study available: European Commission, "Europeans and Nuclear Safety. Fieldwork: September–October 2009, Publication: March 2010," *Special Eurobarometer* 324 (2010), https://ec.europa.eu/commfrontoffice/publicopinion/index.cfm/ResultDoc/download/DocumentKy/55216.
69. E.g., Bangladesh, Belarus, Turkey, Vietnam, Jordan, and United Arab Emirates, where Russia seeks to promote its technology through concessional financial packages and fuel services.
70. Schneider et al., *The World Nuclear Industry*, 47.
71. E.g., *Nuclear New Build: Insights into Financing and Project Management* (Paris: OECD-NEA, 2015), https://www.oecd-nea.org/ndd/pubs/2015/7195-nn-build-2015.pdf.
72. Markku Lehtonen, "NEA Framing Nuclear Megaproject 'Pathologies': Vices of the Modern Western Society?," *Nuclear Technology* 207, no. 9 (2021): 1329–50, http://dx.doi.org/10.1080/00295450.2021.1885952.
73. Lehtonen, "NEA Framing Nuclear Megaproject 'Pathologies.' "
74. Markku Lehtonen interview with an OECD NEA official, Boulogne, July 8, 2019.
75. Bruno Latour, *Aramis, or the Love of Technology* (Cambridge, MA: Harvard University Press, 1996).

Contributors

EDITORS

Arne Kaijser is emeritus professor of history of technology at KTH (Sweden). His main research interests concern the historical development of infrastructural systems and their interaction with society at large. His research has focused on a variety of energy, transportation, communication, and water infrastructure systems in Sweden, the Nordic countries, and Europe. Kaijser was the president of the Society for History of Technology (SHOT) in 2009–2010. He has been an active participant in the Tensions of Europe network and was project leader for a European Science Foundation funded project on European critical infrastructures.

Markku Lehtonen is Marie Curie-Sklodowska Fellow at Universitat Pompeu Fabra (Spain), associate researcher at the Groupe de sociologie pragmatique et réflexive (GSPR) at the Ecole des Hautes Etudes en Sciences Sociales (EHESS) in Paris, and associate faculty at the Science Policy Research Unit of the University of Sussex, where he worked as research fellow from 2005 to 2015. He was a researcher at the ESSEC Business School (Cergy-Pontoise, France) in 2016–2018 and at the Institut Francilien Recherche Innovation Société (IFRIS) at the Université Paris-Est Marne-la-Vallée in 2012–2014. Lehtonen holds a PhD in environmental economics (University of Versailles, France) and an MSc in environmental studies (University of Helsinki, Finland). He conducts research in the areas of energy, environmental, and sustainability policies, on topics such as controversies, appraisal, trust, and mistrust concerning nuclear-sector megaprojects in European countries.

Jan-Henrik Meyer is researcher at the Max Planck Institute for European Legal History and associate researcher at the Leibniz Center for Contemporary History (Germany). Meyer's research combines an interest in the history, politics, and law of European integration with environmental history. His PhD on the European public sphere inquired into how citizens and the media viewed the emerging European Union, corroborated it, and challenged its legitimacy.

Within the HoNESt project, he examined the role of societal actors and international organizations and the relations between nuclear energy and society in Europe, and led the work package on dissemination. His current research traces the emergence of European environmental law in legal scholarship, legislation, and jurisdiction between the 1970s and the 1990s. He is also co-leader of a research project on the West German government's public engagement, the "Citizens' Dialogue on Nuclear Energy (Bürgerdialog Kernenergie) 1974–1983."

Mar Rubio-Varas is director of the Institute for Advanced Research in Business and Economics and full professor of economic history at the Universidad Pública de Navarra (Spain). She obtained her PhD from the London School of Economics (UK) and her BSc in economics from the University Carlos III of Madrid. She was a Fulbright fellow at the department of economics at the University of California at Berkeley (US) in 2001–2002. Her research interests focus on the long-term relationships between energy consumption and economic growth, including energy security and independence, and transitions to a low-carbon economy. Rubio-Varas has published extensively in the fields of economic history and environmental and energy economics. Within the HoNESt project, she led the work package linking history and social sciences and the Scientific Secretariat.

AUTHORS

Stathis Arapostathis is associate professor at the National and Kapodistrian University of Athens (Greece), School of Sciences. He has a PhD in history of technology from University of Oxford (2006), an MSc in history and philosophy of science and technology from the National and Kapodistrian University of Athens (1999), and a BSc in physics from Aristotelio University of Thessaloniki (1996). Arapostathis specializes in the study of public discourses of science and technology relating to environment and toxic regimes; regulation/law and technosciences; legal patent disputes and intellectual property politics; research and technological infrastructures; and the constitution of technoscientific expertise and its role in shaping public policies of science and technology.

Robert Bud is emeritus keeper at the Science Museum in London (UK) and was research keeper at the same institution until 2018. Bud's research has focused on the history of applied science from the fall of the Bastille to the raising of the Iron Curtain. In 2008–2009 he worked with Professor Peter Hennessy on the lecture series on the politics of energy subsequently published by the British

Academy. His publications include monographs on the history of biotechnology and of penicillin. He is a past council member of both the Society for the History of Technology and the British Society for the History of Science.

Pieter Cools is a postdoctoral researcher at the Center for Research on Environmental and Social Change, department of sociology, University of Antwerp, Belgium. His research addresses topics of participation, state-civil society relations, and different types of knowledge and expertise in various policy areas, applying diverse research approaches. Cools obtained his PhD in 2017 and MSc in 2012, both in sociology, from the University of Antwerp. After a PhD thesis on social innovation in social policy, he worked on public engagement in radioactive waste management and nuclear power in two EU-funded projects, HoNESt and Modern2020. Currently he works on public participation in urban planning in Antwerp.

Matthew Cotton is professor of public policy at the Teesside University (UK) and was senior lecturer in human geography and environmental studies at the University of York (UK) from 2016 to 2020. His research focuses on the social and ethical dimensions of environmental management and technology governance, with particular interest in the environmental justice dimensions of environmental policy and planning decisions around energy and waste infrastructure, climate change, and natural resource management. He has published on issues of nuclear power, shale gas extraction, plastic waste, climate adaptation, petrochemicals, onshore wind, electricity transmission and distribution systems, and radioactive waste management.

Ann Enander is professor emeritus at the Swedish Defence University. She specializes in risk perception and communication, crisis preparedness, and emergency management at the local, regional, and national levels. Her work includes analyses of many of the major critical events affecting Scandinavia. Enander's interest in the interactions between nuclear energy and society originate from her studies of the effects of the Chernobyl accident in Sweden and also later studies of public perceptions relating to warning systems and emergency planning around the Swedish nuclear power plants.

Josep Espluga is associate professor at the department of sociology of the Autonomous University of Barcelona (Spain) and senior research fellow at the university's Institute of Public Policies and Government. His research focuses on the relationships between health, work, environment, and territory, with particular attention to the social perception of technological risks and environmental conflicts.

Paul Josephson is professor of history at Colby College (US) and visiting professor at Tomsk State University (Russia). Josephson, a specialist in twentieth-century big science and technology and in environmental history, is the author of fourteen books, including *Red Atom* on Soviet civilian nuclear programs and most recently *Traffic* and *Chicken*. He has visited the former Soviet Union dozens of times since 1984 for research, including research institutes and power stations in Kyiv, Rivne, and Kharkiv, Ukraine, and in Moscow, St. Petersburg, Murmansk region, and Siberia, in Russia. He is currently working on a global environmental history of the nuclear age.

Tatiana Kasperski is associate researcher at the Pompeu Fabra University in Barcelona (Spain), where she previously was a Marie Curie Fellow. She is a specialist on nuclear power in the former Soviet Union, and author of *Les politiques de la radioactivité: Tchernobyl et la mémoire nationale en Biélorussie contemporaine* (Paris: Editions Petra, 2020). She earned her PhD in political science in 2012 at Sciences Po in Paris. She has since worked on several projects on public engagement with nuclear power, including most recently within "Atomic Heritage Goes Critical: Waste, Community and Nuclear Imaginaries," a project led by the University of Linköping, Sweden.

Wilfried Konrad is senior researcher at DIALOGIK, nonprofit institute for communication and cooperation research in Stuttgart, Germany. He carries out research on energy-related topics such as smart grids, carbon capture and storage, and energy use at home. Previously, Konrad was a researcher and project leader at the Institute for Social Research in Frankfurt, Germany, and at the Institute for Ecological Economy Research in Heidelberg, Germany. His interest in the history of nuclear energy and society originates from research in the context of the German *Energiewende*, which included the investigation of the social acceptance of nuclear power in comparison with other energy technologies.

Beatriz Medina is partner and research consultant at Water, Environment and Business for Development. She holds an MSc in environment and health communication (Pompeu Fabra University, 2015) and an MSc in environmental sciences (University of Alcalá de Henares, 2006). She worked as a research associate at the department of sociology at the Autonomous University of Barcelona in 2015–2019. Medina conducts her research and consultancy work in the areas of socioenvironmental studies and science communication, as well as engagement processes and the analysis of social perceptions in environmental issues. She has extensive research experience in projects funded by the European Union and international development agencies in Europe, Africa, Asia, and Latin America.

Ana Prades is senior researcher at Ciemat (Research Center for Energy, Technology and the Environment), a public research body depending on the Spanish Ministry of Science, Innovation, and Universities, where she started her professional activity in 1990. She is currently the head of the Ciemat's Sociotechnical Research Center in Barcelona. Her main research interests are risk perceptions and communication, and their implications for the design and implementation of environmental policies. Prades has been involved in research projects on public engagement methods, lay reasoning and understanding of new technologies, social acceptance of energy technologies, and innovative knowledge brokerage tools to support policymaking and policy implementation in the field of sustainable consumption.

Albert Presas i Puig is associate professor of history of science at the Pompeu Fabra University in Barcelona, and previously a research fellow at the Max Planck Institute for the History of Science in Berlin. His research focuses on science, culture, and power; science in the European periphery; science and Francoism; and the history of nuclear energy. He has been involved in research projects funded by various national and international institutions. Presas i Puig was the lead coordinator of the HoNESt project.

Helmuth Trischler is head of research at the Deutsches Museum, Munich, professor of modern history and the history of technology at Ludwig Maximilian University in Munich, and director (with Professor Christof Mauch) of the Rachel Carson Center. Trischler's main research interests are knowledge societies and innovation cultures in international comparison; science, technology, and European integration; transport history; and environmental history. He is the author of thirty-six books and edited volumes and some hundred articles, and coeditor of a number of book series, including *Umwelt und Geschichte* (Göttingen: Vandenhoeck & Ruprecht) and The Environment in History: International Perspectives (Oxford and New York: Berghahn Books).

Index